DAIRY CATTLE BREEDS

Dr. H. H. Kildee (1884–   ), Dean
Emeritus of Iowa State University.
He guided and inspired students and
breeders of dairy cattle, judged shows,
served on committees of true type,
unified score cards, and type classification, and was a type classifier with
Holsteins and Jerseys. His investigations of grading up common cows
with purebred dairy bulls inspired
wide improvement of dairy cattle.

Dr. C. H. Eckles (1875–1933), teacher and pioneer research worker in dairy husbandry, dairy products, and
dairy cattle nutrition. His influence continues to spread
through his publications, the work of his students, and
the 142 graduate students who hold many key positions
in research, teaching, production, and industry. He was
acclaimed one of the "Ten Master Minds of Dairying"
and affectionately called "The Chief" by associates.

DAIRY CATTLE BREEDS

ORIGIN AND DEVELOPMENT

RAYMOND B. BECKER

UNIVERSITY OF FLORIDA PRESS
GAINESVILLE / 1973

ACKNOWLEDGMENTS

The author and publishers hereby express their thanks to the following, whose generous contributions have made possible the publication of this work:

ALVAREZ JERSEY FARM, INC., Jacksonville, Florida
AMERICAN BREEDERS SERVICE, INC., DeForest, Wisconsin
 (Subsidiary of W. R. Grace & Company)
ELBERT CAMMACK, Geneva, Florida
FLORIDA BROWN SWISS CLUB
FLORIDA GUERNSEY CATTLE CLUB
FLORIDA JERSEY CATTLE CLUB
FLORIDA HOLSTEIN-FRIESIAN ASSOCIATION, INC.
MR. and MRS. JAMES HITE, Summerfield, Florida
V. C. JOHNSON, Dinsmore, Florida
JUDSON MINEAR, Palm City, Florida
NORTHERN OHIO BREEDERS ASSOCIATION, INC., Tiffin, Ohio
TOM C. and JULIA G. PERRY, Moore Haven, Florida
DONALD D. PLATT, Orlando, Florida
C. W. REAVES, Gainesville, Florida
JOHN SARGEANT, Lakeland, Florida
THE WALTER SCHMID FAMILY, Tallevast, Florida
J. K. STUART, Bartow, Florida
CARROLL L. WARD, JR., Astatula, Florida
CARROLL L. WARD, SR., Christmas, Florida
MR. and MRS. WALTER WELKENER, Jacksonville, Florida

PREFACE

When Dr. H. H. Kildee taught "dairy breeds" as a separate course, no textbook was available. Inspiration from this course led me to continue studying. Dr. C. H. Eckles and C. S. Plumb advised on sources of original materials. Libraries were searched in the United States. Dr. Sir John Hammond gave assistance at the School of Agriculture in Cambridge. Dr. A. C. McCandlish arranged appointments in Scotland. Konsulent K. M. Andersen assisted with Danish materials and travel to original sources. Dean E. L. Anthony supplied an unpublished dissertation on the Red Danish milk breed. Breed secretaries and Dr. H. H. Hume wrote letters of introduction which gave entree in England, the Channel Islands, and Europe. Intensive study was conducted at antiquarian bookshops, cattle shows, breed offices, libraries, museums, and on farms in Europe. Curators aided in measuring fossil *Bos* skulls in Cambridge, Cal-

v

cutta, Leeuwarden, London, and Zurich. The British Museum of Natural History supplied photographs of *Bos* skulls and provided access to the library on mammalian paleontology.

Graphs, maps, and vignettes heading selected chapters were prepared from historical material by our son George F. Becker. Dean H. H. Kildee and Dr. C. H. Eckles granted use of their photographs. Dean Kildee selected some key illustrations. Rand McNally Company provided copyrighted maps on which to superimpose information. F. Windels permitted the use of two illustrations from *Four Hundred Centuries of Cave Art*. A photograph of *Bos primigenius* was bought from the Danish National Museum. Dr. J. U. Duerst's photograph of *Bos longifrons* was loaned by the Zootechnische u. Veterinar Hygienisches Institut in Bern. Dr. M. V. A. Sastry photographed two skulls in Calcutta. The Milk Industry Foundation gave two pictures of rock paintings from the Frobenius-Fox expedition. Secretary H. G. Shepard supplied original copies of early ideal Jersey type. Hugh Bone copied the original photograph of the first public milking trial in Ayr. R. W. Hobbs provided the photograph of his eight Dairy Shorthorn cows.

M. S. Prescott of the *Holstein-Friesian World* presented photographs of Solomon Hoxie, Spring Brook Bess Burke 2d, and Wisconsin Admiral Burke Lad. Mrs. Laura Baxter sent the engravings of Kitty Clay 3rd and Kitty Clay 4th. The John Gosling meat cutting demonstration and Dr. S. M. Babcock were photographed by the staff of Iowa State College and the University of Wisconsin. Some pictures were obtained from Robert F. Hildebrand and Harry A. Strohmeyer, Jr. Ralph Sneeringer of the University of Florida copied some photographs with permission.

The chapters on the respective breeds were reviewed critically by the following authorities. Ayrshire—John Graham, David Gibson, Jr., Doris E. Chadburn, and G. A. Bowling; Brown Swiss—John Graham, Fred S. Idtse, W. Engeler, and R. W. Stumbo; Dutch Belted—C. H. Willoughby; Guernsey—H. C. Le Page and Karl B. Musser; Holstein-Friesian—Dr. J. M. Dijkstra and H. W. Norton, Jr.; Jersey—H. C. Shepard and Lynn Copeland; Milking Shorthorn —Arthur Furneaux, W. E. Dixon, and Jesse B. Oakley; Red Dane— K. M. Andersen, K. Hansen, Ejner Nielsen, and Dean E. L. Anthony.

Dean H. H. Kildee, Dr. I. R. Jones, our daughters Mrs. Elizabeth J. Mitchell and Mrs. Ann M. Herrick, and Robert A. Herrick reviewed the entire manuscript. Sincere appreciation is expressed to them and to many other persons, here and overseas, who contributed the historical information.

Through the active interest of Extension Dairyman C. W. Reaves, Director of Special Programs Albert F. Cribbett, and Dr. E. T. York, Vice President for Agricultural Affairs at the University of Florida, a number of interested dairy people contributed to The SHARE Council, University of Florida Foundation, Incorporated. Their loyal cooperation enabled the University of Florida Press to produce this volume. To those many authorities and to all others participating in the production, the author expresses his humble appreciation and thanks.

Grateful acknowledgment and thanks are due to my wife Harriet and to our children. Their patience, encouragement, and cooperation were most helpful.

INTRODUCTION

This book is organized into four sections. Chapters 1–4 describe the geological origin of the genus *Bos* and domestication and early development of common cattle. Man possibly came from east-central Africa to the regions where cattle roamed wild. He hunted cattle for food during the Pleistocene Age and into historical times. Capture and domestication predated written history. Early artists pictured the chase, and later some tame cattle, on rocks and on the walls of caves. Neolithic peoples brought small domesticated cattle from western Asia into Europe, following watercourses where travel was easiest. They brought some cultivated cereals and reached the British Isles and Channel Islands over land connections. Breakdown of feudal tenure and enclosure of lands allowed owners to select bulls to mate with their cows. Better crops and feed stored for winter use were corequisite with selection in improvement of cattle.

Fairs, markets, and agricultural shows rewarded and inspired men with good animals.

Chapters 5–20 trace the gradual development of breeds. A few individuals initiated private herdbooks to keep reliable pedigrees. Solomon Hoxie believed that a herdbook should record conformation or production of individual animals "upon which a science of cattle culture could be based." Associations of breeders developed programs to measure achievements and granted recognition to breeders who qualified for them. Improvement of cattle once was largely an art, dependent on the observing eye and analytical mind of a few leading breeders. Mendel's laws of inheritance and later discoveries added science to art, increasing the rate of improvement. Improved dairy cattle served as the foundation stock in the United States.

Heredity is estimated by biometricians to contribute less than 20 percent to milk producing capacity; 80 percent is a factor of environment (breeding efficiency, disease control, management, nutrition, and other agencies). Such contributions are assembled in part in chapter 21.

The Summary, based on breed chapters, constitutes chapter 22. What does the future hold for further improvements among dairy cattle? The germ plasms of animals possess several types of hereditary genes. Desirable characters have been segregated and disseminated from seedstock such as Penshurst Man O'War, Jane of Vernon, May Rose 2d, and other improvers. Some undesirable recessive genes have been traced even to seedstock animals of the highest qualities. Such genes in heterozygous form were present unrecognized through many generations. They can crop out among some of the progeny from matings between heterozygous parents. Plant breeders have developed disease-resistant varieties by applying known methods to their foundation stocks used in pollinations. The science of improved cattle breeding lies in the future, with methods known at present. The plant breeders' methods can be duplicated by cooperation among dedicated breeders, using the tool of artificial breeding in order to obtain proofs and application to develop better strains of dairy cattle. Examples of such accomplishments have been cited for Ayrshires, Friesians, and Holstein-Friesians, and in several breeding references.

CONTENTS

xi

CHAPTER 1

GEOLOGICAL ORIGIN OF CATTLE

THE HISTORY of the origin and development of animal life is fragmentary as obtained by paleontological studies. Although countless numbers of animals lived, remains of only a few were preserved in fossil form. Some animals drowned in floods, became mired in some bog, or were eaten by predatory animals or man; their bones became covered and preserved from the elements. Erosion, extreme drouth, excavation, or dredging revealed those few specimens to man, but conditions to preserve bones existed in limited areas. Therefore many specimens disintegrated, leaving possibly only a tooth or some study bones. Furthermore the value of these remains may not have been recognized by their discoverers; often the fossils were not turned over to an agency interested in their significance. This imperfect means represents the tools with which to interpret past ages.

GEOLOGICAL AGES

One needs to know measures of time to realize the significance of origins, migrations, and descent of species. The ages of fishes, reptiles, and mammals are characterized by movements and deposition of earth with entrapped remains of life of each period. Typical exposed deposits have been explored and their fauna described. Rates of sedimentation, climatic changes, potassium-argon ratios, and rate of disintegration of radio-carbon-14 have served as methods of estimating time and are subject to further investigation. No attempt will be made to assign years to these periods, but quoted estimates may be repeated (Table 1.1).

The earliest fossil remains of mammals are chiefly those of the marsupials. Such remains are found in rocks of the Triassic and Jurassic periods in Australasia, where the marsupials were protected from encroachment by higher mammals. Some higher placental animals appeared in the Oligocene and Miocene periods, but many did not appear until the Pliocene. Migrations occurred over a long period; their time and direction were affected by geographic and climatic barriers such as mountains, deserts, seas, and icecaps during glacial periods. Man destroyed wild species during the Old Stone and New Stone Ages, and became a disseminator of domesticated animals in the New Stone Age.

Fossil remains of cattle (genus *Bos*) include teeth, skulls, and other bones distributed in parts of southern and western Asia, Europe, and northern Africa. Great Britain and the Channel Islands, which were connected with the continent by land, contributed to early records of cattle. Wild cattle (true genus *Bos*) did not appear in the western hemisphere.

THE MIOCENE AGE

Investigations into the origin of cattle lead into mammalian paleontology, based on few preserved specimens. Personal viewpoints affect the conclusions, which are subject to reinterpretation when additional discoveries may be made in Asia, Africa, or Europe.

TABLE 1.1
GEOLOGICAL AGES RELATED TO ORIGIN OF SPECIES

Era	Major periods	Typical life
Cenozoic	Quaternary	
	Present	Man developed culturally; agriculture; improved livestock.
	Recent	Man an artisan; early domestication of animals in the eastern hemisphere; lake dwellings; late New Stone, Bronze, and Iron ages.
	Pleistocene	Glacial periods; man a huntsman; early cave deposits, valley gravels; *Bos primigenius* and other large mammals; migrations over land connections. Old Stone Age.
	Tertiary	
	Pliocene	Fossil man in east-central Africa; *Leptobos* and other ruminants; mountain upheavals.
	Miocene	Grassy plains; many mammals; early ruminants and horses.
	Oligocene	Increased forests and some coal formation; apes, early ruminants.
	Eocene	Placental mammals with hoofs and grinding molar teeth; some coal formation.
	Paleocene	Many ancient mammals.
	Mesozoic	
	Cretaceous	Broad-leaf forests increase; some coal formation; birds, snakes; last of dinosaurs.
	Jurassic	Toothed birds; more mammals; dicotyledonous plants.
	Triassic	Land plants, dinosaurs, reptiles, primitive mammals; gypsum and salt deposits.
	Paleozoic	
	Permian	Reptiles, insects suited to less humid environment.
	Carboniferous	Forests of the coal measures; amphibians, sharks, crinoids.
	Devonian	Trees, ferns, marine fishes appear.
	Silurian	Land plants, early insects.
	Ordovician	Snails, molluscs, sponges, corals, freshwater fishes.
	Cambrian	Early fossil marine life; trilobites appear.
Precambrian	Proterozoic	Primitive marine forms appear rarely, entrapped by sedimentation.
	Archeozoic	Igneous rock, metamorphosis occurring.

J. Cossar Ewart reported to the Scottish Cattle Breeding Conference that:

> At the end of the Miocene Age, the immense area between the Ganges and the Jumna [rivers], now occupied by the Siwalik Hills, consisted of boundless well-watered plains. That they were fertile will be evident in that they supported a large number of mammals, including three-toed horses, pigs, sheep, goats and antelopes, also buffaloes, bison, and of especial interest *Leptobos,* the oldest and in many ways the most primitive known member of the ox family.

Some writers, however, did not regard *Leptobos* as a true member of the ox family but rather as an older form from which the true ox may have descended.

The Siwalik Hills (Fig. 1.1) are a former ancient flood plain extending along the Himalayan foothills in East Punjab into United Provinces in northern India. Hollow-horned mammals, including three species of true oxen (*Bos*) which were ancestors of domestic cattle, appear to have originated in this region.

THE PLIOCENE AGE

Pilgrim classed the fauna of the Pinjor zone (in the upper Siwalik Hills) at the headwaters of the Bunnah River as belonging to the lower part of the middle Pliocene Age. Fossil camel, *Hemibos,* horse, *Leptobos,* and others occurred here. Overlaying boulder conglomerate also yielded fossil buffalo, camel, hippopotamus, horse, rhinoceros, and swine, as well as *B. acutifrons, B. planifrons,* and *B. platyrhinus.* He concluded, "Then the first appearance of true *Bos* is in the Upper Pliocene of the Siwaliks, while *Leptobos* and *Hemibos* precede it in the Middle and Lower Pliocene."

THE PLEISTOCENE AGE

The earliest fossil remains of true cattle were found in the lower Pleistocene deposits in the Siwalik Hills of north central India below the Himalayan mountains (Fig. 1.2). This earliest true ox was discovered by Hackett in the Narbada Valley in 1874, and was named *B. acutifrons* Lydekker. The skull has a sharp ridge from the poll down to the middle of the forehead. The horns were nearly 10 feet from tip to tip, and extended outward and upward.

The next younger gravel deposits of the Narbada Valley in this region yielded remains of *B. namadicus* (or *B. planifrons* Lydekker), as shown in Figure 1.3. *B. planifrons*, an extinct Indian ox, was of slender build, with horns of the bull set low on the skull. It was described first by Falconer and Cautley, and called also *B. taurus macroceros* Duerst, or *B. palaeogarus* by Rutimeyer. This wild species was contemporaneous with early man in India during the Old Stone Age. Its remains were present also in the lowest levels excavated at Anau in Turkestan by Duerst in 1904. The species which Duerst found in the higher deposits at Anau were smaller and more refined and had shorter horns. He described it as *B. taurus brachyceros*. Northern India was a center from which

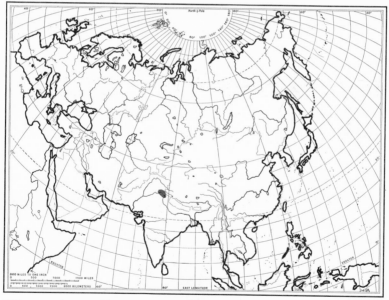

FIG. 1.1. The Siwalik Hills extend along the southern Himalayan foothills in East Punjab and United Provinces between headwaters of the Jumna and Ganges rivers. The genus *Leptobos* and other hollow-horned mammals developed here during the late Miocene and early Pliocene ages. (Use of the copyrighted background map by permission of the Rand McNally Company.)

hollow-horned mammals (*Cavicornia*) disseminated. Bovines appeared first in the sub-Himalayas.

The small cattle (*Hemibos*) are related to the existing anoa of the Celebes. Long-skulled forms such as the ancestral ox (*Leptobos*) appear to be similar to the species *L. etruscus* in the Val d'Arno in the "Recent Pliocene Fauna" of Europe. The true ox (*Bos*) appeared in Europe after the beginning of Pleistocene times —the second faunal stage.

The Swiss paleontologist Professor L. Rutimeyer regarded *B. acutifrons* and *B. namadicus* as the Asiatic and probably older forms of *B. primigenius*; Richard Lydekker considered them distinct, but suggested *B. namadicus* as a descendant of *B. acutifrons*. *B. planifrons* Lydekker, with shorter horns and flattened frontal bones, may have been the female of *B. acutifrons*.

FIG. 1.2. *Bos acutifrons* was the largest known wild ox. The bone horn cores spanned 86.5 inches even though broken off where yet over 3 inches in diameter. (Photographed by Dr. M. V. A. Sastry, Geological Survey of India.)

FIG. 1.3. *Bos namadicus*, or *Bos planifrons*, was contemporary with early man during the Pleistocene Age in the Punjab province of India. (Photographed by Dr. M. V. A. Sastry.)

Lydekker wrote of the Siwalik fauna in the Pliocene:

> Originally discovered in the outer ranges of the typical Himalayan area, the Siwalik fauna has been traced towards the northwest into Punjab, Kach, Sind and the northwestern frontier of Baluchistan; the beds from the two latter areas being lower in the series than those from the typical Siwalik Hills, and containing an older assemblage of forms, although several are common to all. . . .
>
> Goats and oxen for the first time made their appearance, the former being represented by species belonging to the typical *Capra*, and to the shorter-horned genus *Hemitragus*. The oxen (*Bos*) included members of all existing groups with upright triangular horns nearly allied to the anoa of the Celebes. . . . Genera like *Hippopotamus, Bos, Capra, Equus* and *Elephas* are unknown previous to the Siwalik epoch, and some of them were evolved at or about that time in the Indian area.

Lydekker considered *B. taurus primigenius* to be the ancestral stock of domesticated cattle. *B. fraseri* has been identified with a skull from the Pleistocene formation in the Narbada valley. The genera *Equus* and *Elephas* existed in North America in an earlier period. Teeth of the camel occur also in some hard rock phosphate deposits in America.

Rutimeyer stated that *B. etruscus* H. Falconer (or *L. elatus*) was found with remains of mastodon, elephant, rhinoceros, and hippopotamus in the late Pliocene deposits of the Astigiana, between San Paula and Dusino, Italy. *B. etruscus*, a specimen of which is in the museum in Turin, Italy, ranged widely in Italy and France.

B. etruscus (male) had wide heavy horn bases and a less prominent poll than other later European species of cattle. Females were hornless. From its anatomy, *B. etruscus* appeared to be related to the banteng or Java ox of the present day—*Bos sondaicus*—but with horns placed low on the skull near the eyes. A Siwalik representative, *B. falconeri*, had a more slender skull and horns of the bulls turned upward more.

THE OLD STONE AGE

Many fossil remains of *B. primigenius* Bojanus (Fig. 1.4), representing the Old Stone Age, have been found over western Asia, northern Africa, and nearly all of Europe and the British Isles.

Waterworn rocks, broken by man for a sharp edge, were called Soan-type artifacts in southern Asia. They occurred with a *B. namadicus* skull during the Pleistocene in the Punjab province of India.

The older river-drift gravel beds along the Somme, Oise, and Thames rivers in France and England yielded fossilized bones of the primitive wild ox. Specimens occurred in the more recent gravels along the river valleys of France. Ludwig H. Bojanus described the fossil remains of the ox in these gravel beds in 1827 and named the species *B. primigenius*. This species roamed wild over all of western Europe and northern Africa.

The crudest flint implements made by man were associated with fossil ox bones in the older river drifts. The first of these implements recognized as the work of early man were discovered by M. Boucher de Perthes in 1847 near Amiens and Abbeville in the Somme River valley of France. J. Wyatt found similar ones near

FIG. 1.4. The great ox *Bos primigenius* Bojanus spread during the Pleistocene Age over Europe and the British Isles. The last specimen died in captivity in 1627. This specimen was taken from a deposit near Athol, Perthshire. M-2245. British Museum.

Bedford, England, along with bones of deer, hippopotamus, horse, mammoth, ox, and rhinoceros.

MAN AS HUNTERS

Man sometimes lived in caves and shelters during part of the next stage of civilization. Primitive man used flaked weapons and tools; he brought home the quarry to his cave where the flesh was eaten and the long bones broken for marrow. *B. primigenius* bones were not disfigured by gnawing, indicating the dog had not yet become man's companion.

An early hunter broke a young bull's lumbar vertebra with a spear. The animal escaped and the bone healed. The bull broke through the ice and drowned when five to six years old. Professor Sven Nilsson excavated it in 1840 from 10 feet deep in the peat bog at Onnarp, Sweden. Another ox drowned in a bog in northwestern Sjaelland, Denmark. Almost the entire skeleton was recovered (Fig. 1.5) on removal of peat. Small flint microliths were embedded in two ribs. A skull dug from Burwell Fen, near Cambridge, England,

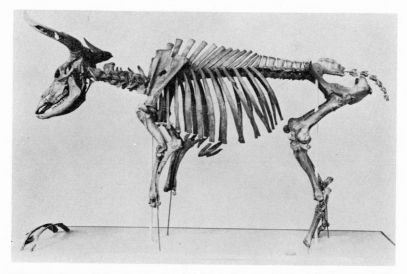

FIG. 1.5. *Bos primigenius* was hunted by early man for food. This animal, the skeleton of which is in the Danish National Museum at Copenhagen, was shot in the flank and two ribs with small flint microliths before it drowned in a peat bog near Vig on the Island of Sjaelland, Denmark.

had a broken celt or flint axe in its forehead (Fig. 1.6). A skull preserved at Bromberg, Prussia, had three spear wounds on the forehead. These findings indicated that man hunted the wild ox *B. primigenius* for food.

B. primigenius stood 6 to 7 feet tall at the withers. A mature cow

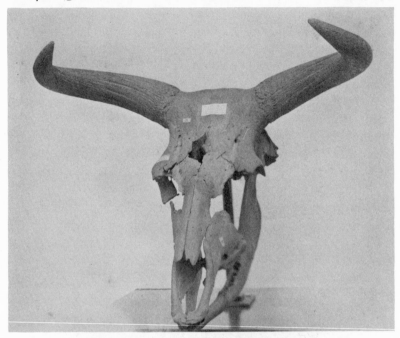

FIG. 1.6. A flint axe penetrated this *Bos primigenius* skull, found in Burwell Fen, near Cambridge, England.

skeleton, which was exhibited at the Technical Agricultural High School in Berlin, was taken from the bottom of a peat bog at Guhlen, Brandenburg, Germany. An almost complete skeleton was in the Sedgwick (Woodwardian) Museum at Cambridge, England.

CAVES AND CAVE PAINTINGS

Caves have yielded considerable evidence concerning cattle. E. O. James mentioned that:

> the Abbés A. and J. Souyssenie and Barden found in a low-roofed cave near the village of La Chapelle-sur-Saints . . . in the department of Cobreze, a Neanderthal skeleton lying in a small pit near the center of the passage . . . stones surrounding the skeleton. Mousterian flints, estimated at two thousand, and the bones of the woolly rhinoceros, reindeer, ibex, bison, cave-hyena, and other cold-loving animals occurred in the deposit, while above the skull were the leg bones of the ancient type of ox, and pieces of quarts, flint, ochre, and broken bones were arranged around the skeleton.

The wild ox was used for food long before the advent of Neolithic man and a smaller kind of domesticated cattle. Keith estimated the Mousterian period at the Würm glaciation of northern Europe at about 40,000 B.C.

Cave paintings were discovered in 1879 by the five-year old daughter of the Spanish Marquis de Sautuola while he was excavating Altamira Cave in the Pyrenees Mountains near Santillana del Mar. Similar cave paintings have been found in France, Spain, Italy, and the Libyan Desert in Africa. Cliff and rock shelters southwest of Tripoli in the Sahara Desert bear paintings of cattle in domestication.

Primitive cave paintings date to the Aurignacian period of the Old Stone Age, estimated at about 20,000 B.C. The older designs are crude outline drawings, but others appear in good proportions. Early designs were colored in red or black on limestone walls and ceilings of ancient caves (Fig. 1.7). Later paintings combined three or more colors. A cave at Pasiega, Spain, contained over 250 paintings and 36 engravings of bison, chamois, deer, horses, ibex, and

stag done chiefly in red. Animals in caves were depicted as pierced
by arrows or spears.

Many paintings showed scenes of the chase (Fig. 1.8), or of cows
and younger animals being killed by spears or arrows—an unlikely
practice if domestication had been known. Such paintings were in
the Albarracia, La Madelaine, Lascaux, Sovigna, and other caves.
One painting (Plate XII, of J. Cabre) pictured in red three cattle
with long horns directed upward and outward. Human beings were
close to three horned cattle in at least one cave painting. No
weapons were in their hands, and the cattle were standing quietly.
Was this intended to indicate domesticated cattle? Ernst Grosse observed that hunting people neglected plants in cave paintings.

THE NEW STONE AGE

Man in Europe was still a hunter and fisherman at the beginning
of the New Stone Age. *Køkkenmøddinger*—shell mounds—along
the North Sea and deposits in cave dwellings contained fossil bones
of animals not domesticated.

FIG. 1.7. A large red dappled cow from a cave painting at Lascaux, France.
(With permission of F. Windels from *Four Hundred Centuries of Cave Art.*
Copyrighted.)

Various stages of cultures were described in the Royal Scottish Museum in Edinburgh:

Aurignacian: The tools—beautifully made end-scrapers. Points and some bone implements with a split base.

Solutrian: People small in number. Chief invention was a slender type of javelin, head shaped like a laurel leaf (made by pressure flaking).

Magdalenian: Bone and flint implements were in use. Lance heads were typical of the earlier Solutrian culture. Cave art was at its height, and drawings of contemporaneous animals, such as bison, reindeer, and mammoth, are found on cave walls and on pieces of bone.

Mesolithic culture, Asilian, kitchen-middens, etc.: At the close of the Paleolithic period a sudden change in climate took place. Milder conditions prevailed, forest reappeared.

Neolithic man lived in huts and began agriculture and the domestication of animals. Pottery making was begun, and his implements were formed by polishing and grinding.

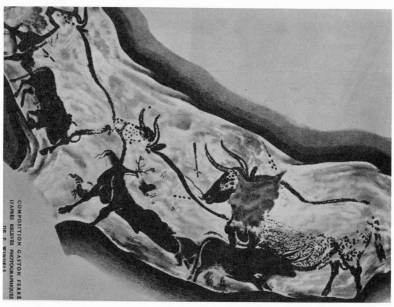

FIG. 1.8. A frieze from the "Hall of Cattle," Lascaux, France, discovered in 1940 near Montignac-sur-Vezere. The *Bos primigenius* bull, at the right in back, has a spear in the muzzle and a throwing stick to the left of its horns. (With permission of F. Windels from *Four Hundred Centuries of Cave Art.* Copyrighted.)

The Museum legend concerning *B. taurus primigenius* stated:

> Although the Urus has been extinct in Scotland for many centuries, it once lived throughout the length of the land. Its remains have been found from Wigtonshire to Caithness in marl deposits, from the floors of lakes which succeeded the Glacial Period, and in peat bogs. That it was hunted by the early settlers in Scotland is shown by its bones occurring in broch and cave deposits. The Urus was a large strong beast standing about six feet high at the shoulder. The horns were very long, and the horn cores were long, curved, and massive.

John Fleming owned a cattle skull 27.5 inches in length and 11.5 inches across the orbits. Richard Owen, describing a skull in the British Museum of Natural History that was found near Blair Atholl in Perthshire, stated that it was a yard long, and that the horn cores spanned 3 feet 6 inches.

In the older *Køkkenmøddinger* or shell mounds along the Danish seacoast, no traces of cereal grains were found. Domestic fowls were absent, but bones of ducks, geese, and swans were common. The stag, roe deer, and wild boar (*Sus scrofa* L.) comprised about 97 percent of the mammalian remains. Bear, beaver, dog, fox, hedgehog, lynx, marten, mouse, otter, porpoise, seal, water rat, wolf, and urus were represented. Traces of a smaller ox also were found. Only the dog was domesticated, according to Professor Steenstrup, a Danish archeologist. Flint implements were plentiful, but metals were absent in these mounds. Zeuner (1963) concluded that settled agriculture preceded domestication of the "crop-robbers" such as cattle, water buffalo, yak, and pig. He regarded domestication of the cow as most significant.

Professor J. J. A. Worsaae considered that during the New Stone Age inhabitants of Denmark possessed tame cattle and horses, and probably some knowledge of agriculture. Relics in the later shell mounds fitted with the early Neolithic Stone Age, when the art of polishing flint implements was known. He stated further:

> The inhabitants of Denmark, and the west of Europe, in the stone-period, are therefore to be designated as forming the transition between the past ancient nomadic races, and the more recent agricultural and civilized nameless tribes. . . . The

inhabitants of Denmark during the bronze-period were the people who first brought with them a peculiar degree of civilization. To them were owing the introduction of metals, the progress of agriculture and of navigation.

Although some *B. primigenius* were domesticated in Europe, this species mainly was hunted as a wild animal. Fossil bones were found in the lower (older) kitchen debris of early lake dwellers in Italy, Switzerland, Germany, and Great Britain. *B. primigenius,* which remained wild during historic times, became extinct in the British Isles before the close of the Bronze Age.

Caesar called the larger bovine "urus," which he mentioned as native in the Hercynian or Black Forest of Germany. Tacitus and Pliny wrote that the horns of these cattle, used as drinking cups, sometimes held as much as 2 *urs* (2 liters). A free translation by Lydekker of *De Bello Gallico,* book vi, chapter xxix (written about 65 B.C.) stated:

> There is a third kind of these animals which are called uri. In size these are but little inferior to elephants, although in appearance, color, and form they are bulls. Their strength and speed are great. They spare neither man nor beast when they see them. . . . In the expanse of their horns, as well as in form and appearance, they differ much from our [domesticated] oxen.

A few aurochs were in the province of Maine about A.D. 550. They were hunted by Charlemagne in forests near Aix-la-Chapelle, Rhenish Prussia, in the ninth century. Records in a Swiss abbey mentioned auroch meat near the close of the tenth century, and crusaders crossing Germany in the eleventh century saw the animals.

Skulls and bones of *B. longifrons* Owen, resembling Island-type Jerseys in size and proportions, were found commonly in the relic beds of lake dwellings, in morasses, and near old forts in Europe and the British Isles, and were associated with stone and bronze implements. The "Niebelungen Lied," an early German legendary poem, cited Siegfried as killing a wisent (European *Bison bonasus*) and four urus near Würm in the twelfth century. Beltz described a chart made about 1284, citing urus between the Duaa and Dnieper

TABLE 1.2

MEASUREMENTS (IN CENTIMETERS) OF SKULLS OF FOSSIL *BOS*, AND THOSE FROM PRESENT-DAY CATTLE

Species and location	Length of skull	Between horn cores	Narrowest part of forehead	Maximum across eye sockets	Horns — Circumference at base of horn cores		Horns — Length of outer curvature		Source or notes
					Right	Left	Right	Left	
Leptobos falconeri									
British Museum[a]	42.4	7.3	18.9	21.8	22.0	22.8	[b]	[b]	Siwalik Hills, Pliocene
Bos acutifrons, F. & Cautley National Museum, Calcutta	[b]	12.7	21.6	24.0	40.0	40.1	125.1[b]	109.4[b]	Upper Siwaliks, Upper Pliocene
Bos namadicus Falconer (F-155)	54.7	18.3	20.5	26.3	34.9	34.3	79.3[b]	104.5[b]	Narbada Valley, Middle Pleistocene
Bos primigenius Bojanus									
British Museum	75.6	26.0	25.4	32.4	35.0		72.4		Blair Atholl, Perthshire, Scotland
		20.5			42.0	41.0	81.3	86.0	Barrington, Pleistocene
	69.8	22.8							Marl pit, Scotland
Cambridge	65.0	20.0	22.7	28.0	32.3	32.5	65.0	66.0	Stone axe in forehead; Burwell Fen
		19.0	26.3		42.5	40.2[b]	90.5	72.5[b]	Barrington, Pleistocene (C. E. Gray)
		19.5	22.2	28.2	33.5				Peat deposit, Scotland
		19.0	24.8	30.5	40.0	39.5	86.0	84.5	Fen land
	61.2	17.0	23.3	30.2	30.3	29.5	58.3	59.5	Burwell Fen, Cambridge. 29.021
University of Copenhagen	70.7	40.4	20.5	30.2	31.4	31.8	56.2	56.7	Peat bog, Denmark. Male
	63.2	19.0	21.3	26.9	26.5	26.2	41.6[b]	46.0[b]	Peat bog, Denmark. Female
University of Lund	61.3	16.6[c]	17.1		17.0	17.0	24.3	22.0[b]	Onnarp peat bog, Sweden. Sven Nilsson.
Bos longifrons Owen[d]									
British Museum	52.6	16.2	18.5	22.9	20.1	19.8	17.5	20.2	Walthamstown, Essex. Male
British Museum	41.3[b]	16.2	15.8	19.9	15.3	15.8	11.5[b]	15.5[b]	Burwell Fen, flat forehead
	49.4	15.3	17.5	22.8	18.5	18.5	20.0[b]	17.0[b]	Reach Fen
	48.6	14.5[c]	17.0	22.8	22.2	22.4	22.4	22.4	Clapton, Essex. M-4097. Male

Bos taurus brachyceros Rutimeyer[a, f]

Source									
		15.9	13.8	17.3	12.0				Ireland
		15.0	14.2	19.8	11.5				Kutterschitz
Zurich	43.2	14.6	15.2	19.7	11.3	11.5	11.5	9.4[b]	C-21
		14.0	17.3	23.0					Ireland
Zurich	44.6	13.5	14.6	19.4	12.4	12.0	10.6[b]	10.5[b]	B-2
		13.0	14.7	18.6	12.6				Peat deposit, Ireland
		12.8	14.4	18.2	13.0				Hostomitz
		12.8	14.3	19.2	11.3				Peat deposit, Ireland. Female
		12.7	15.0	19.6	14.0				Anatolia. Female
		12.3	13.7	17.0	11.0				Tschonschitz
		12.2	13.4	13.7	9.2				Cave at Lascaux, France
		12.1	13.4	17.4	9.2				Thames River. Male
		11.9	13.7	18.4	10.5				Peat bog, Walthamstown, Essex
Cattle breeds, present									
Cambridge	50.0	16.6	19.5	23.5	25.8	25.8	37.5	39.9	Ayrshire male, over horn shells
Cambridge	45.0	14.0	15.8	20.5	15.3	15.3	25.0	24.0	Ayrshire female, over horn shells
Zurich	50.7	23.0	20.4	25.6	24.0	23.9	26.5	25.0	Brown Swiss male, over horn shells
Zurich	49.2	18.9	18.1	23.0	16.0	16.0	17.0	18.0	Brown Swiss female, over horn shells
Dinsmore Farms	50.7	13.8	19.8	26.0	26.7	27.0	27.0	28.1	Guernsey male, Florida
Florida Station	49.5	14.3	15.8	22.5	dehorned	dehorned			Guernsey female, Florida
Florida Station	51.5	12.4	15.3	21.5	dehorned	dehorned			Holstein female
Florida[e]	47.9	34.6	15.5	24.4	18.8	19.1	18.8	18.1	Average of 4 males
Florida Station	43.1	13.1	14.0	20.7	11.8	11.7	15.6	15.8	Average of 9 females
Range Station	54.3	18.3	19.5	24.0	29.3	30.4	34.5	35.0	Guzerat Brahman male
Range Station	49.5	15.3	17.2	19.8	20.5	20.9	27.6	28.7	Guzerat Brahman female

a. British Museum of Natural History, London.   b. Part broken or worn.   c. Prominent poll.

d. Owen considered skulls with smaller horn measurements to be females of the species.

e. Three Jersey males and 9 Jersey cows were from the Florida Agricultural Experiment Station. One male was from Highview Farm courtesy of Carlos Griggs. Ages of Ayrshires and Brown Swiss were not available; all others were mature cattle of the present breeds.

f. *Bos taurus brachyceros* Rutimeyer and *Bos longifrons* Owen are synonymous; usages depend on countries.

rivers and Carpathian Mountains. The species finally became extinct there in the seventeenth century.

Lydekker, Whitehead, and James Wilson thought White Park cattle of the British Isles descended from more or less domesticated early, not wild, aurochs. Since some White Park cattle were polled, this suggested some relation to early Norse cattle of the polled species *B. akeratos*. Roman cattle, as well as *B. longifrons,* may be in the ancestry.

Drinking horns made from the outer horn shells of the great wild ox are in many European museums. The Friesch Museum in Leeuwarden had 13 such drinking horns, some carved or ornamented and others mounted in brass and silver. The largest had an inside diameter at the base of over 4 inches and exceeded 24 inches in length of outer curvature. Smaller ones were of similar proportions. Two horns bore dates of A.D. 1397 and A.D. 1571.

A painting made about A.D. 1500, and found by British zoologist Hamilton Smith in an antique shop at Augsburg in 1827, represented a rough-haired maneless bull with large coarse head, thick neck, and a small dewlap. Its horns turned forward and outward, and were light colored with black tips. The hair was sooty black with a light ring around the muzzle. (Morse reproduced Baron Herberstein's woodcut after Nehring, in the USDA Bureau of Animal Industry 27th annual report.) Herberstein stated that the urus and European bison lived within historic times. A free translation of Nehring's account of Herberstein's *Rerum Moscovitearum Commentarii,* published in 1549, stated:

> Of the wild animals in lands belonging to Lithuania, is one which they call "suber." It is called "bison" in Latin, while Germans call it aurochs. Closely related to it is another "lur," or Latin "urus." We Germans call it bisent incorrectly, for its form is that of a wild ox. Its color is nearly black, and a grayish stripe along the back.

Editions of 1551 and 1556 contained pictures of both urus and bison. An edition of 1557 mentioned that forest cattle (*Boves sylvestris*) differed from domestic cattle only in their black color and white stripe along the back. Herberstein went to Moscow several times and saw both urus and bison.

Wrzeeniewski (1878) wrote that these wild cattle lived in the woods of Jakterwka until the seventeenth century. The last known specimen died in 1627 in the Zoological Garden of Count Zamoisky.

Other fossil species related closely to *B. primigenius* are *B. trocheceros*, *B. frontosus*, *B. brachyceros* or *longifrons*, *B. namadicus*, *B. brachycephalus*, and *B. typicus*. Morse believed them so nearly related that some and perhaps nearly all could be varieties descended from it. All species of *Bos* which lived wild in Pliocene and Pleistocene eras in Europe are extinct. Domesticated cattle presumably are descended mainly from *B. primigenius*, *B. longifrons*, *B. frontosus*, and *B. trocheceros*.

The taurine group is represented also by *B. taurus mauritanicus* Thomas, probably identical with *B. episthonomus* of Pomel, in Algeria and Tunis until historic times. This may be a variety of *B. primigenius*, distinguished by a shorter forehead, horns curved more downward and less forward, with larger and more slender legs. *B. indicus* and others were in Asia and parts of Africa. These zebu cattle have been distributed widely in the warm zones.

Skulls of various species of genus *Bos* differ in size and shape. Measurements of typical skulls of *B. primigenius* and of *B. longifrons* are shown in comparison with those of present dairy breeds in Table 1.2.

REFERENCES

Adametz, L. 1898. Studien uber *Bos* (*brachyceros*) *europaeus*, die wilde Stammform der Brachyceros-Rassen des europaischen Hausrindes. *Z. Landwirtsch.* 46:269–320.

Arendander, E. D. 1898. Studien uber das ungehornte Rindvieh im nordlichen Europa unser besonderer Beruchsichtigung der nordschwedischen Fjellrasse, nebst Untersuchingen uber die Ursachen der Hornlosiskeit. *Ber. Physiologisch. Lab. Versuchanstalt Landwirtsch. Inst. Univ. Halle* 2(13):172.

Babington, Charles G. 1864. On a skull of *Bos primigenius* associated with flint implements. *Antiquarian Commun.* 2:285–88.

Beltz, R. 1898. *Bos primigenius* in Mittelalter. *Globus* 73(7):116–17.

Bojanus, Ludwig H. 1827. De Uro Nostrate Eiusque Sceleto Commentatio. *Verhandl. Kaiserlichen Leopoldinisch-Carolinischen Akad. Naturforsch.* 13(2):413–78.

Cabre, Juan. 1915. El rupestra en Espana. *Memoria No. 1.* Madrid.

Curtiss, Garniss H. 1961. A clock for the ages: Potassium-argon. *Natl. Geographic Mag.* 120:590–92.

Dawkins, William B. 1866. On the fossil British oxen. Part 1. Bos urus, Caesar. *Quart. J. Geol. Soc. London* 22:391–402.

Degerbol, M. 1963. Prehistoric cattle in Denmark and adjacent areas. *Roy. Anthropol. Inst. Occasional Paper 18*, pp. 68–79.

Duerst, J. U. 1908. Animal remains from the excavations at Anau, and the horse of Anau in its relation to the races of domestic horses. *Explorations in Turkestan. Exposition of 1904.* Vol. 2. Carnegie Inst., Washington, D.C. Pp. 341–44.

Ewart, J. Cossar. 1925. The origin of cattle. *Proc. Scottish Cattle Breeding Conf.* Oliver & Boyd, London. Pp. 1–46.

Falconer, Hugh. 1859. Descriptive catalogue of the fossil remains of Vertebrata from the Siwalik Hills, the Narbudda, Perim Island, etc., in the Museum of the Asiatic Society of Bengal. Calcutta.

Fredsjo, A., S. Janson, and C. A. Moberg. 1956. *Hallristningas i Sverige.* Oskarshamns-Bladets Boktryckeri.

Herberstein, Sigmund. 1549. *Rerum Moscoviticarum Commentstii.* Basil. Later eds., 1551 and 1556.

Hughes, T. McKenny. 1894. The evolution of the British breeds of cattle. *J. Roy. Agr. Soc. Engl.* 5(ser. 3):561–63.

James, E. O. 1927. *The Stone Age.* Sheldon Press, London.

Keller, Conrad. 1902. *Die Abstammung der Haustiere.* Zurich.

Klindt-Jensen, Ole. 1957. *Denmark before the Vikings.* Praeger, New York.

Lydekker, Richard. 1898. *Wild oxen, sheep, and goats of all lands, living and extinct.* London.

Morse, E. W. 1910. The ancestry of domesticated cattle. USDA *Bur. Animal Ind. 27th Ann. Rept.*, pp. 187–239.

Nilsson, Sven. 1849. On the extinct and existing bovine animals of Scandinavia. *Ann. Mag. Nat. Hist.* 4(ser. 2):256–69.

Owen, Richard. 1860. *Paleontology, or a systematic survey of extinct animals and their geological relations.* Edinburgh.

Piggott, Stuart. 1961. *The dawn of civilization. The first world survey of human cultures in early times.* McGraw Hill, New York.

Rutimeyer, L. 1862. Die Fauna der Pfahlbauten in der Schweiz. Neue Denkschriften der allgemeinen. *Schweiz. Ges. gesamten Naturw.* 19:1–248.

Werner, Hugo. 1902 *Die Rinderzucht.* 2nd ed. Berlin.

Whitehead, G. Kenneth. 1953. *The ancient White Cattle of Britain and their descendants.* Faber & Faber, London.

Wilson, James. 1909. *The evolution of British cattle and the fashioning of breeds.* Vinton & Co., London.

Windels, F. 1952. *Four hundred centuries of cave art.* (Foreword by Abbe Breuil.) Montignac, France.

Worsaae, J. J. A., and William J. Thomas. 1849. *The primeval antiquities of Denmark.* J. H. Parker, London.

Wrzesnioski, August. 1878. Studien zur Geschichte des polnischen Tur (Ur, Uru, Bos primigenius Bojanus). *Z. Wiss. Zool.* 30:493–555.

CHAPTER 2

DOMESTICATION OF CATTLE

Domestication of cattle was man's greatest exploitation of the wild animal kingdom, according to F. E. Zeuner. Much evidence from the later period of prehistory—during, and following Neolithic times—has been gathered during the past century. Evidence has come from excavations, cave paintings, rock engravings, and the study of the origin of Aryan languages. Development of a system of timing by radiocarbon-14 brought some systematization to previously isolated observations. Two factors affect dependability of such time estimates: biological contaminations and the distinction between remains of wild and domesticated oxen in the same region. Since domestication preceded written history, the exact time and place of this event pends further discoveries. Reed believed that *B. longifrons* cattle were domesticated about 6000 B.C., probably in

Headpiece: Vignette of lake-dweller's hut.

21

the Zagros Mountains and their grassy forelands (hilly flank areas), where cereal farming and village settlements had begun.

An advance in civilization was associated with domestication first of the dog as a hunting companion, then of goats and sheep, and later of cattle. Mucke theorized that a primitive people who made little use of hunting weapons had been involved in domesticating animals. Zeuner grouped cattle with crop trespassers. He believed that such proximity was one reason for early domestication. J. U. Duerst concluded from excavations at Anau, a delta-oasis in Turkestan:

> The agricultural stage of human development (crop growing) must also have preceded the state of cattle breeders, but through the accomplished domestication of ruminants, men obtained freedom of motion for traveling with cattle for good pastures, and commenced a nomadic life. This must be the real explanation of the origin of the wandering people, which Mucke cannot explain, and he consequently considers a priori that nomadic peoples were nomadic before the domestication of cattle. . . . Consequently the first domestication of cattle must have been by a settled people such as the Anau-li were. . . . Importation of metals from India came at a later date.

R. Pumpelly believed that wild cattle were driven by thirst during drouth to the better-watered oases. J. U. Duerst excavated mounds of ancient settlements at Anau and found the remains of domesticated cattle were at a higher level than the earlier levels containing wheat and barley. The wild species *B. namadicus* was in the lowest level at Anau. This species was contemporaneous as a wild animal with early man in India during the Old Stone Age. Though hunting weapons were lacking, he found no enclosures for holding cattle.

Shalidar cave and the nearby Zawi Chemi village sites long were occupied. Excavations of the cave floor down 14 meters to limestone bedrock yielded several human skeletons and evidence from four strata of soil (and fallen limerock slabs). The lower part of layer B-1, colored by decayed organic matter (from animal droppings), contained grinding stones to prepare acorns or grass seeds as food, and a hafted stone sickle or cutting tool. This layer was dated by carbon-14 at about 8,650 years ago. The presence of snail shells, and suggested storage pits or basins in the discolored soil, indicated use of gathered and stored foods. Sheep had been domesticated. The next lower soil layer had less color. Flint hunting weapons were present, but there were no hand milling stones, querns, or disintegrating baskets or fabrics.

R. J. Braidwood excavated three sites eastward of the Tigris River, representing different periods. The Palegawra cave yielded many flint blade-tools and some unworked animal bones. Most of the bones were from wild horses, deer, and gazelles, but some were of sheep, goats, and pigs possibly killed by chance among known wild animals of the region. Many fragments of milling stones suggested attempts at reaping wild grains for food, but no grains were found in the cave. At Karim Shahir, a later site, about half the bones were of sheep, goats, and pigs. A mound excavated at Jarmo dated perhaps a thousand years later. Four flint blades of a sickle were found in position, with scattered barley and wheat grains. Many bones were of sheep, goat, pig, dog, and some large cattle. These excavations suggest the progression of herding to keep meat available for food.

Evidence of the earliest recognized domesticated cattle was found at Banahill and on the Diyala plains in northern Iraq. C. A. Reed estimated the time at some 7,000 years ago. Domesticated cattle were known to be at Thessaly, Greece, on a site dated by carbon-14 at 5550 B.C. Excavations at Tall Arpachiyah eastward of the Tigris River revealed that cattle were in domestication there long before 2900 B.C. M. E. L. Mallowan estimated the Tall Halaf culture there at about 4500 B.C. Decorations of pottery showed long-horn cattle. A model head, dug from a stratum almost down to virgin soil, had incurved horns that pointed almost directly forward. The poll was wide, and the forehead was of medium height. Four metal objects that were found included lead and a copper chisel among the pottery and many stone tools. A seashell on the site was at least 1,000 miles from the Indian Ocean.

THE ARYANS

The Aryans appeared first as a hunting people and then as a crop-gathering people. Roots of their language included some of pastoral pursuits. Names of money and booty were derived from words referring to cattle in several languages of Aryan origin. Lord or prince, *Gopatis*, originally meant guardian of the cattle. Words expanded to mean district or country, or even the earth, once meant pasturage. Keary wrote:

> The evidence of language points to the belief that the ancient Aryans had only made some beginnings of agriculture . . . for among the words common to the whole Aryan race there were very few connected with farming, whereas their vocabulary is redolent of the herd, the cattle-fold, the herdsman, the milking-time. Even the word daughter (Greek—*Thurster*; Sanskrit—*duhitar*) means the milker and that seems to throw back the practice of milking to a very remote antiquity.

The Aryan branch who wrote Sanskrit, according to Sayce, were

> nomad herdsmen, living in hovels . . . which could be erected in a few hours, and left again as the cattle moved into higher ground around the approach of spring, or descended into the valley when winter approached. . . . Cattle, sheep, goats and swine were all kept; the dog had been domesticated, and in all probability the horse.

The Parsees, who followed the religion of Zoroaster, possessed as their Bible and prayer book the Avesta or Zend-Avesta, which is comprised of several parts. The "songs of praise" paid reverence to the ox.

> In the ox is our strength,
> in the ox is our speech,
> in the ox is our victory,
> in the ox is our nourishment,
> in the ox is our clothings,
> in the ox is our agriculture,
> which furnishes to us food.

The Aryan diety Indra was spoken of as a bull in the Vedic

hymns; the clouds still more commonly were the cows of Indra, and the rain their milk. The wicked *Panis* (evil beings of fog and mist) were mentioned in the Vedas as stealing the cows from the fields and hiding them in caves, from which they were recovered later.

In Sanskrit, *Gopatis* or patriarch meant lord of the cattle; morning, calling of the cattle; and evening, the milking time. *Pecunia*, Latin for money, was derived from Sanskrit *pecus*, which originally referred to cattle. The English word *fee* was from the Aryan word for cattle. *Owiefech*—Anglo-Saxon for movable property—referred to living cattle, and immovable property was dead cattle. Cattle were the principal medium of barter or exchange.

EARLY ARYAN CATTLE

Cattle were in domestication and were regarded highly many centuries before the first permanent written history of the Aryan race. In early times the Aryans occupied an area north of the Hindu Kush or Caucasus border and west of the Boler Tagh mountain ranges of west central Asia. They possessed cattle, horses, and "little cattle" (goats and/or sheep). Their religion and history were passed down by word of mouth in the form of chants, hymns, and prayers that mentioned their leaders, faith, and problems. Limited moisture and scarcity of arable land eastward of the Caspian Sea and the Sea of Aral led them to develop some irrigation from the Oxus (now the Amu Darya River) and Jaxartes River. Their herds and lands were raided and overrun by tribes from the northward, as related in the first four *Gathas*—Odes to Zarathustra (Zoroaster)—in the Avesta. The Aryans spread out to new lands, taking cattle with them. The Indo-Aryan branch settled in India, and Irano-Aryans migrated into Persia and westward. Writings of the Indo-Aryans are recorded in the Vedic hymns.

The Aryans differed from the Semites of that period, the former having changed to a settled agriculture. Their Turanian neighbors were nomadic tribes whose territory surrounded the Aryans. The early influence of the Turanians disappeared in Europe before the advance of the Celts and other Aryan branches who came westward slowly, bringing domesticated cattle.

INDUS VALLEY CIVILIZATION

Sir John Marshall directed excavations at Mohenjo-Daro in the Indus Valley between 1922 and 1927 in settlements of a pre-Aryan people, since established to date between 3050 to 2550 B.C., down to 1500 B.C. These people used implements of stone, bronze, and copper; they cultivated barley, wheat, date palms, and cotton and had domesticated zebus (humped cattle), short-horned cattle, buffaloes, camels, dogs, elephants, sheep, and swine. Oxen were yoked to wheeled vehicles. Beef, mutton, pork, poultry, fish, and turtles were among their foods. Milk and vegetables were presumed to have been other important parts of their diet.

The city had substantial homes, paved streets, a public bath, and sewers. Remains of humped cattle were abundant at every level. A short-horned species of humpless cattle was represented among the terra cotta intaglios discovered, but none of their bones were identified. The intaglio terra cotta seals unearthed depicted 408 bulls of several species. *B. gaurus* was on 17 seals, and *B. indicus* was portrayed definitely on 27 seals. Many bulls were not humped. Because the side view showed only one horn, short-horned bulls depicted were called "unicorn" by Marshall. Frederichs thought these animals to be the aurochs, or *B. primigenius* and *B. namadicus* species, based on the seal-amulets.

Some of these cattle had excellent conformations; 53 had relatively level rumps, while 328 had definite slope to the rumps. Short sloping rumps are common among humped Indian cattle today (Brahman or zebu cattle). Ernst Mackay mentioned a figurine of a Brahman bull as a fine example of such workmanship. Copper plates bore designs of cattle, one being a zebu or humped. Crocodiles, dogs, elephants, rhinoceros, sheep, and tigers also were represented on the seals.

The city was destroyed and its people killed in the streets by the Aryans when they invaded India from the northwest about 1500 B.C. The language of these pre-Aryan people had not been deciphered when Sir John Marshall's report was published in 1931, and it was still undeciphered in 1964.

The University of Pennsylvania Museum, British Museum, and

Department of Antiquities at Bagdad cooperated in excavating the ancient city of Ur and vicinity. C. Leonard Woolley, who led the expeditions, wrote of this early civilization: "It is not beside the point that Dumusi 'the Shepherd' ranks amongst the kings who reigned before the Flood, or that the traditional title of the Sumarian ruler was Patasi, 'the tenant farmer' of the God; the Al'Ubaid society was one of shepherds, farmers and fishermen, as we can tell from the remains."

The Royal Cemetery was dated between 3500 and 3200 B.C. The golden head of a bull, a silver donkey from the pole of Queen Shubad's chariot, and the "ram caught in the thicket" were objects that dealt with domestic animals. Bulls' heads of copper were found, which had wide polls and incurved horns similar to some British cattle. These objects were used for ornamentation or worship. Ox carts were found in the earliest royal graves at Ur. Oxen were used for plowing and working on the threshing floor. Remains of an ox were attached to the king's wagon in death pit P.G. 1789: "This ox was about the same height as a Chartley bull, a long-horned breed representing approximately the average size of European domesticated cattle." This was about 5,000 years ago. Indian influence was brought about by trade. Woolley wrote: "In the second phase of the Royal Cemetery decadence is visibly setting in. The animal scenes are still there, though with certain modifications—the hill creatures, the spotted leopard and the smooth-horned highland bull have been replaced by the water buffalo, and instead of the naked beardless hunter, comes one wearing the flat cap of the north on the bearded figure of the mythological Gilgamesh."

A temple excavated at Al'Ubaid bore an inscribed tablet to the reign of A-anni-padda, second king of the first dynasty of Ur. A row of copper statues of oxen stood on the floor along the wall. A copper frieze (Fig. 2.1) had a row of oxen in high relief, shown in the act of rising.

Higher up was a second frieze of mosaic, figures in shell or limestone . . . set against a background of blackstone; there are rows of cattle and a fresh version of the familiar scene in which men milk their cows outside the reed-built byre, but here there are also men, clean-shaved priests, who strain the milk and pour it into great stone jars; it is the farm of the goddess Ninkursag, and her priests store the divine milk which was the food of her foster-son the king.

History and tradition state that Ur of the Chaldees was among the oldest cities established by the Sumarians, who had acquired the art of writing by using a metal stylus on clay tablets. A considerable library maintained at Ninevah (Ashurbanipal) in the seventh century before Christ helped to establish time back another 1,500 years. From these reports, the earliest definite date was 3100 B.C.,

FIG. 2.1. An inlay frieze from the temple of Ninkhursag Al'Ubaid, at Ur of the Chaldees, about 2700 B.C., portrays a milking scene and caring for the milk. (Courtesy of the University of Pennsylvania Museum.)

when Mes-Anni-Padda, first king of the first dynasty of Ur, ascended the throne. (Woolley later re-estimated this date at about 2700 B.C.) This people worshipped gods, including the Moon God ("the young bull of heaven"), who was a great landowner. His tenants paid rent in cattle, sheep, goats, barley, oil, rounds of cheese, pots of clarified butter, and bales of wool for which the scribes made duplicate receipts on clay tablets, one of which was filed in the records of the Ziggurat temple. This temple, with a chapel to the moon, antedated the birth of Abraham in the same city by 400 years.

Daily offerings of butter, cheese, and dates to major dieties were recorded on six tablets in the Ningal temple, as reported by H. H.

Figulla. Milk, bread, white beans, flour, honey, and salt were mentioned.

A change in the river channel withdrew irrigation water from the canals about the second century before Christ, and doomed the city. It is of note that cows, goats, and sheep provided milk from which cheese and sour cooking butter were made at that early time.

The Greeks were believed to have appeared in Greece, or at least in Asia Minor, about 1900 B.C., and were probably preceded by the Latin branch of the Aryans, as well as by the Celts in northern Europe. In Greek mythology, Hermes (Mercury) stole the cattle of Apollo that were feeding on the Pierian mountains, and hid them. A Vedic hymn mentioned "those who sleep by the cattle. . . ."

Keary held that:

> Possession of cattle was a guarantee against want, and an inducement to a more regular and orderly mode of living. . . . The importance attached to cattle . . . is evidenced by the frequent use of words in their origin relating to cattle, in all the Aryan languages, to express many ordinary incidents of life. Cattle occupied a prominent place in Aryan mythology (the Vedic hymns), titles of honour, names of divisions of the day, divisions of land, for property and money.

Races of people who used the Sanskrit language were the Iranic (Persians) and Armenians. Races of Aryan stock in Europe included Greeks, Latins, Celts (Gauls and Britons), Teutons, Slavs, Lettics, and Albanians.

Ancestors of the Parsees down to the end of the Sassanian dynasty ruled over the people of Anau. Duerst believed from remains of wild cattle in the lowest excavations at Anau, and of a somewhat different domesticated type at higher levels, that this may have been the region where cattle were first domesticated. An early civilization excavated in the Indus Valley by Marshall possessed domesticated cattle at an early period. The advanced civilization in the Indus Valley may push back the time of domesticated cattle to an early time in this part of Asia.

CATTLE AND DAIRY PRODUCTS IN THE HOLY BIBLE

The Hebrews were an agricultural people owning camels, cattle, horses, and sheep. The Bible contains many references to cattle, butter, cheese, and milk. Cattle were mentioned first in the version of the Creation (Gen. 1:24–26):

> And God said, Let the earth bring forth the living creatures after his kind, cattle, and creeping thing, and beast of the earth after his kind: and it was so.
> And God made the beast of the earth after his kind, and cattle after their kind, and every thing that creepeth upon the earth after his kind: and God saw that it was good.
> And God said, Let us make man in our image, after our likeness: and let them have dominion over the fish of the sea, and over the fowl of the air, and over the cattle, and over all the earth, and over every creeping thing that creepeth upon the earth.

Some Hebrews led a nomadic life about 3875 B.C. (Gen. 4:20): "And Adah bare Jabal: he was the father of such as dwell in tents, and of such as have cattle."

At the time of the flood (2349 B.C.), Noah took pairs of each kind of animal into the ark, and the remainder perished (Gen. 7:23): "And every living substance was destroyed which was upon the face of the ground, both man, and cattle, and the creeping things, and the fowl of the heaven; and they were destroyed from the earth; and Noah only remained alive, and they that were with him in the ark." Four hundred years later (1920 B.C.) the cattle were distinguished by species in Genesis 12:16: "And he entreated Abram well for her sake: and he had sheep, and oxen, and he asses, and manservants, and maidservants, and she asses, and camels."

When Abram, Lot, and their families and their followers went out of Egypt in 1918 B.C. (Gen. 13:2–11):

> And Abram was very rich in cattle, in silver, and in gold. . . .
> And Lot also, which went with Abram, had flocks and herds, and tents. And the land was not able to bear them, that they might dwell together: for their substance was great. . . . And there was a strife between the herdmen of Abram's cattle and the herdmen of Lot's cattle. . . . And Abram said unto Lot,

Let there be no strife, I pray thee, between me and thee, and between my herdmen and thy herdmen; for we be brethren.

Abram suggested that they separate. Lot chose the well-watered plain of Jordan to the east, while Abram went in the opposite direction. Then in Genesis 15:7, God spoke to Abram: "And he said to him, I am the Lord that brought thee out of Ur of Chaldees, to give thee this land to inherit it." Woolley excavated a temple frieze showing a milking scene at Ur, city of Abraham.

The famous narrative of early cattle breeding was the agreement between Laban and Jacob (Gen. 30:28–43) whereby Jacob received all broken-colored animals as pay for herding Laban's cattle. Jacob presented cattle to his brother Esau (Gen. 32:15) and Esau took them into Canaan (Gen. 36:6). Jacob's son Joseph (Gen. 41:17–27) interpreted Pharaoh's dream of seven fat oxen devoured by seven lean oxen as foretelling seven years of famine, against which Pharoah stored grain for this period of adversity. Jacob traded cattle and lands for food in Egypt during the drouth (Gen. 46:6–32). God promised Moses "a land flowing with milk and honey" as a home for his chosen people (Exod. 13:5).

Moses mentioned burned offerings of cattle several times in the book of Leviticus. Moses's scouts reported that the Promised Land was flowing with milk and honey. This description was repeated several times (Num. 13:27; and Deut. 11:9; 27:3; 31:20; and 32:14). The latter stated: "Butter of kine, and milk of sheep, with fat of lambs, and rams of the breed of Bashan, and goats, with the fat of kidneys of wheat; and thou didst drink the pure blood of the grape."

Joshua succeeded Moses as leader in 1451 B.C. Concerning settlement in the land (Josh. 21:2): "And they spake unto them at Shiloh in the land of Canaan, saying, The Lord commanded by the hand of Moses to give us cities to dwell in, with the suburbs thereof for our cattle."

When Jael begged of Sisera in 1296 B.C. (Judg. 4:19): "Give me, I pray thee, a little water to drink; for I am thirsty. And she opened a bottle of milk, and gave him drink." Also "He asked water, and

she gave him milk; she brought forth butter in a lordly dish"
(Judg. 5:25).

The mother of David directed (1 Sam. 17:18): "And carry these
ten cheeses unto the captain of their thousand, and look how thy
brethren fare, and take their pledge." David met and slew the Phil-
istine giant, Goliath, with a smooth pebble from the brook, directed
from his sling.

At a later time (2 Sam. 17:27–29): "And it came to pass, when
David was come to Mahanaim, that Shedi . . . brought beds, and
basins, and earthen vessels, and wheat . . . and parched pulse. And
honey, and butter, and sheep, and cheese of kine, for David, and
for the people that were with him, to eat: for they said, The people
is hungry, and weary, and thirsty, in the wilderness."

In 1014 B.C., Solomon mentioned milk among his valued foods
(Song of Sol. 5:1): "I am come into my garden, my sister, my
spouse: I have gathered my myrrh with my spice; I have eaten my
honeycomb with my honey; I have drunk my wine with my milk:
eat, O friends; drink, yea, drink abundantly, O beloved."

Isaiah (7:32) mentioned in a prosperous time: "And it shall come
to pass, for the abundance of milk that they shall give, he shall eat
butter: for butter and honey shall every one eat that is left in the
land."

Cattle, milk, butter, and cheese were valued highly by the Hebrews from the earliest written history. The word "butter" was
changed by the translators in the new revised version of the Bible
to "curds." However, cheese, butter, and clarified butter oil were
known at an early period.

ROCK PAINTINGS

Childe described the rock paintings in Spain, stating that "on the
cave walls and in adjacent shelters their inhabitants have painted in
a conventional manner wild animals and episodes of the chase, but
also domesticated cattle, sheep, goats, swine and equids, and pastoral scenes and even an agricultural diety holding a sickle; sledges
and, in the north, wheeled carts are also depicted."

Burkitt reported on a group of paintings of the later Aeneolithic

Age (Spanish Group III): "An extremely interesting art-group that occurs in rock shelters belonging to the late Neolithic and Copper Age periods has been studied in the Spanish peninsula. . . . That the folks who made these paintings practiced the domestication of animals is shown by a very charming example found at Las Canferras de Penarrubia in the Sierra Morena of an animal led by a halter."

M. de Morgan found that oxen were used to till the soil in Egypt at an early date. He also found enclosures where the animals were penned at night. Egyptian monuments indicate that humped cattle —*B. indicus*—were in domestication as early as the twelfth dynasty, 2100 B.C. In Mesopotamia and Arabia, cattle were in domestication at the same time as in Egypt. Regular trade routes passed through the region. Adametz believed in 1920 that the time of domestication in Egypt had not been determined. He thought the earliest Egyptian "Hamiten" race was descended from *B. primigenius* although their withers were developed strongly. Zebus were in Babylonia about 2000 B.C., and were taken to Arabia from there. The horns of early Egyptian cattle were more slender than were those of zebus. Since the oldest goats and sheep came to Egypt from Babylonia, Adametz believed that cattle had been brought over the same route across southern Arabia.

Rock paintings in southeastern Libya (Figs. 2.2 and 2.3) show cattle in domestication at an early period.

Lieutenant Brennans of the French Camel Corps observed many paintings in 1933 and later, on rock walls of overhanging cliffs and in caverns (once human shelters) in the Tassili of the Ajjers. The region is an eroded sandstone plateau in the Sahara Desert southwest of Tripoli. Henri Lhote and associates transcribed these paintings for the Museum of Man in Paris. Few wild oxen were pictured. One fresco showed herdsmen defending their cattle against raiders with bows and arrows. The largest herd numbered 65 animals, accompanied by herdsmen. They were portrayed in red, brown, and yellow colors, some with white markings. Many had wide, upturned horns similar to those in Egyptian sculptures. Their horns were longer than those of early *B. longifrons* in Europe. A domesticated dog sometimes was pictured. The "Bovidean" period indicated a migration of Neolithic man into the region westward of Egypt at around 3530 B.C.

DOMESTICATED CATTLE

Many writers of prehistory pointed to evidence that the Aryan people introduced domesticated cattle from western Asia into Europe. Remains of *B. longifrons* predominated in relic beds of the early Neolithic settlements. The oldest post-glacial settlers of the fertile Danube Valley and its tributaries possessed short-horned cattle, turbary sheep, and a few pigs. They ascended the valley westward,

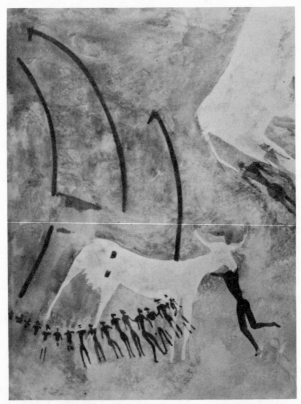

FIG. 2.2. Rock paintings discovered at Ain Dua in southeastern Libya by the Frobenius-Fox expedition indicate that the cow was domesticated more than 6,000 years ago. (From *Milk Industry News*, Volume 2, Number 1, 1938. Courtesy of Milk Industry Foundation.)

where one site was timed by radiocarbon-14 at 4000 B.C. They spread southward into Switzerland and Italy, westward down the Rhine, and across Belgium and France to the Channel and British Isles (Fig. 2.4).

Migrations were traced by the peculiarly shaped polished flint implements and a crude beaker-type of clay pottery, as they moved westward over the steppes of southern Russia, into Hungary, Galicia, Silesia, the Rhineland, Belgium, Normandy, and the Channel and British Isles. These people chose to settle on loess soils and near streams. They were agriculturists and fishermen as well as owners of flocks and herds. Barley, flax, millet, and wheat were introduced in their migrations to newer lands. Childe, in reviewing evidence discovered in many locations, mentioned that inhabitants of the Grecian mainland lived in island villages, hunted deer and other wild life, and possessed domesticated cattle, sheep, and swine. The polished shoe-last celt, a typical implement, was really a hoe used in Neolithic agriculture in the Danube Valley. It was used as a weapon as well.

Another people invaded Eastern Thessaly in the second period (2600–2499 B.C.). They made clay figurines, and added models of cattle to the small human images.

FIG. 2.3. Paintings of cows in what is presently the Libyan desert were made about 4000 B.C., or earlier. This one pictures ancient tribesmen worshiping a cow. (Courtesy of Milk Industry Foundation.)

Childe presumed that the megalith builders introduced domesticated cattle into England, since bones of cattle, sheep, and swine have been recovered from burial barrows. Cultivated grains have not been connected with this people in Britain.

DISCOVERY OF LAKE DWELLINGS IN SWITZERLAND

The winter of 1853–54 was dry and cold. Little snow fell in the Alps, and the water level in many lakes became the lowest on record. Local people built a wall along the new waterline on the edge of Lake Zurich between Ober Meilen and Dollikon. While removing mud from the lake bottom onto the reclaimed area, they found quantities of piling, animal horns, and some implements. A. Aeppli of Meilen believed these specimens to be of human workmanship, and called Dr. Ferdinand Keller's attention to them. Thus the Swiss lake dwellings were recognized.

Over 200 lake dwellings have been found since in Switzerland, mainly representing the Stone and Bronze Ages, but a few settle-

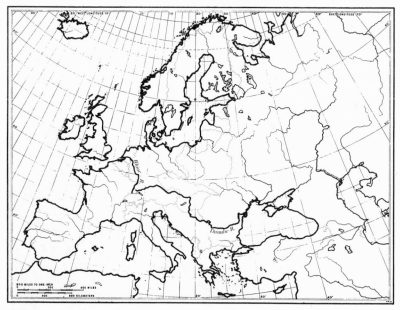

FIG. 2.4. Neolithic man migrated from western Asia up the Danube River and down the Rhine, bringing *Bos longifrons* Owen as a domesticated animal. (Copyrighted background map by permission of Rand McNally Company.)

ments continued in Roman times. The dwellings were built of wattle and clay daub, on platforms erected on poles driven into the lake bottom. Quartz and flint arrows, stone axes and scrapers, crude pottery, bone and wooden weapons, and pieces of flax fabrics were found at Robenhausen and Wangen.

L. Rutimeyer identified remains of 10 fishes, 4 reptiles, 26 birds, and 30 quadrupeds—dog, goat, horse, pig, sheep, and two species of oxen. Bones of the stag and ox exceeded those of other species combined. The stag outnumbered the ox in specimens from the earlier settlements at Moosseedorf, Robenhausen, and Wauwyl. The

TABLE 2.1

RELATIVE FREQUENCY OF MAMMALIAN REMAINS FROM SWISS LAKE DWELLINGS

	Moossee-dorf[x]	Wauwyl[x]	Roben-hausen[x]	Wangen	Meilen	Concise	Bienne
Bos primigenius	2	2	3	1		2	
Bos bison[a]	1	1	4	?			
Bos taurus primigenius	2	?	5	?	2	5	2
Bos taurus brachyceros	5	5	2	5	5	2	5
Bos taurus frontosus		1				2	2

Key:
  x—began in the Stone Age.
  1—denotes a single specimen.
  2—indicates remains of several individuals were recovered.
  3—specimens were common.
  4—specimens very common.
  5—specimens present in great numbers.
a. *Bos bison* must be an error of identification. *Bison bonasus* still lives in woods of Poland, while *Bison latifrons* has become extinct.

reverse was true in later settlements on the western lakes—Wangen and Meilen. Bones of swine were next in abundance. Sheep remains increased in late settlements. Bear, wolf, urus, bison, and elk appeared to have been taken occasionally. Rutimeyer gave Sir John Lubbock a table of animal remains recovered from the lake bottoms, part of which are listed in Table 2.1.

Rutimeyer used *B. brachyceros* as the name for *B. longifrons*. *B. bison*, the present American bison, was absent; *Bison bonasus* still lives in woods in Poland, while *Bison latifrons* is extinct. Horse remains were scarce in lake dwellings before the Bronze Age. *B. tro-choceros* was found at Concise. It had not been identified in earlier pileworks. He believed these specimens of *B. primigenius* and *B. bison* (or europas) were wild, and that the lake dwellers possessed four principal species of domesticated oxen. The first of these in the early pileworks resembled the urus or *B. primigenius,* and no doubt was descended from it. This species now is represented best by cattle in Friesland, Jutland, and Holstein. The second, *B. trocho-*

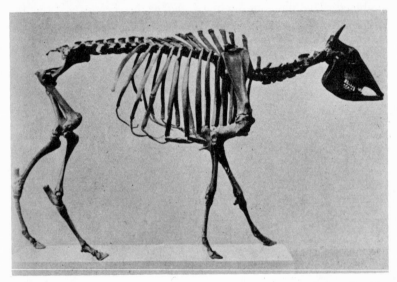

FIG. 2.5. This skeleton of *Bos longifrons* Owen was recovered from the lake-dweller site in a peat bog at Schussenried. (From the museum at Stuttgart. Photograph by J. U. Duerst.)

ceros, resembled a fossil form observed in the diluvium of Arezza and Siena, described by F. von Meyer. It had not been found in the Stone Age settlements. Rutimeyer regarded it as scarcely distinguishable from the urus and observed that its peculiarities were developed principally in females. The third, *B. frontosus,* occurred sparingly in older pileworks, became more frequent in Bronze Age villages, and prevails now in northern Switzerland as the Simmentaler breed. Rutimeyer considered the latter also derived from the urus. The fourth was *B. longifrons,* or *brachyceros* (short horns), as shown in Figure 2.5. *Brachyceros* had been applied previously by

Dr. Gray to a different African ox. *B. longifrons* was abundant in the pileworks. It was not wild in Europe. The brown cow of Switzerland descended mainly from it.

The food of the pileworks dwellers included six-row barley, three species of wheat, and two species of millet. Oats were brought during the Bronze Age. Wild fruits, fish, and flesh of wild and domesticated animals were used. Lubbock believed that milk was an important item of their diet. Pottery colanders, to separate curds from whey, were found in dredgings of the Swiss lake dwellers.

Rutimeyer commented on *B. longifrons* bones from these dwellings: "The race which clearly predominated through the whole Stone Age and was found . . . in the formations which we . . . reckon among the oldest in Wangen and Moosseedorf, I may safely call the Peat Race, or the Peat Cow. Its chief characteristics . . . apart from the skull, is the small length and height of its body, and the exceptionally short but remarkably fine and delicate limbs."

Richard Owen and McKenney Hughes also commented on the small size of this species. James Wilson concluded that the Celtic (British) strain of *B. longifrons* probably was predominantly black in color.

Nilsson described *B. longifrons* in Sweden as

the smallest of all the ox tribe which lived in the wild state in our portion of the globe. To judge from the skeleton, it was 5 feet 4 inches from the nape to the end of the rump bone, the head about 1 foot 4 inches, so that the entire length must have been 6 feet 8 inches. From the slender shape of its bones, its body must rather have resembled a deer than our common tame ox [of Sweden?]; its legs at the extremities are certainly somewhat shorter and also thinner than those of a crown deer (full antlered red deer).

Prehistoric People in Italy

Canon Isaac Taylor wrote that prehistoric peoples invaded Italy in succession, and brought new cultures to the Po Valley. The Iberian savages came as hunters, lived in caves, and possessed no pottery. They were followed by the Umbro-Latin race who built huts and pile dwellings. The latter race possessed cattle and sheep, made canoes, invented the wagon, and gradually acquired knowledge of bronze. The Latins spoke an Aryan language, and reached Europe probably not more than 6,000 or 7,000 years ago, with domesticated dogs, cattle, and sheep. The Latins erected pile dwellings in the lakes of northern Italy, Germany, and Switzerland. Some lake dwellings were occupied continuously from the Stone Age, through the Bronze Age, and into the early Iron Age. The people stored acorns, hazelnuts, and water chestnuts. Later they began to grow barley, wheat, and flax. They learned to spin and weave fabrics, tan leather, and even to make boats.

Some small lakes in northern Italy became peat bogs. People digging peat from such a moor at Mercurage, near Arona, discovered successive layers of such a settlement. The deep layer yielded bones of the wild boar and stag, with a few of domesticated cattle and sheep. There were stores of hazelnuts, acorns, and water chestnuts along with flint tools and crude pottery, but no metal. The upper relic bed contained bones of the ox and sheep. The settlement was destroyed before the agricultural stage had been reached.

As population increased and spread, villages were erected on dry land, the remains of which formed small knolls or *terre mare* (marl beds), the successive strata of debris extending over parts of the Stone and Bronze Ages. Nearly 100 such mounds were known, from which have come such objects as strainers for preparing honey, hand mills for grinding grains, and dishes perforated with holes "which were probably used for making cheese." No iron, gold, silver, or glass were found. At some period in the Bronze Age, the Umbrians were overwhelmed by an invasion of the Etruscans from the north. All of their settlements were destroyed before the advent of the Iron Age, which probably commenced in Italy about the ninth or tenth century before Christ.

Klatt summarized 325 references on various kinds of domesticated animals. He concluded that the polled character could be a mutation among domesticated cattle, and that differences in dimensions of skulls, horn cores, and other bones might have resulted from selections of individual breeding animals.

REFERENCES

Braidwood, L. 1959. *Digging beyond the Tigris.* Abelard-Schuman, New York. Pp. 261–81.

Braidwood, R. J. 1958. Near Eastern prehistory. *Science* 127:1419–30.

Burkitt, M. C. 1921. *Prehistory.* Cambridge Univ. Press, Cambridge.

———. 1949. *The Old Stone Age, a study of paleolithic times.* 2nd ed. Cambridge Univ. Press, Cambridge.

Childe, V. G. 1925. *The dawn of European civilization.* (Rev. ed., 1939.) London.

Curwen, E. Cecil. 1931–32. Prehistoric agriculture in Britain. *J. Bath and West and South Counties Soc.,* 6(ser. 6):7–20.

———. 1938. Early agriculture in Denmark. *Antiquities* 12(46):135–53.

Dawkins, W. Boyd. 1880. *Early man in Britain.* Macmillan, London.

Duerst, J. U. 1899. *Die Rinder von Babylonien, Assurien und Agypten und ihr Zusammen hang mit den Rindern der Alten Welt.* Berlin.

Dunbar, Carl O. 1960. *Historical geology.* 2nd ed. Wiley, New York. Pp. 415–16.

Figulla, H. H. 1953. Accounts concerning allocations of provisions for offerings in the Ningal-Temple at Ur. *Iraq* 15:171–92.

Hilzheimer, Max. 1933. Unser Wissen von der Entwicklung der Haustierwelt Mitteleuropas. *Vierteljahrsch. Naturforsch. Ges.* 78:218–24.

Isaac, Erich. 1962. On the domestication of cattle. *Science* 137:195–204.

James, E. O. 1927. *The Stone Age.* Sheldon Press, London.

Keary, C. F. 1912. *The dawn of history and introduction to prehistoric study.* Rev. ed. Scribner & Sons, New York.

Keller, Ferdinand. 1878. *The lake dwellings of Switzerland and other parts of Europe.* Trans. by J. E. Lee. 2nd ed. Longmans, Green & Co., London.

Klatt, B. 1927. Entstehung der Haustiere. *Handbuch der Vererbungswissenschaft.* Vol. 3. Berlin.

Lhote, Henri. 1959. *A la Decouverte des Fresques du Tassil* [The search for the Tasseli frescoes]. Trans. by Alan Houghton Broderick. Dutton, New York.

Lubbock, Sir John. 1889. *The origin of civilization and the primitive conditions of man.* 5th ed. Longmans, Green & Co., London.

Lydekker, Richard, et al. 1901. *The new natural history.* Vol. 2. Mammals. Merrill & Baker, New York.

Mackay, Ernst. 1935. *The Indus civilization.* Luzac, London.

Mallowan, M. E. L. *Twenty-five years of Mesopotamian discoveries (1932–1956).* Pamphlet Collection, Univ. of Minnesota Library.

Mallowan, M. E. L., and J. G. Rose. 1935. Excavations at Tall Arpachiyah, 1933. *Iraq* 2:104, 105–78.

Marshall, Sir John. 1931. *Mohenjodaro and the Indus civilization.* London.

Mucke, J. E. 1898. *Urgeschichte des Ackerbaues und der Viehzucht.* Greifwald.

Myers, J. L. 1911. *The dawn of history.*

Natl. Geographic Soc. 1967. *Everyday life in Bible times.* Supplementary maps.

Obermaier, H. 1916. *Fossil man in Spain. Memoir No. 9.* Yale Univ. Press, New Haven, Conn.

Pike, Albert. 1924. *Irano-Aryan faith and doctrines as contained in the Zend-Avesta (1874).* Standard Printing Co., Louisville, Ky.

———. 1930. *Indo-Aryan dieties and worship, as contained in Rig Veda (1872).* Standard Printing Co., Louisville, Ky.

———. 1930. *Lectures of the Arya (1873).* Standard Printing Co., Louisville, Ky.

Pumpelly, R. 1908. *Explorations in Turkestan. Expedition of 1904.* Carnegie Inst., Washington, D.C.

Reed, Charles A. 1959. Animal domestication in the prehistoric Near East. *Science* 130:1629–39.

Sayce, A. H. 1891. The primitive home of the Aryans. *Smithsonian Inst. Rept. 1890*, pp. 475–87.

Solecki, Ralph S. 1963. Prehistory in Shanidar Valley in northern Iraq. *Science* 139:179–93.

Wheeler, Sir R. E. Mortimer. 1947. *In ancient India, No. 3.* Praeger, New York.

———. 1959. *Early India and Pakistan.* Praeger, New York.

Woolley, Sir C. L. n.d. *The development of Sumerian art. Ur excavations.* 2 vols. Clarendon Press, Oxford.

———. 1958. *History unearthed.* E. Benn, London.

Zeuner, F. E. 1963. *A history of domesticated animals.* Hutchinson, London.

CHAPTER 3

THE BRONZE AGE AND EARLY HISTORY

CULTURAL STAGES spread slowly with waves of migration from the East, or as commerce increased along channels of trade and barter. As the culture of the Danubian settlers moved slowly during the Neolithic period (New Stone Age), so the Bronze Age cultures progressed as tribes that possessed improved tools and weapons of bronze migrated. Men of the Bronze Age were warriors and agriculturists, moving onward to new fields with their families and domesticated animals.

Copper and gold were among the early metals used by man. In addition to tools and weapons, these metals were shaped into ornaments and objects of worship. The bull's head from the Royal Cemetery at Ur was among the finest specimens. The earliest bronze implement found at a campsite along the Danube River migration route was timed by radiocarbon-14 at about 2300 B.C. The first bronze alloy, which consisted of 1 part of tin to about 3 to 9 parts of copper, was harder than copper. Perhaps this was discovered during the Indus civilization, in northern Persia (Iraq) or in western Asia. Bronze was known to the early Chaldeans and Egyptians, and there were mines of copper in Israel.

BRONZE AGE ROCK ENGRAVINGS

Bronze was brought westward about 800 B.C. and was found along with flint, stone, and bone implements in the upper strata of many Swiss lake dwellings. Bronze pieces included perforated dishes believed to have been used in draining whey from curds in making cheese. Such dishes have been found also in Bronze Age sites in Italy.

Information is limited on the status of cattle in Europe during the early Bronze Age. Neolithic artists drilled plowing scenes dot-by-dot on the schist rocks at high altitudes near Monte Bego in the Maritime Alps. These engravings showed bulls with exaggerated horns. Sometimes one, two, and even four or five oxen were pulling a wooden plow, guided by one and sometimes two men. These scenes were viewed from above (Fig. 3.1). Similar rock engravings of a plowing scene occur near Tanum, Sweden, in which oxen were viewed from the side (Fig. 3.2).

Two gold cups (Fig. 3.3) found in a grave at Vaphio near Sparta in 1889 were dated by Helen Hardner at 1600 to 1500 B.C., but formerly they were thought to have been made by an artist of the Mycean period about 150 B.C. A hunting scene on one cup showed three wild cattle, one tangled in a net. On the other, a man held a wild ox by a thong fastened about the left hind leg. Three other oxen appeared quiet and domesticated. These scenes presumably represented wild bulls, capture, and domestication. The oxen on the cups were thought to be likenesses of European uruses. The cups are displayed in the National Archeological Museum in Athens.

VOYAGERS IN BRITAIN

The Bronze Age culture was brought to the British Isles by immigrants from the Rhineland. They brought an improved agriculture, and mined Cornish deposits of tin for bronze manufacture.

FIG. 3.1. Engraving of a ploughing scene at Fontanalba in the Maritime Alps during the Bronze Age. (Photograph from the Association for the Study, Protection and Illustration of the Valley of Marvels. Courtesy of Secretary General Henry Musso.)

FIG. 3.2. A ploughing scene with oxen was drilled dot-by-dot into gray and pink granite rock near Tanum, Sweden during the Bronze Age. (Courtesy of Dr. Åke Fredsjø, Keeper of Antiquities.)

J. Cossar Ewart (University of Edinburgh) studied fossil remains of cattle and concluded:

> The examination of Neolithic and Bronze Age deposits proves that for about 18,000 years there have been living in Europe three kinds of tame cattle, viz: polled cattle (Swedish Fjall Breed, after Arenander), cattle with short horns and cattle with long horns. . . .
>
> There is no evidence of existence of a wild ox of the *longifrons* or *brachyceros* type. Writers of cattle with rare exceptions allege that the long-horned cattle of Western Europe are mainly, if not entirely, descended from the *Bos primigenius,* a variety of which (the urus or aurochs) Caesar came across in the Hercynian forest.

Importance of cattle during the Bronze Age was signified by worship of them as idols, mentioned in early Biblical history (Exod. 32:4). Religious life and worship of early Britons was in the hands of Druid priests. One religious ceremony consisted of cutting mistletoe from the sacred oak and subsequently sacrificing two white bulls fastened by their horns to the sacred trees. The ceremony was followed by feasting and rejoicing.

Early Phoenician and Greek voyagers went westward to Spain and even to Britain in search of metals. The Carthaginians sent their captain Himilco on a voyage that took him along the coast of Britain some time between 570 B.C. and 470 B.C. The poem of "Festus Avienus" mentioned Himilco as the discoverer of Land's End in Cornwall, England. He told of tin and lead, and wrote of native Britons: "They migrate the sea in barks built, not of pine or oak, but strange to say, made of skins and leather."

Two branches of the Celtic race reached England, the Goidels arriving first. They found the Iverians (Druids) in possession, and amalgamated with them as a people. The Brythons, who used woven cloth for clothing, also settled in the British Isles several centuries later. Windle stated that "During their occupancy in the fourth century before Christ . . . a syndicate of merchants of Massilia [modern Marseilles] fitted out an expedition . . . under a learned Greek mathematician, Pytheas, who twice visited Britain. He mentioned that 'the natives collect the sheaves in great barns,

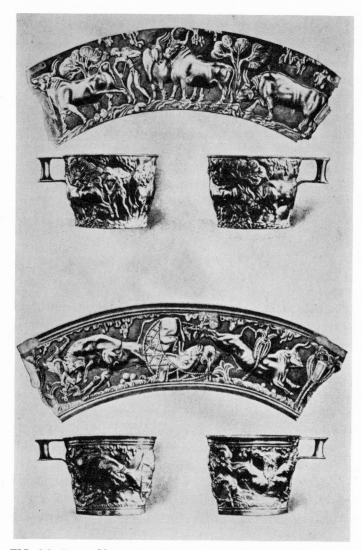

FIG. 3.3. Two gold cups found in the beehive grave at Vaphio near Sparta, Greece in 1889. These figures are believed to illustrate the capture, taming, and domestication of the wild ox. Workmanship is that of about 1500 B.C.

and thrash out their corn [grain] there.'" Lake dwellings were in use, both crannogs and pile dwellings. The Brythons cultivated wheat, and possessed cattle and sheep.

Strabo, the Greek geographer, described the people of Cassiterides (islands of tin), stating: "Walking with staves, and bearded like goats, they subsist by their cattle, leading for the most part a wandering life. And having metal of tin and lead, these and skins they barter with the merchants for earthenware, and salt, and brazen vessels. Formerly the Phoenicians alone carried on this traffic by Gadeira [Gibralter], concealing the passage from every one." Strabo mentioned that the Gauls lived mainly on milk and all kinds of flesh, especially that of swine.

ROMANS IN BRITAIN

The Romans under Caius Julius Caesar invaded Britain in 55 B.C. The Britons drove their cattle inland, attempting to leave the invaders without food. These early domesticated cattle were said by McKenny Hughes and others to have been *B. longifrons* of the Neolithic period, since *B. primigenius* had been destroyed as a wild species in Britain before the Bronze Age.

The Britons claimed to have migrated from Belgium, which also was inhabited by Celts following the Druid religion. The Britons were cultivators who had many cattle, treated their land with manure, and used the plow to produce grain and other crops. The Roman invasion was followed by continuous occupation by armies, retainers, and settlers who introduced horticultural and agricultural plants and brought some cattle. The last Roman garrison was withdrawn in A.D. 142.

McKenny Hughes studied remains of cattle in Great Britain, especially from the peat near Reach Lode north of Cambridge. The latter cattle had smooth polls, long, straight horn cores, and compared favorably with cattle pictured on Roman coins and early relics of Asia and Egypt. The new kind, supposedly introduced during Roman occupation, modified the small Celtic short-horns and contributed to the ancestors of later improved breeds. Ewart mentioned that hornless cattle skulls discovered at Newstead, an old Roman center in Berwickshire, had Roman origins. He also found a modern type of *B. acutifrons* among them. James Wilson and others believed that the Park Cattle of England descended in part from large white cattle introduced during Roman occupation.

INVASIONS FROM NORTHERN EUROPE

Angles from Schleswig-Holstein in southern Denmark, Jutes from Jutland, and Saxons from the northern Netherlands introduced some cattle during the fifth century. Norsemen brought polled cattle of light dun color from the Scandinavian peninsula to coastal settlements.

About 40 years after the Roman forces withdrew, the Saxons were invited to repel invasion by the Picts and Scots from the north A.D. 447–449. They became aggressive settlers and landowners. Windle described a landowner's estate under the Saxon occupancy, with its rampart, ditch, and a palisade or thick hedge on the former. The estate lands were tilled by villeins and theows (slaves), or rented out. The farming operations, according to Windle, were as follows:

the communal officers took charge of the village ploughs and the beasts which drew them were the property of the villeins, the size of whose holdings determined the number of animals required. . . . The smallest holding of land . . . was a bovate . . . this word derived from the Latin *bos*, an ox. . . . Double this amount was a virgate, the normal holding of the villein, who must supply two oxen to the team. (A hide or virgate equals 4 virgates, or a full team of eight oxen.)

It will now sum up these facts as to the village if we take one example of a manor—that of Westminster. THE DOMES-DAY BOOK records that the villa ubi sedet Ecclesia Sci Petri (the Abbey) the abbot of the same place holdeth 13½ hides. There is land for 11 plough teams (8 oxen each). To the demesne belong 9 hides and 1 virgate, and there are 4 plough teams. The villeins have 9 plough teams, and one more might be made. There are:

9 villani with a virgate each;
1 villanus with a hide; (containing 4 virgates);
9 villani with a half virgate each;
1 cottier with 5 acres;
41 cottiers rendering a shilling each for their gardens;

> There are meadows for 11 plough teams;
> Pasture for cattle of the village;
> Woods for 100 pigs.

There are 25 houses of the abbot's soldiers and of other men who render 8s. per annum or £10 in all. In the same villa Rainardus holds 3 hides of the abbot. There is land for two plough teams, and they are there in demesne, and one cottier. Wood for 100 pigs. Pasture for cattle. Four arpents of vinyard newly planted. All of these are worth 60s. This land belonged to the Church of St. Peter.

MARCO POLO'S OBSERVATIONS

The dependence of people on cattle was observed by the Venetian Marco Polo (1254–1324). He described *B. indicus* in Persia thus:

> The beasts also are peculiar; and I will tell you of their oxen. They are very large, and all over white as snow, the hair is very short and smooth, which is owing to the heat of the country. The horns are short and thick, not sharp in the point; and between the shoulders they have a hump some two palms high. There are no handsomer creatures in the world. And when they have to be loaded they kneel like the camel; once the load is adjusted, they rise. Their load is a heavy one, for they are very strong animals.

He wrote concerning coastal India: "The food of the people is flesh, and milk, and rice. The people of the province do not kill animals nor spill blood; so if they want to eat meat, they get the Saracens who live among them to play the butcher."

Tibetans endowed their wives with cattle, slaves, and money according to their ability. Tartars moved with the season to find pasturage, living on milk and meat which their herds supplied, and on wild game. Koumis (a fermented beverage) was made from mare's and cow's milk. After making butter, buttermilk was dried in the sun. Polo wrote, "They also have milk dried into a kind of paste to carry with them; and when they need food they put this in water, and beat it up until it dissolves, and they drink it." Their animals were branded, except sheep and goats, which were herded.

Many people in Kublai Khan's domain in northwestern China used flesh and milk as food. On his return journey, Marco Polo noted at the Port of Aden that livestock subsisted in part upon small fish, either fresh or dried.

REFERENCES

Burkitt, M. C. 1925. *Prehistory.* 2nd ed. Cambridge Univ. Press, Cambridge.

Childe, V. Gordon. 1935. *New light on the most ancient East.* K. Paul, Trench, Trubert & Co., London.

Degerbol, M. 1963. Prehistoric cattle in Denmark and adjacent areas. *Roy. Anthropol. Inst. Occasional Paper 18,* pp. 68–79.

Dickson, Adam. 1782. *The husbandry of the ancients.* 2 vols. Edinburgh.

Ewart, J. Cossar. 1925. The origin of cattle. Cattle breeding. *Proc. Scottish Cattle Breeding Conf.* Owen & Boyd, Edinburgh. Pp. 1–46.

Hughes, T. McKenny. 1894. The evolution of the British breeds of cattle. *J. Roy. Agr. Soc. Engl.* 5(ser. 3):561–63.

————. 1896. On the more ancient breeds of cattle which have been recognized in the British Isles in successive periods, and their relation to other archeological and historical discoveries. *Archaeologia* 55:125–58.

Munro, Robert, et al. 1895. *The British lake village near Glastonbury.* Taunton.

Polo, Marco. 1875. *The book of Marco Polo, the Venetian concerning the kingdoms and marvels of the East.* Trans. and ed. by Col. Henry Yule. 2nd ed. Directors of Old South Works, London.

Reed, C. A. 1961. Osteological evidence for prehistoric domestication in southwestern Asia. *Z. Tierzucht. Zuchtungsbiol.* 76:31–38.

Taylor, Canon Isaac. 1891. The prehistoric races of Italy. *Smithsonian Inst. Rept. 1890,* pp. 489–98.

Wilson, James. 1909. *The evolution of British cattle.* Vinton & Co., London.

Windle, Bertram C. A. 1897. *Life in early Britain.* Putnam, New York.

LAC, CASEVS, BVTYRVM.

CHAPTER 4

CATTLE IMPROVEMENT BEGINS

A GAP IN KNOWLEDGE of cattle improvement extends into the thirteenth century (before 1253) when Walter of Henley, bailiff of Christchurch manor (Canterbury?) wrote from experience. Concerning the area in estate and in pasture: "And if you have land on which you can have cattle, take pains to stock it as the land requires. And know for truth if you are duly stocked, and your cattle well guarded and managed, it should yield three times the land by the extent."

He preferred oxen to horses because of cost, and oxen could be fattened for slaughter in the end. Further:

> Sort out your cattle once a year between Easter and Whitsuntide—that is to say, oxen, cows, and herds—and let those that are not to be kept put to fatten; if you lay out money to

Headpiece: A sixteenth-century dairy farm scene in Europe. (From Mattioli, 1598.)

fatten them with grass you will gain. And know for truth that bad beasts cost more than good. Why? I will tell you. If it be a draught beast he must be more thought of and more spared, and because he is spared the others are burdened for his lack. And if you must buy cattle buy them between Easter and Whitsuntide, for then beasts are spare and cheap. . . . It is well to know how one ought to keep cattle, to teach your people, for when they see that you understand it they will take the more pains to do well. . . . and let your cows have enough feed, that the milk may not be lessened.

How much milk your cow should yield.

If your cows were sorted out, so that the bad were taken away, and your cows fed in pasture or salt marsh, then ought two cows to give a wey of cheese and a half gallon of butter a week. And if they are fed in pasture of wood, or in meadows after mowing, or in stubble, then three cows ought to yield a wey of cheese and half a gallon of butter a week between Easter and Whitsuntide without rewayn. . . .

And if you see it with regard to the three cows that ought to make a wey, one of these cows would be poor, from which one could not have in two days a cheese worth a halfpence; that would be in six days three cheeses, price three halfpence. And the seventh day should keep the tithe and the waste there may be.

THE FEUDAL SYSTEM

The feudal system developed gradually from the mid-Roman empire, the Frankish empire, and later. Free men and small landowners "commended" themselves to strong estate owners and nobles, to whom they rendered service in return for protection. The feudal system was strong in the tenth to twelfth centuries. Large manors and estates sometimes were held subject to the will or whims of the ruler. Estates became hereditary under later feudal codes, with no uniform practices. Some kings became strong and despotic. The barons and leaders of the Roman Catholic Church in England opposed such practices. The "Act Declaring the Rights and Liberties of the Subject and Settling the Succession of the Crowne" in 1689 declared illegal the absolute right of the Crown. This act reserved much authority to an elected Parliament. William and Mary agreed to this Bill of Rights. The bill recognized rights and privileges of common people and free men.

The feudal system of land tenure and servitude, strengthened in England under William the Conqueror and his successors, broke down gradually with establishment of a strong elected parliamentary government.

Enclosure Acts and Selection

The open-field system of culture existed in Great Britain until replaced by nearly 4,000 Enclosure Acts passed by Parliament between 1760 and 1884. Separate acts directed commissioners to divide fields and distribute land among those who had held it jointly. Hedges and roads were built. These changes allowed owners to manage livestock and regulate breeding practices. New cropping systems and rotations could be adopted by farmers. Turnips had been introduced in 1644 and lucerne (alfalfa) later. In feudal times, animals had grazed together under a herdsman, or ran at large. Male animals were used in common, a practice that continued largely even early in the eighteenth century, with little improvement other than by selection.

Selection undoubtedly played a part in improvement of cattle for centuries. Roman agricultural writers mentioned cattle. Cato the Censor (234–149 B.C.) wrote that grazing cattle were more profitable than agriculture, and Columella (about A.D. 50) wrote that probably every farm had grass for some cows, goats, or sheep. Adam Dickson summarized Varro (116–27 B.C.), Columella, and Palladius (author of "De Re Rustica" in the fourth century A.D.) in the words: "The rustic writers are very particular in their direction about buying cattle, among these there is one mention by almost all of them; . . . that the ground to which they are brought, be of the same kind with that on which they are bred." Red and brown cattle were mentioned as more valuable than black cattle.

Leonard Mascall, in *The Government of Cattell,* in 1596 (editions also in 1600 and 1633), used *sort* rather than *breed* when describing characteristics of oxen to buy: "Oxen are according to the region and the country where they are bred; for as there is a diversity of grounds and countries, so likewise there are diversities of bodies, and diversities of natural courage: and likewise diversitie in hairs and horns of them. For those oxen in Asia be of one sort, and those in France of another sort; so likewise here in England of another sort."

FEED CROPS INTRODUCED

Sir Richard Weston (1591–1652) introduced trefoil clover into Surrey County and started a crop rotation founded on clover, flax, and turnips. Timothy Nourse (1700) contended that "grass rais'd by foreign seeds" ought to be permitted since great numbers of cattle were raised that way, and consequently more corn (grains): "Now the more corn and cattle are rais'd, the cheaper must all provisions be, which is generally look'd upon to be a benefit to the publick."

IMPORTANCE OF SIRES

Gervaise Markham wrote the popular *Cheape and Good Husbandry*, the eighth edition of which was printed in 1653. He recognized the importance of the sire: "I think fittest in this place, where I intend to treat of horned Cattell and Neat, to speak first of the choyce of a fair bull, being the breeders principallest instrument of profit." He mentioned the best English cattle for the market being bred in Darbyshire, Lancashire, Lincolnshire, Somersetshire, Staffordshire, and Yorkshire, where they were mainly black. Lincolnshire cattle also were pied with more white than other colors, and had small crooked horns.

Markham wrote:

> Now to mix a race of these and the black ones together is not good, for their shapes and colours are so contrary, that their issue are very uncomely: therefore, I would wish all men to make their breeds either simply from the one and the same kind, or else to mix York-shire . . . with one of the black breeds, and so likewise Lincoln-shire with Somerset-shire, or Somerset-shire with Glocester-shire. . . .
>
> Now for the Cow, you should choose her of the same country with your Bull, and as near as may be of one colour, only her bag or udder would ever be white, with four teats and no more, and her belly would be round and large, her fore-head broad and smooth, and all her other parts such as are before shewed in the male kind.
>
> The use of the Cow is to fold, either for the Dary, or for breed: the Red Cow giveth the best milk, and the black Cow

bringeth forth the goodliest Calf. The yong Cow is the best for breed, yet the indifferent old are not to be refused. That Cow which giveth milke longest is best for both purposes, for she which goeth, longest dry loseth halfe her profit, and is lesse fit for teeming; for commonly they are subject to feed, and that straineth the Womb or Matrix.

DUTCH-BREED FOR MILK

J. W. Gent described cattle during 1669 (2nd edition in 1675) thus:

Of cows and oxen. These worthy sort of beasts are in great request with the husbandman, the Oxe being useful at his cart and plough, the Cow yielding great store of provision both for the family, and the market, and both a very great advantage to the support of the trade of the kingdom.

Concerning their form, nature and choice, I need say little, every *Countryman* almost understanding how to deal with them.

The best sort is the large *Dutch Cow* that brings two calves at one birth, and gives ordinarily two gallons of milk at one meal.

As for their breeding, rearing, breaking, curing of their diseases, and other ordering of them; and of *Milk, Butter* and *Cheese*, etc., I refer you to such authors that do more largely handle the subject than this place will admit of.

J. Mortimer (1721) agreed almost wholly with Markham, but added: "The hardiest are the Scotch; but the best sort of cows for the pail, only they are tender and need very good keeping, are the long-legg'd, short-horn'd Cow of the Dutch-breed, which is to be had in some places of Lincolnshire, but most used in Kent, many of these cows will give two gallons of milk at a meal."

The "Dutch-breed" was much sought after, according to John Lawrence, in 1726, and at higher prices than other cattle. He commented on abuse of commons:

But the encouragement is, the many pernicious commons we have, which for the flush of milk in the few summer-months, makes the poor buy cows, to starve them in winter, and to spend so much time in running after them, as would earn twice the worth of their milk by an ordinary labour; whereas, if the commons were enclosed, some would feed them well all the summer, and others would yield hay for them in the winter; whereby there would be always a tolerable plenty of milk; from which spring many more considerable dairies.

Lawrence told of a cow belonging to the Vicar of Stanford upon Avon that yielded twelve pounds of butter every week for two or three months, and that made tolerable cheese for the family afterwards.

According to R. Bradley in 1729:

> In the choice of cows, those with the following marks are most worthy our esteem. To be high of stature, long-bodied, having great udders, broad foreheads, fair horns and smooth, and almost all other tokens that are required in the bull; especially to be young; for when they are past twelve years old they are not good for brood. But they live many times much longer, if their pasture be good, and they keep from disease. . . . In some places they have common bulls and common boars in every town. . . .
>
> Near London, or other very populous places, the milk of cows will yield a sufficient profit in the pail; but in places remoter from the market, it is either disposed of in the dairy for making butter, if the feed is such as is rich and hearty, and consists of pure grass, which is sweet and free from weeds.

He remarked that an "over-plus of milk" in winter was profitable for butter.

IMPROVED STRAINS NAMED

Toward the end of the eighteenth century, the better strains or varieties of cattle began to be known by names of the counties where they originated. The names "kind" and "sort" gave way gradually to "breed" as cattle took recognizable distinguishing characteristics typical of these areas and as agriculture became more specialized. Thomas Hale recognized this situation in 1758, when he wrote:

> Our bulls differ only in their size, according to the counties in which they are bred. The various parts of this kingdom afford so different pasturage for cattle; that when they are brought into other places, they are called after the name of that whence they came. The Lancashire breed is large, and Welch are smaller, and the Scotch least of all. In Staffordshire

they are commonly black, and in Gloucestershire red; and they have the like differences in other counties.

The husbandman should be acquainted with the several breeds, that he may suit his purchase to his land.

The large kinds are bred where there is good nourishment, and they require the same where they are kept, or they will decline: the poorer and smaller kinds which are used to hard fare, will thrive and fatten upon a moderate land. . . .

The husbandman should have one of these considerations in view, in stocking his land, then using them principally for breed, for milk, or for work; and accordingly as either of these is his principal aim, he is to make his purchase: one breed being fitter for one of these uses, another for another. . . .

Whatever breed he chuses, he should keep entirely to it; that is, the bull and cows should all be of the same kind; for it is a general and true observation, that a mixed race does not succeed so well. . . .

The cow being chiefly intended for the dairy, care is to be taken in purchasing a right kind, for there is a vast difference in the profit of this animal, according to the breed.

They have large cows in all those counties where they breed large oxen, but the size is not all that the husbandman is to consider; the quantity of milk is not always proportioned to the biggness of the beast; and that is to be his chief regard.

Welch and Scotch cows will do upon the poorest pastures: and they will suit some who cannot rise to the price of better kinds. They yield a good quantity of milk if rightly managed; but the fine kinds are the Dutch and Alderney: these are like one another in shape and goodness, but the Alderney cow is preferred, because the hardier.

The Dutch breed have long legs, short horns, and a full body. They are to be had in Kent and Sussex, and some other places where they are carefully kept up without mixture and will yield two gallons at a milking; but in order to do this they require good attention, and good food.

The Alderney cow is like the Dutch in the shortness of her horns, but she is stronger built, and is not so tender. She requires rich feeding, but is not liable to so many accidents, and is equal to the other in quantity and goodness of her milk.

Of which ever kind they be observe the following rules in their choice. Let them have the forehead broad and open; the eyes large and full, and except the Dutch and Alderney breeds, let the horns be large, clean and fair. . . .

Of whatever breed the cow be let her neck be long and thin; her belly deep and large. See that she have a large, good, white and clean looking udder, with four well grown teats.

Let the bull be of the same breed: and let them be of as large a kind as the pastures will support. But it is better to have a cow of a smaller breed well fed than one of the best in the world starv'd.

The red cow it is said gives the best milk, and the black cow is best for her calf; but this is fancy. The red cows milk has been long famous; and a calf of a black cow is accounted good to a proverb; but the breed is the thing of consequence not the colour.

The cow that gives milk the longest is the most profitable, to the husbandman; and this is most the case with those which are neither very young, nor advanced into years.

CHANGES IN ENGLISH AGRICULTURE

Lord Ernle (R. E. Prothero) described this period of change in English agriculture:

The house and homestead of the peasantry (under the manorial system) were originally the only permanent enclosures, and the only property which they could be said to hold in separate ownership. The rest of the village land was held and farmed in common. It consisted of three portions—arable land, meadow, and pasture. Areas were allotted each for cultivation of wheat, barley, and fallow. . . . After the crops were cleared, the fences were removed; common rights were revived; and the cattle of the village wandered promiscuously over the whole.

The meadows were assigned (likewise) to use of individuals from Candlemas to Midsummer Day, and the remainder of the year were pastured in common. Beyond the arable and meadow lands lay the roughest and poorest land which afforded timber of building, fencing, or fuel, mast and acorns for swine, rough pasture for the ordinary live stock, and rushes, reeds and heather for thatches, ropes, beds, and a variety of other uses. . . . In 1764, out of 8,500 parishes (in England), 4,500 were still unenclosed, open-field farms, cultivated upon a cooperative system of agriculture. . . . Sheep are kept for their wool rather than their mutton, and cattle are valued for their milking qualities or their power of draught. . . . On the cow-downs the common herdsman tends the cattle of the community. They begin to feed on the downs in May, and continue to graze there till the meadows are mown, and the crops are cleared from the arable fields. Then they are turned in upon the aftermath, the haulm, and the stubble. . . . The rams and the parish bull are provided.

In the 16th century, agriculture in England became more profitable, enclosures were made, and the rights of common were greatly restricted. . . . Gardening was taken up again late in the 17th century. Deep drainage began to be discussed. From the Flanders of the 17th century, Sir Richard Weston brought turnips and red clover, and Arthur Young called him a greater benefactor than Newton. . . . Perennial rye grass was introduced. White clover was introduced in 1700, and timothy and orchard grass came to England from America in 1760. The 18th century saw revolutions in English farming. One came when Lord Townsend established the Norfolk system (or rotation): wheat, turnips, barley, clover & grass. One half of the land was constantly under grain crops and the other under cattle grazing. Sheep and cattle were fattened on the land on turnips, increasing barley yields.

Trowell (1739), Hale (1756), and Anderson (1775) advised several pastures, and suggested they be grazed in rotation to get greater returns from the land. John Mills wrote *A Treatise on Cattle* in 1776, "showing the most approved methods of breeding, rearing, and fitting for use, horses, asses, mules, horned cattle, sheep, goats and swine with Directions for the proper Treatment of their several Disorders." Other writers included bees and rabbits along with all four-legged cattle. Mills regarded the ox as the most valuable of horned cattle. Also, Mills wrote, "Formerly the wealth of man consisted chiefly in his herds of black cattle . . . for it is only by the cultivation of lands, and the abundance of Cattle that a state can be maintained in a flourishing condition," except for gold and silver. He described the ox by colors and by sorts and sizes in different countries. Richness of milk varied with some sorts. Three classes of cattle were recognized: short-horned, longhorn, and polled or muley cattle.

Ten years later, George Culley, a farmer at Fenton, Northumberland, published *Observations on Live Stock,* containing hints for choosing and improving the best breeds of the most useful kinds of domestic animals. He also included among the livestock rabbits, swine, deer, and poultry. Culley recognized *breeds* of cattle. He pointed out "a very common mistake in endeavoring to unite great-milkers with quick-feeders." He mentioned that neat cattle in the Azores were long-horned and in every respect the same as the Lancashire breed, only less in size. Further, "Mr. Bakewell, from the superior manner in which he has distinguished himself in the breeding of cattle and sheep . . . pointed out some of the principal advantages over those methods that were in greatest repute in his day." Culley also described the short-horned breed of cattle, still in many places called the Dutch breed. He mentioned that Mr. Michael Dobinson brought bulls from Holland to the River Tees area.

Lifting Days

The backward condition of agriculture with regard to livestock and feed supplies in many areas was exemplified by "lifting days" in the spring. A large painted wall panel at the Agricultural Exposition at Bellahoj in 1938 depicted a "lifting day" in 1788 in Denmark when villeinage or serfdom was discontinued. Cows surviving the winter, but too weak to stand, were lifted and helped onto pasture by groups of farmers going from farm to farm. Dr. A. C. McCandlish's (1890–1938) grandmother told him of the practice in Wigtonshire during her childhood. Insufficient or unbalanced feed for the cattle caused them to lose condition during the winter and early spring. These limitations of feed, which prevented animals from developing to their inherited capacity, seriously retarded improvements in livestock. The need for lifting days disappeared when land was enclosed and crop production and farming methods were improved.

The Fairlie Rotation

Interest in improvement of agriculture, including livestock, developed gradually. Innovations in farming spread slowly, as people hesitated to adopt new ideas and practices. As early as 1730, Alexander, Earl of Eglinton, began to lay out roads, plantations, and ditches on his estate in the county of Ayrshire. He opened quarries and encouraged his tenants to progress. He brought a prominent farmer from another district as an example. When leases expired, fields were enclosed and a rotation set up under the new leases whereby some land was placed in sod, and only definite amounts cultivated. This was known as the Fairlie system of rotation, devised by William Fairlie.

Alexander established a Farmers' Society, over which he presided for a number of years. The idea of farmers' societies spread among landowners, and had a great influence on increasing agricultural knowledge. The movement made slow progress among tenants, who looked upon those innovations as a means to increase their rents.

IMPROVEMENTS THROUGH BREEDING

Robert Bakewell (1725–95) was among the first prominent improvers of cattle, sheep, and horses in the British Isles. He was born at Dishley Grange, in Leicestershire, England, where his grandfather and father were tenant farmers and able breeders of livestock. His ancestors included men prominent in the church, diplomacy, and the learned professions.

Bakewell succeeded to the tenancy of Dishley Grange in 1760 because of infirmities of his father, who died in 1773. The farm consisted of 440 acres, of which 110 acres were cultivated. It carried 60 horses, 400 sheep, 510 cattle of all ages, and some swine. Bakewell's experiments in pasture irrigation and improvement, travels, and purchases of breeding stock were so costly that he was declared bankrupt (in the *Leicester Journal*, December 27, 1783). However, his animals were not sold, and an appreciable estate was left to a nephew when Bakewell died in 1795.

Bakewell selected two "Canley" cows from Sir William Gordon of Carrington. They were of an improved strain developed by Sir William Greeley, and later by a Mr. Webster of Canley. Bakewell bought a bull from Westmorland. The use of these animals is illustrated in the breeding of sires that he leased out and used later in his own herd. The pedigree of the bull Shakespeare shows how Bakewell tried to perpetuate desired characteristics of selected animals.

"Twopenny" was named from the prophesy of a visitor that as a calf he was not worth two pence. The bulls, D and Shakespeare,

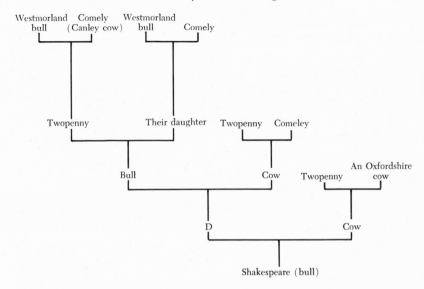

Shakespeare (bull)

were calved about 1772 and 1778, respectively, and were popular sires of the Longhorn breed in their time.

Bakewell studied livestock closely. Professor Low wrote about 1842 that Bakewell took over the management of the ancestral home about 1755, and that in breeding cattle, horses, sheep, and swine:

> He sought for the best animals of their respective kinds, and coupling those together, endeavored to develop in the highest degree, those characteristics which he deemed good. He appeared to have disregarded, or made light of, size in all the animals which he reared, and to have looked mainly to those characteristics of form which indicate a disposition to arrive at early maturity, and become readily fat. He acted to the fullest extent upon the principle that the properties of the parents are communicated to their descendants. This led him to attach the highest importance to what is termed blood, or breeding from individuals the descendants of those of approved qualities. A maxim of his was, that "like begets like"—in principle nothing new, but never, perhaps, acted upon in breeding to the like degree before.

George Culley, a longtime acquaintance of Bakewell's, wrote of a trip to Friesland for Dutch or Flemish mares to improve the English stock. Bakewell was disappointed when a Dutch farmer refused a high price for one mare. Further, in *Observations on Livestock,* Culley cited:

> The kind of cattle that were most esteemed before Bakewell's day, were the large, long-bodied, big-boned, coarse, gummy, flat-sided kind, and often lyery or black-fleshed. On the contrary, this discerning breeder introduced a small, clean-boned, round, short-carcassed, kindly-looking cattle, and inclined to be fat; and it is a fact, that these will both eat less food in proportion, and make themselves sooner fat, than the others; they will in truth pay more for their meat in a given time, than any other sort we know of in the grazing way. His sheep are still more excellent than his cattle.

Bakewell's fame as a breeder spread and attracted many visitors to Dishley—royalty and nobility from England, France, Germany, and Russia, as well as writers and many breeders of livestock. Culley and Arthur Young visited him. Young described a 10-day visit in 1786, as follows: "In breeding his bulls and cows, (and it is the same with his sheep), he entirely set at nought the old ideas of the necessity of variation from crosses; on the contrary, the sons cover the dams, and the sires their daughters, and their progeny equally good, with no attention whatever to vary the race. The old system, in this respect, he thinks erroneous; and founded in opinion only, without attention either to reason or experience."

Bakewell founded the Dishley Society or Tup Club in 1783 to protect and spread the Dishley or New Leicester sheep which he bred. Many rams were leased out for a season, which made it possible for him to return for his own use those that transmitted the desired characteristics to their offspring. This occurred nearly 40 years before foundation of the first herdbook for cattle.

His methods can be summarized under four main guidelines: (1) use judgment of type, with an ideal in mind; (2) secure the best stock available for the purpose; (3) breed closely, and prove the transmitting ability of individual animals, using the desirable ones to the maximum; (4) eliminate undesirable animals from the breeding herd. The principles upon which Bakewell depended in producing his improved stock were "fine form, small bones, and a true disposition to make fat readily." Many prominent breeders since have followed Bakewell's practices of endeavoring to breed the best to the best, with careful selection, and to make the maximum use of desirable sires.

King George turned "Farmer George" and became a correspondent of Bakewell. A host of distinguished names were enrolled under the banner of the "Farmer's Friend," and stamped the period as active in advancing the science and practice of agriculture. The Longhorn breed of cattle is still a minor breed in England. Bakewell's example stimulated improvers of other breeds. Although not the *first* improver, the widespread following of his example established Bakewell as the first *great* improver of livestock.

FAIRS AND MARKETS

Incentive toward improvement of livestock stemmed partly from early fairs and markets. When a group of animals was assembled for sale, the better ones often attracted buyers before the poor animals were taken.

Fairs are of ancient origin. Franchises or charters to hold them at a given time or place were granted by the king or feudal lord. The charters often stated the commodities to be offered. Fairs were held in China in ancient times, and were operating in Champagne and Brie in A.D. 427. In 660 a cattle market was chartered in Utrecht, Holland, where butter and cheese fairs also were held. "The Horse Fair," painted by Rosa Bonheur, pictured that famous event in France.

During Anglo-Saxon times, market transactions were concluded before the "reeve" or before acceptable witnesses for legal security. Buyers and sellers congregated at fairs and markets, which tended to stabilize prices. Fairs—often larger and gayer than markets—increased greatly in the British Isles after the Conquest (1066). Between 1199 and 1483 over 2,800 fairs and markets were chartered in England.

Fairs and markets had an educational value. Discriminating buyers at the fairs selected cattle for breeding purposes. Charles Colling saw some desirable calves at Darlington market, and traced their sire which was bought later by his brother and another breeder. This bull became famous as Hubback (319), a foundation animal of the Shorthorn breed. "The Durham Ox," a steer fattened by Charles Colling to weigh 3,024 pounds, was displayed at the markets by Colling and two subsequent owners until the bull was past 10 years old. Colling fattened the freemartin "The White Heifer that Travelled" to 2,300 pounds, and exhibited her over the country. These events occurred prior to most livestock exhibitions—around 1800 to 1813.

Some "Alderney" cattle were distributed in England through the markets, and others were sold through advertising in current papers.

THE AGRICULTURAL SOCIETIES

Organized exchange of ideas for improvement of agriculture was fostered by formation of "The Honourable Society of Improvers in the Knowledge of Agriculture in Scotland" in Edinburgh on June 8, 1723. Membership included the Dukes of Athol, Hamilton, and Perth, the Marquises of Lothian and Tweedale, eighteen Earls, and representative barons, knights, and gentry of the country. The members were impressed by the low state of manufactures in Scotland, and by "how much the right husbandry and improvement of ground is neglected, partly through the want of skill in those who make a profession thereof, and partly through the want of due encouragement for making proper experiments." Secretary Robert Maxwell published the history of this society in 1743 before a revolution disrupted the country in 1745. A committee of 25 people had the duty of dividing into sections to stimulate investigations of agriculture. These sections were instructed to " . . . mark down their thoughts thereupon in writing, to be revised by the Committee." Also to " . . . correspond with the most intelligent in all the different customs in the nation concerning their different ways of managing their grounds, that what may be amiss may be corrected, and what is profitable initiated."

Members of the Committee were asked to "send up the different ways of the management of their farms, and to form small societies of gentlemen and farmers in their several counties." Advice given on fattening and tending cattle was "Be sure to prepare a careful hand to attend feeding of them, for upon this depends the whole success of the attempt."

The "Select Society," founded in Edinburgh by Allan Ramsey on May 23, 1754, with 15 members interested in philosophical inquiry, grew to over 100 members within a year. Resolutions passed on March 23, 1755, "for the encouragement of Arts, Science, Manufactures, and Agriculture" included " . . . the Society for the above purposes takes the name of 'The Edinburgh Society for Encouraging Arts, Sciences, Manufactures, and Agriculture in Scotland.'" A notice in the Edinburgh newspapers on April 10, 1755, mentioned premiums offered for competitions, including a Gold Medal for "the best dissertation on vegetation and the principles of agriculture." Farmers who wished to assist the Society were invited to notify the secretary. Discussions included estate management, highway construction, length of land leases, rent, taxation, converting moor into arable or good pasture ground, and sowing grass seeds without either lime or manure.

Prizes of £10 were given in December 1756 for the best stallion, and £4 for feeding and selling to the butcher the greatest number of calves at least 6 weeks old. Alexander Ramsey believed these to be the first prizes awarded in Scotland for livestock. Two prizes for cow-milk cheese, and one for salt butter encouraged dairy enterprise in 1756. The Select Society and Edinburgh Society were discontinued in 1765 due to limited finances.

About 50 persons met at Fortune's Tentine Tavern in Edinburgh on February 9, 1784. They organized "The Highland Society of Edinburgh" with John, Duke of Argyll as president, four vice-presidents, and twelve committeemen. Objects of the Society approved on March 12, 1784, included:

An inquiry into the present state of the Highlands and Islands of Scotland, and the condition of their inhabitants.

An inquiry into the means of the improvement of the Highlands by establishing towns and villages; by facilitating communication through different parts of the Highlands of Scotland; by roads and bridges, advancing agriculture, and extending fishieries, introducing useful trades and manufactures; and by an exertion to unite the efforts of the proprietors, and call the attention of the Government towards the encouraging and prosecution of those beneficial purposes.

A Royal Charter was issued on May 17, 1787, which changed the name to "The Highlands Society of Scotland at Edinburgh." Parliament granted £3,000 in 1789, the interest from which was available to pursue stated objects of the Society.

The first step toward cattle improvement was the Gold Medal award (valued at 7 guineas or about $35) "for the best Bull from two to five years old, proper for improving the breed of Highland cattle, and the property and in possession of any proprietor or tenant in Argyllshire—the bulls to be shown at Connell, Parish of Kilmore, on the 20th October (1789)."

Premiums also were offered for bulls owned by tenants having at least 40 cows, and for bulls being in herds that numbered at least 30 cows. The judges were instructed to "pay particular attention to the shape of the bulls, and not to the size, as it was meant to encourage the true breed of Highland cattle." Heifers also were recognized in 1807.

Formation of local societies was proposed and approved at the general meeting in 1792.

The Society's first show was held for three days at Edinburgh in December 1822. Prizes were offered for pairs of oxen in four classes: Shorthorn, Aberdeenshire, West Highland, and a class for "Angus, Fife, and Galloway oxen, or any other breed." Sheep and pigs were recognized in 1823, and breeding cows in 1824. Two Ayrshire oxen and a Shorthorn-Ayrshire ox were recognized in the 1825 show.

Many general agricultural societies were organized soon after. Dublin Society, founded in 1731, supervised agricultural trials under an official experimentalist. The Society for the Encouragement of Arts, Manufactures, and Commerce organized in London in 1754 and began publication of transactions in 1783. Empress Catherine of Russia directed the Free Oeconomical Society to be formed and set up experimental plots near St. Petersburg (Petrograd) under a priest trained on Arthur Young's estate.

AMERICAN FAIRS AND AGRICULTURAL SOCIETIES

The American Philosophical Society was organized at Philadelphia in 1743. Although not an agricultural society, its proceedings included agricultural articles. The Strawberry Fair, authorized by an Act of 1723, was held at Childsberrytown in Berkeley County, Virginia, in the spring and autumn "for exposing for sale horses, cattle and merchandise." This fair and later ones were periodic markets.

Elkanah Watson—"Father of Agricultural Fairs"—was an early leader and organizer and an especially enthusiastic promoter of fairs. Watson procured two Merino sheep in 1808, which he exhibited on the public square in Pittsfield, Massachusetts, near his estate. People interested in these better animals inspired the idea of organizing a different type of agricultural society. Neighbors helped him to found the Berkshire Cattle Show in 1810 and the Berkshire Agricultural Society in 1811. Animals competed for money prizes, and a half-mile parade featured band music and mounted marshals. The show expanded each year to include different types of exhibits—manufactured woolen cloth, a household department, more livestock and farm produce, with a promotional address. New societies were formed with support by state legislatures to hold practical agricultural fairs. Interest declined after Watson died, but by the 1840s it increased again.

The earliest American agricultural society may have been the New York Society for Promoting Art, operating in 1766 and awarding premiums for essays on specific subjects. The New York Society for the Promotion of Agriculture, Arts, and Manufactures dates from 1751. The South Carolina Society for Promoting and Improving Agriculture and other Rural Concerns was projected in 1784. They advocated that plantation owners set aside some land for experiments with animals, implements, and plants. The Philadelphia Society for Promotion of Agriculture, founded in 1785, publicized the production of newly introduced dairy cows in its memoirs.

The Massachusetts Society for Promoting Agriculture, founded in 1792, published reports of its deliberations. They also imported and placed improved breeds of stock with members under an agreement that they be maintained pure and their kind multiplied.

They introduced Merino sheep, Percheron horses, and Ayrshire, Devon, Shorthorn, and other cattle breeds.

In North America, fairs have been adapted also toward agricultural development, education, and promotion by agricultural societies, breed organizations, fair associations, and governmental agencies. Such activities have been expanded to include consignment, promotional, reduction, and dispersal sales of purebred livestock.

REFERENCES

Anderson, James. 1775. *Essays relating to agriculture and rural affairs.* Edinburgh.

Bradley, Richard. 1729. *The gentleman and farmers' guide for the increase and improvement of cattle.* London.

Brown, Robert. 1811. *Treatise on rural affairs.* Edinburgh.

Culley, George. 1786. *Observations on live stock.* (4th ed, 1807) G. G. J. & J. Robinson, London.

Darwin, Charles R. 1859. *Origin of species.* Appleton, New York.

Dickson, Adam. 1788. *The husbandry of the ancients.* 2 vols. Edinburgh.

Garrard, George. 1805. *An atlas of different varieties of cattle of the British Isles and cattle from the East Indies.* (A portfolio.) London.

Gent, J. W. 1675. *Systema agriculturae, being the mystery of husbandry discovered and layd open by J. W.* 2nd ed. London.

Hale, Thomas. 1756. *A compleat body of husbandry.* 2 vols. London.

Huberman, Leo. 1936. *Man and his worldly goods.* Harper, New York.

Lamond, Elizabeth. 1890. *Walter of Henley's husbandry.* London.

Lawrence, John. 1726. *A new system of agriculture. Being a complete body of husbandry and gardening in all parts of them.* London.

――――. 1809. *A general treatise on cattle, the ox, the sheep, and the swine.* C. Whittingham, London.

Leouzon, Louis. 1905. *Agronomes et Eleveurs.* J. B. Baillerie & Sons, Paris.

A Lincolnshire Grazier. 1833. *The complete grazier.* 6th ed. Baldwin & Cradock, London.

Low, David. 1842. *On the domesticated animals of the British Isles.* Longmans, Green & Co., London.

Markham, Gervaise. About 1631. *Cheape and good husbandry.* 8th ed., 1653. J. Harison, London.

Mascall, Leonard. 1596. *The government of cattle.* (Eds. to 1633.) London.

Mills, John. 1776. *A treatise on cattle.* J. Johnson, London.

Mortimer, John. 1707. *The whole art of husbandry, or the way of managing and improving land.* J. H. Mortlock, London.

Neely, Wayne C. 1935. *The agricultural fair.* Columbia Univ. Press, New York.

Pawson, H. Cecil. 1957. *Robert Bakewell.* Crosby Lockwood & Son, London.

Prothero, R. E. (Lord Ernle). 1888. *English farming past and present*. Ed. by Sir A. D. Hall, 1936. London.

————. 1892. Landmarks in British farming. *J. Roy. Agr. Soc. Engl.* 3(ser. 3):11.

Sebright, Sir John Saunders. 1809. *The art of improving the breeds of domestic animals*. London.

True, Rodney H. 1925. The early development of agricultural societies in the United States. *Amer. Hist. Assoc. 1920 Ann. Rept.* Washington, D.C. Pp. 293–305.

Young, Arthur. 1786. A ten day tour to Mr. Bakewell's. *Ann. Agr.* 6:452–502.

CHAPTER 5

AYRSHIRES IN SCOTLAND

FOUNDATION of the Ayrshire breed on native cows dates back between two and three centuries in the districts of Carrick, Cunningham, and Kyle in Ayr (see Fig. 5.1). The county, which extends in a crescent along 70 miles of coast on the Firth of Clyde, is 4 to 28 miles wide and covers 1,149 square miles. The altitude varies between sea level and 2,298 feet; mean temperature ranges from 35° to 65° F.; and annual precipitation averages 35 inches. The cool moist climate favors grasses, cereal grains, and root crops. Dairying is the leading agricultural enterprise. Half the land was in grass in 1925, mostly on clay and heavy loam soils. Turnips, rutabagas, and potatoes were the main tilled crops.

Headpiece: Vignette of Ayrshire cow.

EARLY CATTLE

Remains of extinct *B. primigenius* occur in Pleistocene deposits, the bottoms of lochs and lakes after the Ice Age. Neolithic settlers who brought *B. longifrons* as domesticated cattle lived in huts, made pottery and polished stone implements, and had a settled agriculture. The Romans brought draft cattle from south of the Alps to Scotland in A.D. 80. Norsemen (Danes and others) raided and eventually lived in the region. Red cattle and the polled character trace to Norse cattle. Some polled cattle were among the Roman introduction into England. Timothy Pont wrote in 1600 that much excellent butter was sent to other sections.

Dutch cattle were imported into the lowlands of eastern England before 1600. Mortimer described them as "good milkers, long-legg'd, and short-horn'd" in 1721. Some crossing probably took place in the foundation of the Ayrshire breed.

ORIGIN OF DUNLOP CHEESE

Barbara Gilmore, who fled to Ireland because of religious persecution, brought the knowledge of cheese making (in 1688) to the farm of Hill in the Parish of Dunlop, as told by R. H. Leitch. People suffered many hardships and privations during the religious controversies. Little was done then to improve cattle. Daniel DeFoe (author of *Robinson Crusoe*) wrote before his death in 1731: "The greatest thing this country wants is more enclos'd pastures, by which the farmer would keep stock of cattle well fodder'd in the winter, and which again, would not only furnish good store of butter, cheese, and beef to the market, but would, by their quantity of dung, enrich their soil according to the unanswerable maxim in grazing that stock upon land improves land." He favored northern Ayr, its pastures and cattle.

LAND ENCLOSURE

According to William Aiton, the first Act of the Scottish parliament concerning land enclosure was passed in the reign of James I, about 1457. In 1695 another Act was passed for division of a Common and consolidation of intermixed properties. Agriculture was unprogressive. Although landowners instituted improved methods on their

FIG. 5.1. The districts of Cunningham, Kyle, and Carrick, which comprise the county of Ayr in southwestern Scotland, were the areas in which the Ayrshire breed of dairy cattle was developed.

own farms, tenants hesitated to follow, suspecting a trick by the landlord to exact more rent-in-kind. During winter, cattle became emaciated and weak for lack of proper feed, and many died. When grass became available in spring, farmers—even after 1800—had "lifting days" to get weak cows onto their feet and out to pastures. Lack of proper feed retarded improvement of cattle.

Marshall Stair introduced horse-hoeing, alfalfa, and St. Foin grass into Wigtonshire after 1728; cultivated turnips, carrots, cabbages, and potatoes; subdivided and enclosed his lands; and drained marshes. His sister did the same in Ayr. Stair died in 1747, his sister in 1770 (100 years of age).

EARL OF EGLINTON

Alexander, 10th Earl of Eglinton, succeeded to his estate about 1730. Aiton wrote of his activities:

> He traversed every corner of each of his extensive estates; arranged the divisions and marches of the farm; laid off roads, plantations and ditches, opened quarries, etc. and by the frequent seeing and conversing with his tenants, and pointing out the improvements proper to be executed, he roused them to industry, rendered them more intelligent, and laid the foundation of their future prosperity. He instituted an agricultural society, and presided over it for many years.
>
> The Earl of Eglinton brought from east Lothian, Mr. Wright of Ormiston, an eminent farmer, who introduced into Ayrshire the proper mode of levelling and straighting land, fallowing, drilling, turnip husbandry, etc. His Lordship also put an end on his estate, to that destructive distinction of croft and field land; and the system of over plowing, which had so long and so improperly been pursued; and prohibited his tenants from ploughing more than one third of the land in their possession. That which has since obtained the name of 'Fairly rotation,' was first introduced by Alexander, Earl of Eglinton, and only followed out by Mr. Fairly after his lordship's death. That important branch of rural economy, the improvement of the breed of cattle, did not escape the attention of that worthy and dignified nobleman. Ploughmen, roadmakers, and people conversant in the dairy, were brought by him, from different parts of Britain. Fencing was begun on an extensive scale, and the face of the country was ornamented and sheltered by many

clumps of trees which he caused to be planted on the eminences. New farm houses were begun to be erected on more liberal plans; the tenants were taken bound to crop only one third of their possessions; to manure the land, sow grass-seed, and every improvement of which the ground was susceptible, was planned and begun to be executed by that enlightened nobleman.

Aiton mentioned that John, Earl of Loudon, raised field turnips, cabbages, and carrots as early as 1756. Fullarton had described the seeding of 3 bushels of ryegrass and 12 pounds of clover per acre, to be cut for hay 1 year and pastured for 5 years. The fodder was to be fed upon the ground, and all manure spread upon it.

George Culley listed the long-horned, short-horn'd, polled or Galloway breed, Kiloes or Scottish cattle, and the Alderney or French breed among the several breeds of cattle in the British Isles in 1786. He remembered Michael Dobinson of Durham who brought bulls from Holland early in his life "and those he brought over, I have been told, did much service in improving the breed." Hale (1756) and Arthur Young (1770) mentioned these Dutch cattle in northeastern England.

THE (OLD) STATISTICAL ACCOUNTS

Rural activities in 893 parishes in Scotland were recorded between 1790 and 1798 in 21 volumes known as *The (Old) Statistical Accounts*. These accounts were prepared largely by local ministers and were edited by Sir John Sinclair, Secretary of Agriculture. Some writers gave little attention to cattle, but butter, cheese, or milk cows were mentioned in at least 16 accounts in southwestern Scotland. Thus scanty development of dairying was recognized in Ayr, leading in the northern district of Cunningham. For the parish of Dunlop, it mentioned: "But the principal produce . . . is cheese. For this it has long been known and distinguished, insomuch that all the cheese made in the country about it . . . goes by the name of Dunlop cheese, and finds a ready market on that account. In 1750, the only enclosures were . . . about gentlemen's seats; and in winter . . . the cattle roamed at pleasure and poached on all the arable land. . . . By 1798 most of the land was enclosed, and cattle were confined."

Of nearby Beith Parish—"They almost universally made Dunlop cheese." Likewise in Dalry—"The breed of cows is much improved from what they were." Similar mention of milk cows was made from several surrounding parishes.

The middle district of Kyle likewise was turning to dairying, with cheese made "after the Dunlop manner, and equally good." Of Sorn, just to the south—"The black cattle consist partly of the ancient breed; but mostly of a mixed breed between that and the Cunningham kind. About two-thirds are milk cows and the rest young cattle, rearing for the same purpose."

Tarbelton (westward of Sorn) kept 1,800 cows, and "a prodigious quantity of butter and cheese is made annually here for sale." Cattle in Symington Parish were "generally of a good milk kind, giving from 10 to 14 Scotch pints per day."

Of Kirkswald Parish—"The dairy was in a most neglected state . . . 40 years ago. . . . Now the milk cows are changed to the better, are put into parks sown down with white and yellow clover, and when they live in the house by night or by day, are fed upon cut red clover. Every steading of farm houses has an apartment by itself for a milk house, and every conveniency suited to it. Good butter and cheese are now exported from the parish to the market of Ayr and Paisley."

COLONEL FULLARTON'S SURVEY

Colonel William Fullarton (1793) was appointed by the Board of Agriculture to survey the county of Ayr. He pointed out in his preliminary report that in Cunningham

> a breed of cattle has for more than a century been established, remarkable for the quantity and quality of their milk in proportion to their size. They have long been denominated the Dunlop breed, from the ancient family of that name, or the parish where the breed was first brought to perfection. . . . Within these 20 years, brown and white mottled cattle are so generally preferred as to bring a larger price than others of equal size or shape, if differently marked. It appears, however, that the mottled breed is of different origin from the former stock. . . . This breed was introduced into Ayrshire by the present

Earl of Marchmont . . . from whence they have spread over all the country.

This breed is short in the leg, finely shaped in the head and neck, with small horns, not wide, but tapering to the point. They are neither so thin coated as the Dutch, nor so thick and rough as the Lancashire cattle. They are deep in the body, but not so long, nor so full and ample in the carcase and hind quarters as some other kinds. They usually weigh from 20 to 40 English stone. . . . It is not uncommon for these small cows to give from 24 to 34 English quarts of milk daily, during the summer months, while some of them will give as far as 40 quarts, and yield 8 or 9 English pounds of butter weekly. The breed is now so generally diffused over Cunningham and Coil (Kyle), that few of other sorts are reared on any well regulated farm. The farmers reckon that a cow yielding 20 quarts of milk per day during the summer season, will produce cheese and butter worth about £6 per annum. . . .

In former times a proportion of Dutch or Holderness cattle had been propagated, and when well fed, yielded large quantities of milk. But they were thin haired, lank in the quarters, and delicate in the constitution, which rendered them unfit for a soil such as Ayrshire's. They were, besides, extremely difficult to fatten, yielded little tallow, and from the spareness of their shapes, incapable of carrying much flesh upon the proper places.

Alderneys and Guernseys have also been occasionally introduced, in order to give a richness and colour to the milk and butter; which they do in a degree superior to any other animal of the cow species.

The term "Ayrshire breed" was used by the Reverend David Ure in 1793, when he wrote "with respect to the origin the common account is, that about a century ago, the farmers in Dunlop were at great pains to improve the original breed of the country, by paying strict attention to the marks which their experience had led them to make of a good milk cow."

Forsyth described Ayrshire cows in 1805 as:

Formerly black or brown, with white faces and white streaks along their backs, were prevailing colours; but within these 20 years brown and white-mottled cattle are so generally preferred as to bring a larger price than others of equal size and shape if differently marked. It appears, however, that this mottled breed is of different origin from the former stock; and the rapidity with which they have been diffused over a great extent of the country, to the almost entire exclusion of the preceding race, is a singular circumstance in the history of breeding. . . . The breed is now so generally diffused over Cunningham and Coil (Kyle), that very few of other sorts are reared on any well regulated farm.

When Colonel Fullarton was transferred to other duties, the Board of Agriculture selected William Aiton, native-born in Ayr, who prepared a new report in 1811. He wrote, "I am old enough to remember, nearly, the commencement of enclosing of land, and the introduction of ryegrass, as a crop in the parrish of Kilmarnock. The popular prejudice and extraordinary clamour, among the tenantry against these innovations was very strong. . . . The tenants were disposed to consider every movement they were required to make on their possessions, as tending only to augment their labour, and increase the rent rolls of the proprietor."

Aiton described the weights and measures used in Ayrshire. A Scotch gallon (8 Scotch pints) was 840 English cubic inches, whereas the English "wine gallon" was the present 231 cubic inches in volume. There were 4 different liquid-measure gallons and 1 dry-measure gallon. Cheese, butter, meat, hay, and straw were sold in Ayrshire by the trone or county weight which contained 24 ounces avoirdupois per pound, 16 pounds making 1 stone. English or avoirdupois weight contained 16 ounces per pound, 112 pounds per hundredweight, and 14 pounds per stone "jockey weight" used for groceries and other merchant goods.

Dutch weight contained 17½ ounces of English avoirdupois weight per pound, 16 pounds per Dutch stone (equalling 17½ English pounds, or 11 2/3 pounds trone or county weight in Ayrshire). Troy weight was used for gold and silver, and by apothecaries.

AITON'S SURVEY OF AYR

Aiton wrote:

Next to the melioration of the soil, the raising of grain, sowing grasses, and planting of useful roots; the rearing of cattle, and turning their produce to the best account, form the more important concern of the husbandman. . . . The age, shape and qualities, as well as the sizes . . . require to be attended to . . . the breed is most improved by selecting males of the best shapes and qualities, and by no means so large as the females which they cover, and that the shapes and qualities, as well as the size of the stock are chiefly governed by the food and manner in which the animal is treated. . . .

The only distinction of breeds to be met with in the county of Ayr, are the Galloway and the Dairy Cows. As both are excellent of their kinds, a particular description of each requires to be given.

Galloway cattle prevailed in Carrick until Aiton's time, when:

the dairy breed has been lately introduced, and is fast increasing in Carrick, still the Galloway cow is most common in that district of Ayrshire. . . .

As the county of Ayr formed a part of the kingdom of Galloway, and was inhabited by the same people [Cambrian Britons], their cattle, and the mode of treating them would continue the same in both countries, so long as the inhabitants remained uncivilized; except in so far as they were affected by soil or climate. But the inhabitants of both countries seem to have begun much earlier than their neighbors, to pay attention to cattle and their produce.

The dairy breed of cows in Ayrshire . . . are in fact a breed of cows that have, by crossing, coupling, feeding, and treatment, been improved and brought to a state of perfection which fits them above all others yet known, to answer almost in every diversity of situation, where grains and grasses can be raised to feed them, for the purposes of the dairy, or for fattening them for beef.

That justly celebrated breed have neither been imported from abroad, nor raised to their present excellence, altogether, by the magical effects of gigantic bulls, brought into the district. For though some alterations may have been affected in their size, shape, and colour, by the introduction of a few cows and bulls, of an improved breed, as I shall have occasion to notice; yet the dairy breed of Ayrshire are in a great measure the native indigenous breed of the county of Ayr, improved in their size, shapes and qualities, chiefly by judicious selection, cross coupling, feeding, and treatment, for a long period of time, and with much judgement and attention by the industrious inhabitants of the county, and principally by those in the district of Cunningham.

It appears from the adage taken as the motto, and quoted above ("Kyle for a Man, Carrick for a Cow, Cunningham for Butter and Cheese, And Galloway for Woo',") that the making of butter and cheese had, at the most remote period of their history, been the chief study and the highest boast . . . of the inhabitants of Cunningham. In prosecuting this species of industry, they could not fail to discover, that the cows who were the most amply supplied with suitable food, would yield the greatest quantities of good milk. Hence another adage of unknown antiquity, common in that district. *"The cow gives her milk by the Mou'."* That discovery once made, it was natural for them to do their utmost to supply that food which so much contributed to the milk they wanted; and the improved feeding so given would, independent of other circumstances, tend greatly to the increase and improvement of the stock of cows. . . .

It was chiefly by these means (selecting calves from the better producing cows), and not by changing the stock, or altogether by lining their cows with bulls of greater size, that the dairy breed of Ayrshire attained their present unrivalled perfection. . . .

Some alteration was probably made on the dairy stock of Ayrshire, by the introduction of a few Dutch or English cows and bulls of a size greatly superior to the native race in that county. . . .

Among other crosses with foreign cows or bulls, I understand, that the Earl of Marchmont, about 1750, purchased from the Bishop of Durham, carried to his seat in Berwickshire, several cows and a bull either of the Teeswater or other English breed, of high brown and white colour, now so general in Ayrshire; and that Bruce Campbell, Esq., then factor on his Lordship's estates, in Ayrshire, carried some of that breed to Sornberg, in Kyle, from whence they spread over different parts of the county. A bull of that stock, after coupling with many cows about Cessnock, was brought by Mr. Hamilton, of Sundrum, and left a numerous progeny in that quarter of Ayrshire. I am of opinion that this bull . . . would increase the size, and alter the colour of his progeny, and the large bones and ill shapes, incident to the calves begotten by a large bull, upon a small cow, would be gradually corrected in after generations.

I have also been told that John Dunlop, Esq., of Dunlop, brought some cows of a large size, from a distance, probably of the Dutch, Teeswater, or Lincoln breeds, and that much of the improved breed of Cunningham proceeded chiefly from their origin. John Orr, Esq., of Barrowfield, brought from Glasgow, or some part of the East country, to Grougar, about 1769, several very fine cows of the colour now in vogue; one of whom I remember cost £6, which was more than twice the price of the best cow then in that quarter. As I lived then in that neighborhood, I had access to know that many calves were reared from these cows, and that their offspring have been greatly multiplied, on the strath of the water of Irvine.

Though I have mentioned those, I do not suppose they were the only instances of cows, of larger and improved breeds, being introduced into the county.

It was probably from some or other of these mixtures, that the red and white colours of the present stock, now so common, were introduced. I remember, about 1778 and 1780, that breed became fashionable, with some of the most opulent and tasty farmers, in the parish of Dunlop and Stewarton; and that from these quarters of the county they gradually spread over the other parts, first of Cunningham, afterwards of Kyle, and now of Carrick, and other districts even out of the county. Till these were introduced, the cows of Cunningham were generally black, with some white on their face, belly, neck, back, or tail. The native breed of cows in Scotland, seem to have been generally black, and except in the improved dairy breed, they are still mostly of a dark or black colour. Hence the term black cattle is still applied to cows of every colour, all over Scotland. . . .

The size of the Ayrshire improved dairy cows varies from 20 to 40 stones English, according to the quality and abundance of their food. If cattle are too small for the soil, they will soon rise to the size it can maintain, and the reverse, if they are larger than it is calculated to support.

The shapes most approved of in the dairy breed are as follows. *Head* small, but rather long and narrow at the muzzle. The *eye* small, but smart and lively. The *horns* small, clear, crooked, and their roots at considerable distance from each other. *Neck* long and slender, tapering towards the head, with no loose skin below. *Shoulders* thin. *Fore-quarters* light. *Hindquarters* large. *Back* straight, broad behind, and the joints rather loose and open. *Carcase* deep, and *pelvis* capacious and wide, over the *hips,* with round fleshy *buttocks. Tail* long and small. *Legs* small and short, with firm *joints. Udder* capacious, broad and square, stretching forward and neither fleshy, low hung, nor loose; the *milk veins* large and prominent. *Teats* short, all pointing outwards, and at considerable distance from each other. *Skin* thin and loose. *Hair* soft and woolly. The *head, bones, horns,* and all *parts* of *least value* small; and the general *figure* compact and well proportioned.

The most valuable quality which a dairy cow can possess is that she yields much milk. A cow in Ayrshire that does not milk well will soon come to the hammer. I have never seen cows anywhere that, under the same mode of feeding and treatment, would yield so much milk as the dairy breed of that district. Ten Scotch pints per day is no way uncommon. Several cows yield for some time twelve pints, and some thirteen or fourteen pints per day. I have heard of sixteen or eighteen pints being taken from a cow every day, but I have never seen so much; and I suspect there must have been some froth, either in the milk, or in the story.

Care and feed of the cows was mentioned.

The winter food of the dairy stock in the county of Ayr, from the time that the grass fails in the autumn, till it rises in the month of May, has been chiefly the straw of oats, or . . . the hay of bog meadows, frequently but ill preserved. For a few weeks after they calve, they are allowed some weak corn and chaff boiled, with infusions of hay; and by way of luxury a morsel of rye-grass or lea-hay once every day; and of late years by some farmers, a small quantity of turnips, in the early part of the winter, and a few potatoes in the spring, have been added.

Such meagre feeding, for so long . . . reduced the cow to a skeleton. When turned out to pasture in the month of May, many of the cows are so much dried up and emaciated, that they appear like the ghosts of cows; their milk vessels are dried up, and it is not till they have been several weeks at the grass that they give either much milk, or of a rich quality.

Every dairy farmer will admit that their cows are much injured by the length of the winter. . . . They can . . . shorten the period, and soften the rigours of winter by providing them such stores of turnips, potatoes, and other green food as will render the cattle comfortable, and preserve them in a milky habit till the return of summer. . . . the high price obtained for rye-grass-hay causes the farmer to deal it out but sparingly to the cows. . . . The food in summer, of the dairy stock in the county of Ayr, is generally pasture. In the best cultivated districts, where clover and rye-grass grow luxuriantly and the pasture is nourishing, the cows fare well and produce much milk.

He described feeding of freshly cut clover in the byre, thus getting "double the quantity" of feed from an area as against grazing it. He advocated dividing a pasture and grazing alternately to prevent seeding. Also, the tax on salt deprived cows "of that necessary article of their life and comfort."

ORIGIN OF THE BREED

Aiton continued seeking the origin of improved Ayrshires. His last account stated:

> They have increased to double their former size, and they yield about four, and some of them five, times the quantity of milk they formerly did. By greater attention to their breeding and feeding, they have changed from an ill shaped, puny, mongrel race of cattle, to a fixed and specific breed, of excellent shape, quality and colour. This change has not been effected by merely expelling one breed and introducing another, but on the far sounder principles of careful crossing and better feeding. . . .
>
> These are all the instances of stranger cattle which have been brought into the county of Ayr, as far as I know at the time . . . or have been able to trace, and I am not aware, that more than a dozen or at most twenty such cows ever came into this district. I am disposed to believe, that although they rendered the red colour with white patches fashionable in Ayrshire, they could not have had much effect in changing the breed into their present highly improved condition.

The greatest number of cows then weighed 24 to 36 stones. Better feed and care allowed the cows to develop to the extent of their inherited ability, but more than feed is necessary to establish a breed. Lack of it restricts development and milk production.

John Speir believed that some qualities of Ayrshires depended strongly on Dutch cattle brought to England probably between 1600 and 1750, before the black-and-white color dominated in the Netherlands. A few cattle were sent to Scotland from the Island of Jersey by Field Marshall Henry Seymour Conway and Lieutenant General Andrew Gordon between 1772 and 1806.

New Statistical Accounts

The presence of some Channel Island cattle in southern Scotland was documented in *The New Statistical Account of Scotland* in 1845. The Reverend David Ure, of Glasgow, stated that dairy cows of Roxburgh to the eastward were "a mixture of the Dutch, French, and English kinds. They are short-horned, deep-ribbed, and of a white and red colour. The Ayrshire breed has now got into the county, and is found to answer exceedingly well."

Colonel J. Le Couteur, of Jersey, mentioned the earlier two shipments of Jerseys to Scotland, and some resemblance between the races.

John Speir described introduction of Highland blood into the Ayrshire breed, as follows:

Between 1800 and 1830 the Ayrshires seem to have grown immensely in public favor. . . . The favorite herd at this early period, and the one which exerted probably greater influence on the breed than any other of this period, was that of Theophilus Swinlees, Dalry. . . . He was born 4th April, 1778, and died 18th April, 1872, at 94 years of age. He had a brother Will who was a Highland cattle dealer. . . . Being a neighbor, . . . the writer received direct many of the notes regarding this particular herd. . . . Theo. Paton often repeated to me the story that the basis of his herd was a cross between an Ayrshire bull and a West Highland heifer. The introduction of any Highland blood into the Ayrshire breed has often been disputed, but as far as this particular instance is concerned there is no room left for doubt. That eminent exhibitor and judge of Ayrshire bulls, the late Wm. Bartlemore, of Paisley, says that this animal was a Skye heifer, and that "The first progeny was a red heifer calf, but the dam in milk exhibited such pre-eminent qualities of teat and udder, that he again mated her for years."

It was about this time that the Ayrshire began to have stronger horns than formerly, and with the points turned upward instead of inwards; but whether or not these changes were gradually brought about by natural selection, or by the influence of the Swinlees breed, as it was generally called, there is little evidence to show. . . . Bulls from that herd . . . were introduced into almost every herd of prominence. The change was, however, very gradual, for as late as 50 years ago

(1859) a large proportion of Ayrshires, as I remember them, had incurved horns. Mr. Hamilton (a noted Ayrshire judge, beginning in 1849) says they were often as much curved inward that the points had to be sawn off to prevent them entering the head.

The terms "sort," "kind," and "breed" were used loosely in connection with cattle of the British Isles in the 1700s. Final official sanction of the term "breed of cattle" appears to trace to a report of the Highland and Agricultural Society of Scotland on January 30, 1835. Several distinct breeds were mentioned in this report. Ayrshires were recognized by this Society as a pure breed in 1836. Local cattle were known by the name of Dunlop, Cunningham, and then Ayrshire successively as numbers increased and they became recognized more widely as a dairy breed.

The many shades of black, brown, red, fawn, and cream leave room for thought as to the source of the fawn, yellow, brindle, and cream colors described with Ayrshires imported into Canada and the United States between 1836 and 1891, and registered in Volume 1 of the *Dominion Ayrshire Herd Book*.

FAIRS AND SHOWS

John Speir stated in 1909 that "the breed as we know it today is in great part the result of the showing season. Ayrshire, and more particularly the Cunninghame and Kyle districts, seem to have been about, if not the very first, to adopt a system of holding competitions and shows."

Probably the earliest livestock competition in Scotland took place in December 1756, when the Edinburgh Society awarded a prize to the best draft stallion, and a premium for the greatest number of calves fed and sold to the butcher. A premium for salt butter, and two for cheese, also were awarded in 1756.

The present Highland Society—now named the Royal Highland and Agricultural Society of Scotland—was founded February 9, 1784, and began to award annual premiums for breeding stock in 1789. Parliament appropriated £3,000, the interest on which was used "for advancing agriculture" and other purposes.

In Ayr, the Kilmarnock Farmers' Club, founded in 1786, held its first cattle show at Kilmarnock in 1793. Gilbert Burns (brother of Robert Burns) discussed improving Ayrshire cattle in 1795, stating: "Although much has been done of late in this country in proper selection of the species to breed from, yet much remains to be done. That particular attention out to be given to the whole appearance of the animal, as well as to its colour and horns. That much attention out to be given in the selection of the cow as well as of the bull." The particular type of animal desirable to breed from was discussed shortly thereafter. Premiums for heifers at the show were added in 1807. A picture of an ideal Ayrshire cow, as approved by the Club, was published in 1811.

The Highland Society rotated location of its show over Scotland, with prizes awarded to Ayrshires first in 1814, then in 1816 and 1821. Forty-nine cows and bulls competed in their show at Glasgow in 1826. Two prize cows at the Highland Show in 1828 are pictured in Figure 5.2. Malcolm Brown (1829) mentioned Ayrshires at the show: "This breed has been greatly improved; yet there remains much to be done, and this can only be attained by a careful and continued attention to the temper, size, shape and qualities of

FIG. 5.2. An artist's portrait of the first and second prize Ayrshire cows at the Highland Show in 1828. Note the short horns and closely attached udder. (Portrait by Howe.)

those intended to breed from together with the greatest care in the treatment of the young stock."

The Highland Society's policy committee decided in 1835 to foster "Shorthorned, West Highland, Polled Angus, Polled Aberdeenshire and Galloway," and the Ayrshire as a dairy breed. This action suppressed Horned Aberdeenshires and the dairy breed of Fifeshire. They advocated upgrading native cattle with bulls of approved breeds.

The Highland show held in Ayr in 1835 had 88 Ayrshire entries,

FIG. 5.3. "Geordie," an outstanding first prize winner in 1838 and 1839, was popular. His progeny exerted a wide influence on the breed. He was of Swinlees stock.

mostly from that county. The top bull was of Swinlee breeding. Bulls receiving prizes had to travel the district and serve cows in a radius of 30 miles, if £20 were subscribed. "Geordie" won first prize at the Highland show in 1838 and at the Ayrshire Agricultural Society and Highland shows in 1839 (Fig. 5.3).

CHANGES IN SHOW IDEALS

A General Agricultural Association of Ayrshire held its first show in 1836, and remained permanently in the Burgh of Ayr in 1852. (This still is the leading Ayrshire show in Scotland, equalled on some occasions by the Royal Highland Show.) Judges began gradually to place much emphasis on fine points—upturned horns, teats not over

2½ inches long, level sole to udder, and style—overlooking milking qualities. The standard of 1853 stated: "Milk vessels capacious and extending well forward, hinder part broad, and firmly attached to the body, the sole or under surface nearly level. The teats from 2 to 2½ inches in length, equal in thickness, and hanging perpendicularly; their distance apart at the sides should be equal to one-third the length of the vessel, and across to about one-half of the breadth." Yet the dairy was held to be the chief purpose of the breed.

FIG. 5.4. Photograph of the first public milking trial, Ayrshire, 1860. Sponsored by the Duke of Atholl (right).

D. Tweedie, breeder and prize-winner, mentioned teats "from 1 to 2½ inches in length, equal in thickness, and hanging perpendicularly" about 1865. The Royal Agricultural Society of England recognized Ayrshires at their show in 1855.

The Duke of Atholl believed that milking ability was neglected, and so he arranged for a milking contest at the Ayr show. The cows yielding most for ten consecutive milkings won prizes on April 25, 1860. Glen Gaur, owned by a Mr. Wallace, averaged 26 pounds 5½ ounces per milking or over 263 pounds in 5 days. The contestants are shown in Figure 5.4.

The fad for level soles went so far that some showmen were said to "board" the udders during the night before the show, and "set" the teats with collodion. A rift occurred among the breeders. Some favored the fine points of show type while others wanted deeper udders, longer teats, and lighter natural fleshing. The scale of points adopted in 1884 still held for teats 2 to 2½ inches long and spaced *exactly so* on the udder. An agricultural writer in 1885 criticized:

> In fact the prize milk cow is like the masher of the period— the one can hardly look over his collar, and the other does not give much milk for fear of injuring the symmetry of her vessel. . . .
>
> And all practical men who want to improve their herds stand aloof, because their object is milk and milk alone. They feel that they are more likely to injure than improve their stock by introducing prize strains.
>
> Let breeders breed for milk, and determine not to keep or breed from a cow that does not yield a certain quantity of milk containing a given proportion of cream. Let them use as stock bulls those only which are the produce of cows that comply with the above conditions.

PRODUCING ABILITY RE-EMPHASIZED

No extreme can exist long on an impractical basis. William Bartlemore—breeder, showman, and judge of Ayrshires—thought that the teat craze was passing away in 1889 because of discrimination at sales in favor of cows with longer teats. The export trade also demanded producing ability. Secretary C. M. Winslow of the Ayrshire Breeders' Association in the United States wrote in 1904: "The dairymen of the United States today want a cow of fair size, a large milker, with comfortable teats." The Scale of Points of 1906 specified, with Canadian collaboration, "length 2½ to 3½ inches and not less than 2 inches, thickness in keeping with length, hanging perpendicularly and slightly tapering, and free flow of milk when pressed." A committee petitioned the Highland Society in 1906 that awards to Ayrshire cattle be given to "a type eminently suited for dairy purposes."

Beginning in 1913 the Ayr show fostered "milk record classes," with £15 for prizes from the Department of Agriculture for Scotland. To enter, first-calf heifers must have produced 7,000 pounds of milk with 3.5 percent fat, and cows 8,000 pounds of milk. Animals with production pedigrees gained prominence in the show and sale ring.

THE "NEW" SHOW

The Ayrshire Cattle Herd Book Society set up a committee in 1918 to establish new standards for judging dairy cattle. They proposed as a scale:

Form, symmetry, and constitution	30 points
Mammary development, to include teats, shape of udder, milk veins, etc.	35 points
Authenticated milk yield in the case of a cow	35 points
Authenticated milking pedigree in the case of bulls and heifers.	35 points

This proposed scale allowed 1 point for each 100 pounds of milk over the minimum (5,000 and 6,500 pounds of 3.8 percent milk) until the 35 points were attained. The "New Show," organized by the Herd Book Society, was held at Ayr on February 9 and 10, 1921, with 203 cattle exhibited—the largest show of Ayrshires to that time. A Canadian judged the show.

Beginning in 1921, cows exhibited at the London Dairy Show were required to enter the milking trials. An Ayrshire cow with an udder that was high and wide in rear attachment, full, well veined, and deeper than the "vessel-type" udder won Supreme Individual Championship and the Spencer Challenge Cup in 1923. A team of six Ayrshire cows won the interbreed championship. The combined effects of these shows reestablished the Ayrshire type on a practical foundation of type and production.

A minimum qualifying standard of production was required of females, according to the lactation, for entry at the Scottish Dairy Show. The Herd Book council added an alternative standard for the 1962 show: "Animals which have produced over 400 pound, butterfat at not less than 3.6 percent in any one lactation." These qualifying standards emphasize milking ability as a requirement in Ayrshire show cows.

Production records published for 42 of 67 Ayrshires competing in the 1-day milking trial at the 1963 Scottish Dairy Show in Glasgow averaged:

Type	Milk	Fat	Solids-not-fat	Production points
	(pounds)	(percent)	(percent)	
19 mature cows	78.9	4.59	8.89	175.1
12 4-year olds	65.3	4.97	9.01	151.9
11 first-calf heifers	52.8	4.43	9.08	118.5

HERD BOOK SOCIETY

The *Ayrshire Herd Book* (Volume 1) containing pedigrees of 17 bulls and 59 cows was compiled in 1877 as a private enterprise by Thomas Farrall of Aspatria, Carlisle. Interest arose soon in establishing an Ayrshire herdbook under an impartial official body. The directors of the Ayrshire Agricultural Association met May 15, 1877, and considered a circular issued by Farrall, particularly the suggestion by David Tweedie of Castle Crawford, Abingdon, that the association initiate action. Farrall's *Herd Book* probably would not be continued. A committee of three was appointed, with The Honorable C. R. Vernon as chairman, to take action on Tweedie's suggestion. Mr. Vernon convened a public meeting on June 26, 1877, in the Ayr courthouse at which the Ayrshire Herd Book Society was organized. The name was extended later to Ayrshire Cattle Herd Book Society of Great Britain and Ireland.

Mr. Vernon compiled Volumes 1 and 2 of the official *Ayrshire Herd Book*, which the society published in 1878 and 1879. The society took charge with Volume 3. Records were included of prizes awarded to animals at shows since 1873. Objects of the new Society were: "(1) To maintain the purity of the Breed of Cattle known as Ayrshire Cattle. (2) To collect, verify, preserve, and publish an Ayrshire Cattle Herd Book, with the Pedigrees of the said Cattle, and other useful information concerning them." The oldest animal in Volume 1 was a cow "Premium 1" born in 1852 and died in 1871. A bull "Bob" was born in 1854; his dam was by "Geordie," a show winner of Swinlee stock.

After Volume 10 of the *Herd Book*, "animals are only admitted whose pedigrees can be traced in previous volumes, either through the Dam or Sire, but any animal not tracing as above may be entered in an Appendix, if approved of by the Committee, without a number, thus qualifying produce for future entry."

The Society granted prizes at designated shows in 1903. Production records became part of the requirements for entering cows in Appendix B of the *Herd Book* in Volume 32 (1909). Winning show prizes ceased to qualify for entry in Appendix B in July 1915. Heifers then had to produce 204 pounds of butterfat, and cows had to produce 264 pounds of butterfat in 12 months before, or 40 weeks after, a calving date. A cow's tattoo marks also were used for identification.

Current production and calving intervals for entry of unregistered females in Appendix B are:

Interval between calvings (months)	Butterfat requirements (in pounds)	
	Heifers	Cows
13	285	323
14	313	351
15	342	380

The butterfat test in any lactation founded upon must be 3.75 percent or more. Purebred Ayrshire females by unknown sires were eligible for entry into Appendix C in 1970. Their heifers became eligible for Appendix B.

One male was registered for 14.1 females in 1960.

Herds free from tuberculosis (attested herds) and those tested for brucellosis were recognized by an insignia from 1936 and 1938 respectively. Elimination of both diseases contributed to extend the average useful life of animals.

PEDIGREE REGISTRATION CERTIFICATE SYSTEM

The Society discontinued publishing the *Herd Book* after Volume 78 in 1953, and substituted the Pedigree Registration Certificate System. Official pedigree registration certificates were fully detailed to four generations for lineage and to three generations for production records. Milk records were furnished by the milk recording associations. All certificates were microfilmed for the Society's archives. The name was shortened to Ayrshire Cattle Society in April 1959.

The Cattle Blood Typing Service opened in Edinburgh in September 1966 to serve all of Britain. By looking at 30 to 40 blood groups of an animal, its dam, and sire or possible sires, correctness of parentage can be determined.

The Society is governed by 77 councilmen from the respective districts comprising one or more counties in the British Isles, based on numbers of Ayrshires in each district. The members hold an annual meeting and elect their councilmen by mail ballot.

FIG. 5.5. Ayrshire cows in Quinton Dunlop's herd near the Firth of Clyde.

A typical herd of Ayrshire cows near the Firth of Clyde is shown in Figure 5.5.

Type Classification

The Society planned to begin type classification of Ayrshire cattle in October 1971 with a demonstration during the annual convention. To arrive at a final score for the animal, the standardized method would break down assessments into sections, such as head and back, shoulders and chest, middle and loin, rump and thighs, feet and legs, shape and size of udder, attachment of udder, teats, veins, quality of udder, giving points to each section.

PRODUCTION RECORDS

Colonel William Fullarton wrote in 1793: "In Cunningham . . . a breed of cattle has for more than a century been established, remarkable for the quantity and quality of their milk in proportion to their size."

Aiton mentioned many cows would yield 3,046 to 3,225 quarts of milk in a season and others not half that amount. William Harley's model dairy near Glasgow averaged 12 quarts of milk per day. The first authenticated production records were from the milking contest sponsored by the Duke of Atholl in 1860. The top five cows yielded between 22 and 26 pounds 5½ ounces per milking for ten consecutive milkings. The duke bought the winning cow, which produced 6,258 quarts (over 15,000 pounds) of measured milk in 12 months.

The yields in a good average Ayrshire herd were estimated by William Bartlemore in the 1880s to be 6,300 to 6,600 pounds of milk per year, with 3.5 to 4.5 percent of butterfat. Cows were fed mainly on pasture, hay, and roots in season.

Early in 1903 the breed historian John Speir proposed systematic milk recording to the Highland and Agricultural Society directors. They appropriated £200 annually for 5 years. Three societies operated during the cheese-making season. The movement spread. The Ayrshire Cattle Milk Records Committee was set up from 1908 through 1913, representing the Highland and Herd Book societies. The 1,265 records to the end of 1907 averaged 6,784 pounds of milk, 3.71 percent and 252 pounds of butterfat. The committee included the agricultural colleges in 1914, the year the name was changed to the Scottish Milk Records Association. Increased grants aided in support until World War II (1939). Local societies and members assessed the costs largely among themselves and the Herd Book Society. Milk recording was extended gradually to a yearly basis.

Milk records were divided into three classes:

	Cows	Heifer
Class I	Not under 250 pounds fat	Not under 200 pounds fat
Class II	Intermediate	Intermediate
Class III	Under 166 pounds fat	Under 133 pounds fat

Records in Class I were published; cows in Class II were regarded of commercial quality, while the number or percentage of Class III animals *retained in a herd* were published. The stigma of retaining Class III cows induced disposal. This plan gave incentive whereby production of the breed might improve. The standards were increased in 1933 to 280 pounds in Class I and under 224 pounds in Class III for cows, and 186 and 149 pounds of fat in those classes for first-calf heifers.

A Special Production Register was established in Scotland in 1950. Registered cows were eligible in Section I on producing 400 pounds of butterfat in 305 days and a living calf within 13 months of previous calving. From 1954, nonpedigree cows were included in the Register. Cows that calved between 13 and 15 months after start of the year's record and produced 450 pounds of butterfat in 365 days were eligible for Section 2. Average production of Ayrshires was reported by the National Milk Records in the fiscal year 1959–60 as 8,563 pounds of milk, 3.82 percent and 327 pounds of butterfat from 107,253 lactations. Butterfat tests have fluctuated between 3.82 and 3.91 percent for the breed average in the last 10 years.

Lifetime production is recognized by Production Clubs. Seventeen great Ayrshire cows have achieved membership in the 200,000-pound Club. Some 373 cows rated in the 150,000-pound Club, and 7,037 cows were listed in the 100,000-pound Club by the spring of 1970.

When James A. Patterson retired in 1967, the Scottish Milk Records Association combined with the Scottish Milk Marketing Board and the office was moved from Ayr to Glasgow.

PROGENY TESTING

Dr. A. C. McCandlish of the West of Scotland Agricultural College began in 1932 to publish comparisons of daughters and dams of living Ayrshire sires. Lists of Approved Dams, which he began, were published first in 1939. Combined activity of the Ayrshire Cattle Herd Book Society, Scottish Milk Records Association, and the agricultural colleges made the production records more useful to Ayrshire breeders. Progeny testing was placed under a Research Committee of the Herd Book Society in 1947.

APPROVED SIRES

The Council of the Herd Book Society adopted requirements for Approved Sires and Approved Dams in 1946, as in the United States. An Ayrshire bull must have had 10 or more tested daughters with complete lactations and each produced at least 8,000 pounds of milk, 3.8 percent and 320 pounds of fat in 305 days on a 2× mature equivalent basis to become an Approved Sire. At least 50 percent of his daughters past 3 years old must have been tested and all first-lactation records average at least 9,000 pounds of milk and 342 pounds of butterfat on a mature equivalent basis. Daughters that died before calving or were in nontesting herds were excluded.

Rules revised in 1961 required first-calf heifers to yield at least 7,500 pounds (actual) of milk testing 3.8 percent and 285 pounds of butterfat. The bull must have rated at least 101 points or more in the Relative Rating Value assessment based on contemporary herdmate comparisons by the Scottish Bureau of Records in Edinburgh or the Livestock Records Bureau of the Milk Marketing Board of England and Wales, with at least 80 percent of all daughters' milk recorded. Records were computed to two milkings daily, in 305 days or less.

Production requirements for Approved Ayrshire Sires in 1966 were increased to the following averages:

first lactation—8,500 pounds milk, 4 percent and 340 pounds fat
second lactation—9,000 pounds milk, 4 percent and 360 pounds fat
third and later lactations—10,000 pounds milk,
4 percent and 400 pounds fat

Some 2,046 Ayrshire bulls qualified as Approved Sires before 1968.

APPROVED DAMS

Before 1961 there were two requirements for Approved Dams: (a) Three or more daughters must produce an average of 9,000 pounds of milk, 3.8 percent and 342 pounds of butterfat in 305 days on a mature equivalent basis. Sixty percent of them must each have yielded 8,500 pounds of milk, 3.8 percent and 340 pounds of butter-fat. (b) Two tested daughters qualified their dam if each produced more than 9,500 pounds of milk, 3.8 percent and 361 pounds of butterfat in 305 days. The index of a son was equal to that of a daughter's record, providing 50 percent of his 3-year-old daughters were tested.

In 1961 the Council revised the requirements so that the *actual* average of all daughters' records be at least 7,500 pounds of milk, 3.8 percent and 285 pounds of butterfat on two milkings daily. Not less than 60 percent of all tested progeny must each produce at least 8,000 pounds of milk and 320 pounds of fat. When a dam had only two daughters, each must yield at least 8,500 pounds of milk, 3.8 percent and 323 pounds of butterfat. With too few daughters, the average of all daughters of her son were equivalent to the records of a daughter. Only first lactations were studied, except at an owner's request. Since 1954 an Approved Dam also must have produced two 305-day records attaining the standard of 9,000 pounds of milk and 3.8 percent butterfat.

In 1966 production requirements of daughters (at least two) were the same as those for Approved Sires, with 4.0 percent fat. Requirements for total solids or solids-not-fat in the milk were held in abeyance by the Research Council since insufficient numbers of these records were available then. Some 4,701 cows earned the Approved Dam rating before 1966.

In 1966 cows were studied for "Approval" only on request. One point was earned for each pound of butterfat that the daughters' average production exceeded the breed standard for first, second, or subsequent 305-day lactations. The production standard continued with 4.0 percent fat. Dams qualifying were designated as Classified Approved Dams.

Solids-Not-Fat in Ayrshire Milk

Attention was given to butterfat contents of milk by the Scottish Milk Marketing Board (organized in 1933) and other agencies since early in World War II. Lower prices were paid for milk which fell below 3.5 percent fat during the winter, and under 3.4 percent fat in summer. The board notified all producers when solids-not-fat fell under 8.5 percent. No further steps were taken in Scotland prior to 1962. The Milk Marketing Board of England and Wales, supported by the National Farmers Union, announced that in October 1962 premium payments would begin for all milk containing over 12.6 percent total solids. A price penalty would apply to milk under 12 percent total solids or 8.4 percent solids-not-fat. The policy was approved by the Ministry of Agriculture.

Heredity is recognized as important with relation to the solids-not-fat contents of milk. The Scottish Milk Marketing Board, Hannah Dairy Research Institute, and the Institute of Animal Genetics in the University of Edinburgh cooperated in 1949 to study milk from about 500 daughter-dam pairs of cows. A pilot program was begun in 1954 whereby solids-not-fat was determined in the bulk milk of selected herds in one milk recording circuit. Control recorders were trained at the Hannah Dairy Research Institute in correct use of the lactometer. Solids-not-fat determinations have been made routinely since October 1957 on bulk milk of each herd. These records are provided for voluntary use of herd owners. In 1962 the Scottish Milk Marketing Board began payments for milk based on fat and solids-not-fat.

The Ayrshire Cattle Society publicized that 319 Ayrshire heifers at all B.O.C.M. Ltd. progeny tests averaged 9.11 percent solids-not-fat and 3.90 percent fat. The 715 Ayrshire cows at all postwar London Dairy Shows averaged 8.95 percent solids-not-fat and 4.15 percent butterfat; the 296 Ayrshire heifers averaged 9.10 percent solids-not-fat and 4.38 percent fat. This emphasized the desirable quality of Ayrshire milk with regard to its contents of solids-not-fat.

The Council of the Ayrshire Cattle Society sent a resolution to all Milk Marketing Boards in June 1962, as follows: "The Ayrshire Cattle Society of Great Britain and Ireland requests all Milk Marketing Boards to provide facilities to all members of Milk Recording Societies for the testing of solids-not-fat at the very earliest opportunity. In particular, it is emphasized that those facilities be made to all breeders of pedigree dairy cattle in view of the importance of these tests in the selection of breeding for total solids." It was asked that all breeders be given equal opportunity to apply such tests with their registered cattle.

IMPACT OF WORLD WAR II

D. Marshall stated that poultry and swine decreased in number under feed rationing in Scotland. Fewer sheep were kept on tillable land. Horse population decreased and use of government tractors increased. Acres of vetch, beans, kale, and cabbages increased for feeding milk cows. Dairy cattle increased from 741,000 in 1939 to 823,000 in 1945, with registered animals increasing. Milk production decreased in the first 2 years, then increased:

	1939	1944–45
Summer milk, gallons	77,046,000	82,977,000
Winter milk, gallons	54,451,000	58,852,000

Milk was rationed for children and invalids, and increased to 0.66 of a pint daily per person—double that in 1935. Eradication of tuberculosis in cattle progressed rapidly.

LICENSING BULLS

The Department of Agriculture for Scotland established licensing of bulls, effective June 1, 1950. For License A, a bull must be eligible or registered in a dairy breed herdbook. His dam and sire's dam were required to meet standard production requirements. Milk records on the Pedigree Registration Certificate of the Ayrshire Cattle Society were accepted for granting a Dairy A bull license. License B can be issued to some bulls from first-calf heifers, with some restrictions. In 1960, 1,014 A and 436 B licenses were granted for Ayrshire bulls.

The Society registered 1,541 males and 26,577 females in 1968. In addition, 477 females qualified in Appendix A and 478 in Appendix B in the grading-up system. This was a ratio of 1 male to 17.9 females entered.

ARTIFICIAL BREEDING AND PROVING BULLS

Artificial breeding in Scotland began before World War II, and with licensed bulls since 1950. A polled Ayrshire bull was kept at one center in 1955. Progeny testing of Ayrshire bulls in artificial use was begun in 1954 at the South Cathkin Testing Center of the British Oil and Cake Mills, Limited, in Lanarkshire. The Scottish Milk Marketing Board and Ayrshire Cattle Society were co-sponsors. The farm tests about 13 heifers by each of 4 Ayrshire bulls at a time. Heifers are loaned, and held under similar feeding and management through an average of 270 days of the first lactation. Rate of milk letdown, total yield, fat, and solids-not-fat percentages are recorded; conformation and udder attachments are observed. Milk from 44 daughters recorded during 1964–65 ranged from 8.78 to 9.44, and averaged 9.11 percent solids-not-fat. About 1 bull in 12 has been considered outstanding, but it is considered equally valuable to detect below-average bulls before allowing them heavy service.

The Board of Records (Livestock Records Bureau) was established under the Department of Agriculture for Scotland in 1955 through its Animal Breeding Research Organization. They receive and analyze all milk records obtained by the milk recording bodies in Scotland. They cooperate with the Scottish Milk Marketing Board, which owns the bulls in artificial service, and with the Ayrshire Cattle Society. The bureau carries out virtually a progeny test by contemporary comparisons in the field. Bull evaluation by contemporary comparisons was extended in 1962 for earlier estimations of their potential value. The earliest heifers by a young sire are identified by tattoos or registration numbers. When these heifers freshen, Milk Records supervisors are contacted for the first 90-day yields together with that of a heifer by another sire calving near the same time on the same farm. Such preliminary comparisons afford early relative indications of the better and poorer sires as concerns peak of lactation of their daughters. The Livestock Records Bureau follows this with the regular contemporary comparison, which includes the results of persistency.

Intercountry shipments of Ayrshire semen must be authorized by a Pedigree Semen Export Certificate by the Ayrshire Cattle Society in the exporting country, and must meet sanitary regulations of respective health authorities.

MERIT BULLS

The revised Constitution and Rules that took effect in 1964 provided for Merit Bull registration certificates to be issued to males whose dams qualified with two consecutive 305-day lactations, as follows:

360 pounds butterfat at 3.8 percent minimum of 5 tests
in first lactation
420 pounds butterfat at 3.8 percent minimum of 5 tests
in second lactation
450 pounds butterfat at 3.8 percent minimum of 5 tests
in third lactation
500 pounds butterfat at 3.7 percent minimum of 5 tests
in fourth or subsequent lactations

This standard recognized that butterfat percentages in milk decreased gradually after cows reached maturity. The above standard was changed to 4.0 percent fat for 1968 and later. Less than 10 percent of Merit Bulls registered before 1967 had dams with two lactations below 4 percent butterfat.

Production records of cows in the 100,000-pound Club, averaging under 3.6 percent butterfat were not published after 1962. The mode of butterfat tests of 1,728 lifetime records of cows in the 100,000-pound Club between June 1963 and the close of 1966 was 3.85 percent, with a range of 3.60 and 5.01 percent fat. The 178 lifetime records of cows in the 150,000-pound Club in 1961–66 also had a mode at 3.85 percent, with a range of 3.26 to 4.70 percent fat.

A blood typing scheme for all Britain was established in Edinburgh in September 1966. The service was available to members on a fee basis. The society also agreed to blood type every two hundredth male registered (at random), as well as all bulls selected for the Society's Young Sire Scheme and cases referred to the Discipline Committee.

AYRSHIRE TRENDS

A trend has occurred away from the traditionally horned Ayrshire of Swinlee type. This was noticeable particularly in eastern Scotland and England where many herds are handled on the yard or open court system. The proportion of dehorned Ayrshires was insignificant 25 years ago. From three polled Ayrshires from the United States in 1948 and a polled bull in artificial service first in 1955, the polled character has spread.

Advertisements or illustrations in *The Ayrshire Cattle Society's Journal* for Winter 1970 showed some trends within the breed. One herd was naturally polled and 47 apparently were dehorned. The polled cow Hartly Design 689695 had been the Supreme Individual Champion and member of the winning Bledisloe Trophy team at the London Dairy Show in 1964. Eight illustrations were of horned Ayrshires. Of the 132 advertisements 35 were mainly announcements. Three herds proudly reported high total milk solids, and 50 with 4 percent average butterfat tests or above. Sixty-three herds were listed as accredited (brucellosis-free), and three were attested (tuberculosis-free), one since 1911. Undoubtedly others were under test, but the brief announcements precluded mention.

A survey in 1967 found that 1.75 percent of Ayrshires being registered were naturally polled. Some 60 to 80 percent of Ayrshires in the United Kingdom are dehorned because of pen or loose housing of milking herds. The Ministry of Agriculture (England) or the Department of Agriculture for Scotland had provided Strain 19 Bang's vaccine free for immunizing against brucellosis for calves 151 to 240 days old, beginning in May 1962.

The Ayrshire Cattle Society's Journal is published quarterly. The *Research Bulletin* formerly contained the list of recently Approved Sires, Approved Dams, cows that qualified for the 100,000-, 150,000-, and 200,000-pound Clubs with milk production.

J. Lawson is the General Secretary and Editor, at 1 Racecourse Road, Ayr., Scotland.

REFERENCES

Anonymous. 1804. Various methods calculated for the improvement of the County of Caithness carried on in the course of the year 1803. *Farmers' Mag.* 5:6.

————. 1877. Proceedings of meeting to organize the Herd Book Society. *N. Brit. Agriculturist* 29(n.s.):426–27.

————. 1949. *The Ayrshire breed brochure from Ayrshire Cattle Herd Book Society of Great Britain and Ireland.*

Aiton, William. 1811. *A general view of the agriculture of the county of Ayr.* A Vepier, Trongate. Glasgow.

————. 1825. *A treatise on the dairy breed of cows and dairy husbandry.*

————. 1830–31. On the characteristics of cattle. *Quart. J. Agr.* 2:485–97.

————. 1833–34. On the dairy cattle of Ayrshire. *Quart. J. Agr.* 4:763–80.

Auld, Rev. Robert, and Alexander Cuthill. 1845. *New Statistical Account of Scotland* 16:50.

Bartlemore, William. Cited by John Speir. 1909.

Brisbane, Rev. Thomas. 1790–98. *The (Old) Statistical Account of Scotland* 9:537.

Buchanan, James. About 1880. In J. P. Sheldon, *Dairy Farming: Being the theory, practice and methods of dairying.* London.

Campbell, A. B. 1845. *New Statistical Account of Scotland* 16:828.

Culley, George. 1786. *Observations on live stock.* G. G. J. & J. Robinson, London.

DeFoe, Daniel. 1738. *A tour thro' the whole Island of Great Britain.* Rev. ed.

Dickie, Rev. Matthew. 1845. *New Statistical Account of Scotland* 5:288, 300.

Dickson, James. 1835–36. On the application of the points by which cattle are judged. *Quart. J. Agr.* 6:546–98.

Dickson, R. W. 1805. *Practical agriculture.* Vol. 2. London. Pp. 994, 1117.

Donald, H. P., D. W. Deas, and A. L. Wilson. 1952. The genetic analysis of the incidence of dropsical calves at birth in herds of Ayrshire cattle. *Brit. Vet. J.* 108:227–45.

Donaldson, Rev. William. 1811. *Trans. Highland Agr. Soc. Scotland.*

Farrall, Thomas. 1876. On the Ayrshire breed of cows. *Trans. Highland Agr. Soc. Scotland* 8(ser. 4):129–47.

———. 1877. *The Ayrshire Herd Book.*

Forsyth, Ro. 1805. *Beauties of Scotland.* 4 vols. Pp. 439, 445–46.

Fowler, A. B. 1933. The Ayrshire breed of cattle: A genetic study. *J. Dairy Res.* 4:11–26.

Fullarton, John. 1845. *New Statistical Account of Scotland* 16:261.

Fullarton, Col. William. 1793. *General view of the agriculture of the county of Ayr, with observations on the means of its improvement.*

Garrard, George. 1800 and later. *Description of the various types of oxen common in the British Isles.* Portfolio. London.

Harley, William. 1829. *The Harlenian dairy system.*

Hogg, Thomas, Jr. 1845. *New Statistical Account of Scotland* 16:225.

Hughes, T. McKenny. 1896. On the more important breeds of cattle which have been recognized in the British Isles in successive periods and their relation to other archeological and historical discoveries. *Archaeologia* 5(ser. 2):125–58.

Lawrence, John. 1809. *A general treatise on cattle.* 2nd ed. C. Wittingham. P. 69.

A Lincolnshire Grazier. 1808. *The complete grazier.* (6th ed., 1833.) W. Jackson, New York.

Low, David. 1842. *On the domesticated animals of the British Isles.* Longmans, Green & Co., London.

McCandlish, A. C. 1918. Black and white Ayrshires. *Ayrshire Quart.* 3(4):24–28.

M'Kinley, Rev. James, David Strong, and Andrew Hamilton. 1845. *New Statistical Account of Scotland* 5:545.

MacNeilage, Archibald. 1901. Famous Ayrshire sires. *Trans. Highland Agr. Soc. Scotland* 13(ser. 4):149–76.

Marshall, David. 1946. Scottish agriculture during the war. *Trans. Highland Agr. Soc. Scotland* 58(ser. 5):1–77.

Martin, W. C. L. 1854. *Cattle: Their history and various breeds.* London. P. 107.

Murray, Gilbert. 1875. *The Ayrshire breed of cattle.* In J. Coleman, *The cattle of Great Britain.* Horace Cox, London.

Pettigrew, Robert. 1845. *New Statistical Account of Scotland* 16:111–17.

Pont, Timothy. About 1600. *Topographical account of the district of Cunningham, Ayrshire.* Pp. 9–16.

Pryde, George S. 1937. *Ayr Burgh Accounts, 1534–1624.* 3rd ser. P. 28.

Ramsay, Alexander. 1879. *History of the Highland and Agricultural Society of Scotland.* W. Blackwood & Sons, Edinburgh.

Reed, O. E. 1924. A trip to Scotland. *Ayrshire Dig.* 10(4):10–11.

Reid, John. 1949. Breeding better cows. In *The Ayrshire breed* (brochure). Ayr. Pp. 29–33.

Robertson. 1829. *Progress of improvements in Ayrshire, more particularly Cunninghame.*

Robertson, K. J. 1962. Speeding up field progeny results. *Ayrshire Cattle Soc. J.* 34(3):122–26.

Sinclair, Sir John. *The (Old) Statistical Accounts.* 21 vols. (Written by pastors and others.)

————. 1812. *An account of the systems of husbandry adopted in the more improved districts of Scotland.* Arch Constable & Co., Edinburgh.

Smith, A. D. Buchanan. 1936. The value of the livestock show. In Ayrshire Agr. Assoc., *The Book of the Association.* Glasgow. Pp. 106–10.

————. 1937. *The Ayrshire breed. Past—Present—Future.* Glasgow. Pp. 106–10.

Speir, John. 1909. Early history of the Ayrshire breed of cattle. In *Report of Milk Records for the Season 1908.* Pp. 271–317.

Sturrock, Archibald. 1866–67. Report of the agriculture of Ayrshire. *Trans. Highland Agr. Soc. Scotland* 1(ser. 4):21–106.

Tweedie, D. 1865. *The Upper Ward of Lanarkshire.* Vol. 4. P. 14.

Ure, Rev. David. 1793. *The history of Ruthglen and East Kilbride.* Pp. 184, 187–89.

Wallace, Robert, and J. A. Scott Watson. 1923. *Farm live stock of Great Britain.* 5th ed. Oliver & Boyd, Edinburgh.

Youatt, William. 1893. *The complete grazier.* 13th ed. William Fream, London.

Miscellaneous Breed Publications

1877–1958. *Ayrshire Cattle Herd Book.* Vols. 1–78.

1936. *The Book of the Association, 1836–1936.* Ayrshire Agr. Assoc., Ayr.

1954. Ayrshire bull proving station, South Cathkin. *Ayrshire Cattle Soc. J.* 27:7,79.

1954. *Ayrshire Cattle Herd Book Society Res. Bull.* 6th ed.

1961. Rules governing Club cows. *Ayrshire Cattle Soc. J.* 33(2):84–85.

Rules governing the Approval of Ayrshire sires.

Rules governing the Approval of Ayrshire dams.

CHAPTER 6

AYRSHIRES IN AMERICA

THE FIRST Ayrshires imported to America supplied fresh milk for the passengers and officers aboard sailing vessels, and were sold on reaching port. An Ayrshire bull and cow were imported to New York by Henry W. Mills about 1822. Dr. White owned an Ayrshire cow in Duchess County, New York, in 1828. The Massachusetts Society for the Promotion of Agriculture and John P. Cushing imported 1 bull and 7 cows in 1837. Their second importation in 1845 numbered 1 bull and 4 cows, and the third included 4 bulls and 11 heifers "mainly of Swinlee stock" in 1858. The Society reported one of their cows produced 10 pounds of butter in a week at 12 years old in January, 4 months after calving, when fed hay and a quart of carrots daily. Another cow yielded 15 pounds of butter per week on common feed, and had four daughters, all first-rate cows. The Society contrasted these Ayrshires with famous "native" cows that left no progeny equal to themselves.

Several importations came between 1860 and 1870. Cows were milked by hand in the United States by men who preferred animals with long teats. When Scottish emphasis changed toward slightly longer teats (2 to 2½ inches in 1884, and 2½ to 3½ inches in 1906), 2,867 Ayrshires entered the country from 1885 to 1926. Seventy-five importers were listed in Volume 1 of the *Ayrshire Record*.

Colonel Zadock Pratt reported yearly production of about 50 "native" cows during 1857 to 1861 averaging between 4,355 and 5,209 pounds of milk per cow. He mentioned that cows could be improved by a single cross to one of the pure breeds.

AYRSHIRE HERD BOOK

The Massachusetts Society required their animals to be kept pure. The Association of Breeders of Thoroughbred Neat Stock assembled the first *Ayrshire Herd Book*. This Association dealt separately with Ayrshire, Alderney, Devon, Hereford, Jersey, and Shorthorn cattle. It was sponsored by an "Alderney" breeder in 1859. Volume 1 had pedigrees of 79 Ayrshire bulls and 217 cows under 130 ownerships.

Volume 2 appeared in 1686 and Volume 3 in 1871. Lewis Sturtevant studied these volumes and found them to contain the following:

	Imported	Tracing to importation	Others
Males	59	482	407
Females	192	872	914

Dissatisfied because lack of records prevented tracing some animals completely to importation, Sturtevant Brothers assembled the *North American Ayrshire Register,* wherein every animal was traced to importation. Volume 1 entered 238 bulls and 521 females, to the importation of 1837. The cow Twinney 500 had "Swinley horns."

J. R. Sturtevant called a meeting of 16 persons "to fix up the herdbook of Mr. Boggs" (the first herdbook). Some 24 persons attended a second meeting in 1875. The old volumes were "brushed aside," and a new *Ayrshire Record* was established with J. D. W. French as editor for the Ayrshire Breeders' Association. They printed Volume 4 before *The Ayrshire Record, New Series, Volume 1* appeared in 1876. The Ayrshire Breeders' Association and North American Ayrshire Register united in 1881 and incorporated legally in 1886 under the name of The Ayrshire Breeders' Association.

The Constitution stated: "We, the undersigned breeders of Ayrshire cattle, recognizing the importance of a trustworthy Herd Book that shall be accepted as final authority in all questions of pedigree, and desiring to secure the cooperation of all who feel an interest in preserving the purity of this stock, do hereby agree to form an Association for the publication of a Herd Book, and for such other purposes as may be conducive to the interest of breeders, and adopt the following Constitution."

The Association received registrations from 2,273 owners in 1969. Some 850 Ayrshire bulls and 11,739 females were registered during 1969, with the numbers in 18 states increasing over those of the previous year.

Operating departments of the Association are Office Management, Records Department, Registrations and Transfers, Promotion, and *The Ayrshire Digest*. A Promotion Director is active.

Officers of the Ayrshire Breeders' Association are elected annually. The nominating committee divided the country into eight districts in 1946, based on registrations and breed activities. Districts were adjusted as breed distribution changed. Registrations for the 1956 revision were as follows:

District	Area	Average of 1953–55 registrations
1	New England states	4,407
2	New York	3,947
3	Pennsylvania, New Jersey, Delaware, Maryland	3,893
4	Ohio, Kentucky, Tennessee, West Virginia, Virginia, North Carolina, South Carolina, Georgia, Florida, Alabama, Mississippi, Louisiana	3,796
5	Michigan, Wisconsin, Illinois, Iowa, Missouri, Arkansas, Indiana	3,729
6	Minnesota, North Dakota, South Dakota, Nebraska, Kansas, Oklahoma, Texas, Montana, Wyoming, Colorado, New Mexico, Idaho, Utah, Nevada, Arizona, Washington, Oregon, California	4,013

A director was elected annually from each district to serve a 3-year term. Officers included a president, four vice-presidents, and the directors. Proxies were specific by custom, avoiding abuse of this privilege. No proxies could be solicited beyond the state or district in which a member lived. A committee approved proxies 7 days before the meeting. An Executive Committee consisted of the President, First Vice-President and six Directors. Elections have been by secret mail ballot before the annual meeting, since 1959. No proxies were voted. Members voted to reduce from 18 to 12 directors in 1968.

SELECTIVE REGISTRATION

A pedigree evaluation plan for male calves was begun in 1945. The program, revised later, included three classes of registration certificates.

A "Preferred" Ayrshire bull registration certificate was issued to a calf (a) by an Approved Sire and from an Approved Dam, or (b) by an Approved Sire and from a dam with 9,750 pounds of milk and 400 pounds of butterfat in the first lactation, and she by an Approved Sire. The average of two or three 305-day records computed to a $2\times$ mature equivalent basis also was recognized for the dam's production requirements.

A "Selected" pedigree certificate was issued (a) to a male calf by an Approved Sire, and from a dam with 9,750 pounds of milk and 400 pounds of butterfat; (b) to a male calf by a "Preferred" pedigree sire, and from an Approved Dam, or a dam with 9,750 pounds of milk and 400 pounds of butterfat; or (c) to a male calf by a sire with a transmitting index of 9,750 pounds of milk and 400 pounds of butterfat, or from an Approved Dam or a dam with 9,750 pounds of milk and 400 pounds of butterfat. In the last instance, the dam's sire must have been Approved or have the transmitting index mentioned. If the dam was Approved, it was not required of her sire. Records were for the first lactation, or average of the first two or three lactations. Lactations were for 305-day $2\times$ mature equivalent basis. Records of dams at comparable ages were used in comparing with daughters.

A "Standard" pedigree certificate was issued if the higher certificate was not rated. During 1970, 40 bulls received Preferred registration certificates, 63 rated Selected pedigree certificates, and 862 were Standard. All production requirements were increased in 1963 to 10,000 pounds of milk and 410 pounds of butterfat, and in 1970 to 12,500 pounds of milk and 500 pounds of butterfat.

IDENTITY ENROLLMENT

In 1970 the Directors of the Ayrshire Breeders' Association recommended two steps for the Identity Enrollment of unregistered Ayrshire females: (a) Females were approved for Ayrshire type and color by a breed representative who inserted an official ear tattoo to be recorded by the Association; and (b) a daughter by a registered Ayrshire bull must qualify with an inspection score of at least 0.825 points and have a standard DHIA record at least equaling the DHIR breed average for the age group.

Based on USDA Sire Summaries, Ayrshire bulls were Production Approved when their daughters qualified on a graded scale with the following:

Predicted difference	Repeatability
(pounds milk)	(percent)
+ 800	25
+ 500	60
+ 100	90

Double Approved sires had at least ten daughters with an average final score of 0.850 or more by the latest classification program.

Most Ayrshires are red and white in varying proportions. Some are mahogany instead of red. A few cattle are "black and white," or very dark mahogany.

AYRSHIRES IN THE SHOW RING

Ayrshire cattle were displayed at the agricultural fairs of New England and New York in 1842 and earlier. They competed in mixed classes with Devon, Durham, Hereford, and native cattle. The New York State Agricultural Society appointed a committee to discuss " . . . those forms, qualities and properties which most conduce to intrinsic value; and also that the distinctive characteristics of each separate breed may be as closely defined as possible." They met October 17–18, 1843. George Randall and others drew up a scale of points for judging Ayrshires, presented January 3, 1844. This "Scale of Points in the Ayrshire Cow" was cited in Volume 1 of the *Dominion Ayrshire Herd Book* in 1853. The 100-point scale gave 12 points to udder, which was to have the following characteristics:

Udder—In this breed is of more especial importance as the Ayrshires have been bred almost exclusively with reference to their milking properties. The great feature of the udder should be capacity, without being fleshy. It should be carried squarely and broadly forward, and show itself largely behind. As it rises upward, it should not mingle too immediately with the muscles of the thighs, but continue to preserve its own particular texture of skin—thin, delicate, and ample in its folds. The teats should stand wide apart, and be lengthy, but not large and coarse.

Ayrshires were displayed at the New York State Fair in 1852, 1854, and 1861. The Massachusetts Society and Massachusetts Board of Agriculture sponsored an exhibition in Boston in October 1857. Nineteen monetary prizes were offered in seven classes for Ayrshires.

The Ayrshire Breeders' Association adopted a Scale of Points for Ayrshire cows in 1885, giving 33 points to udder and teats from a total of 100 points. It provided for "teats from 2½ to 3 inches in length, equal in thickness, the thickness being proportional to the length." The udder was to be capacious, and sole nearly level. A Scale of Points for bulls was adopted in 1887.

Conflict arose in 1893 between American preference for milking cows with convenient size of teats and the Scottish "vessel-type" Ayrshire at the World's Columbian Exposition in Chicago. The Canadian judge preferred Canadian cows with upturned horns and vessel-type udders. "He wholly ignored milking points," according to *The Country Gentleman*. After this incident, Robert Wallace of Auchenbrain, Ayrshire, wrote: "I frankly admit that far too many of our Ayrshires have been judged from a wrong standpoint, our judges going wholly for beauty of form without giving, I may say, any heed to what is likely to be the most useful in the dairy. I have left off following the fancy, and am going in for what is likely to produce milk."

Ayrshire standards changed because of the Columbian Exposition. American breeders recognized that beauty might be added to pure utility.

The scale of points of the Ayrshire Breeders' Association in 1901 was worded similarly to the Canadian and Scottish scales of 1884 even to the udder and teat descriptions (from 2½ to 3 inches in length). The Swinlee type exerted little influence on Ayrshires in the United States before 1900. The older type in America was modified eventually. Scottish-bred Ayrshires exhibited at the Pan-American Exposition (Buffalo, 1901) and Louisiana Purchase Exposition (St. Louis, 1904) attracted attention.

NATIONAL DAIRY SHOW

The National Dairy Association held its first show in Chicago on February 15–24, 1906. Ayrshires, Dutch Belteds, Guernseys, Holstein-Friesians, and Jerseys were exhibited. Brown Swiss were admitted in 1907 after that breed association voted to become a dairy rather than a dual-purpose breed. The association sponsored the National Dairy Show, which was the court of last resort in dairy type. Regional shows at Springfield, Massachusetts; Waterloo, Iowa; and Portland, Oregon, were an intermediate level above state fairs. The National moved from region to region—Springfield, Massachusetts, to San Francisco over 40 years. It was interrupted by foot and mouth disease (1915), a depression (1932–34), and World War II. The Association, which had been a center of many national events, disbanded in 1946.

The Ayrshire Digest stated that not a single "vessel-type" udder was seen in the 1939 shows. "According to present-day standards, there is beauty in the popular udder of this period, even though it is built for utility." A similar change occurred in Scotland following the "new show" in 1921, with emphasis on udder quality and utility rather than vessel-type udders.

ALL-AMERICAN CONTEST

An All-American Contest, open to winners of first or second prize at county or district shows, or the top four places at state or larger shows, was initiated in 1960. Canadian animals competing in the United States shows were eligible also. Owners made nominations and submitted a glossy photograph and the show record. Females of milking age were required to be on production test before the shows. A nominating committee selected six animals from each class for placing by the All-American Judging Committee. A Junior All-

FIG. 6.1. The ideal Ayrshire cow was prepared by a True Type committee and two artists in 1924.

American Contest under the same rules limited nominations to owners under 21 years of age.

TRUE TYPE

The Ayrshire Breeders' Association appointed John Cochrane and A. H. Tryon as a True Type Committee in 1921. They studied many photographs and employed Robert F. Hildebrand and Robert F. Heinze to present their ideas in a painting. The picture portrayed True Type. The Ideal Ayrshire Cow possessed mammary development, dairy conformation, ruggedness, capacity, general

lines, and style of the breed. The udder was a compromise between vessel-type and the deep udders of some heavy milkers
(Fig. 6.1). A True Type model revised in 1970 had increased stature and was dehorned.

A scale of points for cows was revised in 1931. The committee re-
gretted insufficient information to revise the bull scorecard. Long-
wearing qualities were emphasized rather than "fancy" points.
Eight points were given to feet and legs, 10 to barrel, and 30 for
mammary development. Teats were to be of "convenient size."
Ideal weight of mature Ayrshire cows was given as 1,000 to 1,400
pounds. Many Penshurst Farm cows exceeded 1,400 pounds. Classification inspectors measured some 2,000 cows in 1942, of which 877
averaged 1,156 pounds, Excellent and Very Good cows tending to
be larger than others.

Type Classification

Type classification with Ayrshires began in 1941, following classification of three other breeds. All cows in the milking herd and bulls
past 18 months old have been scored since 1947. Association inspectors rated registered Ayrshires as:

	(points)
Excellent	More than 90
Very Good	85–89
Good Plus	80–84
Good	75–79
Fair	70–74
Poor	Fewer than 70

Mature cows should tape 1,200 pounds and heifers under 36
months at least 900 pounds to be eligible for a 90-point score. Average scores of Ayrshires classified in 1970 were:

Classification score	Bulls		Cows	
	(number)	(percent)	(number)	(percent)
90–100	7	12.1	126	5.2
85–89	39	67.3	1413	58.3
80-84	11	18.9	823	34.0
75–79	1	1.7	51	2.1
Under 75			10	0.4

The average score of 2,477 Ayrshires in 93 herds in 1970 was 0.855. Average production of cows increased with type ratings, with small differences between groups.

The classification form contained 10 divisions under the break-down system, including head and neck, shoulders and chest, middle and loin, rump and thighs, feet and legs, general quality, breed characteristics. The mammary system was divided into shape and size, fore and rear udder attachments, teats, veins, and quality. Suspensory ligament was considered. Each part was scored on a 100-point basis; the scores were totalled and divided by 10 for total score. A score could be raised or lowered at successive reclassifications. Distinct breaking away of the udder was faulted according to age of the animal, as also were winged shoulders or faulty feet and legs. The inspector recorded abnormalities such as wry-tail, wry-face, pop eye, cropped ears, or other hereditary characters. The classification was recorded on the registration certificate by the Association.

The classification system was improved in 1967 by adding descriptive terms for which each of the ten divisions was faulted on the basis of a 100 score. Descriptive terms of faults were:

Stature
    A. too small
    B. legs too short
    C. lack of stretch
    D. general coarseness
    E. lack of symmetry

Dairy Characteristics
    A. thick and/or patchy
    B. coarse
    C. lack of refinement
    D. thick thighs
    E. lack of breed character

Head and neck
    A. head too short
    B. head narrow
    C. Roman nose

  D. neck short and thick

Shoulders and chest

  A. shoulders thick

  B. shoulders open or loose

  C. crops weak and chine not prominent

  D. chest shallow or constricted

Middle

  A. slab sided

  B. shallow

  C. short

  D. round ribbed

Loin and rump

  A. loin narrow

  B. loin weak or flat

  C. low thurls

  D. low pins

  E. rump short

  F. tail high or coarse

Feet and legs

  A. rear legs sickle

  B. rear legs crooked

  C. rear legs post

  E. pasterns weak

  F. feet shallow heel

  G. feet splayed

  H. feet toe out

Mammary system and appearance

  A. too large or bulgy

  B. too small

  C. short fore

  D. short rear

  E. pendulous

  F. quartered

  G. tilted

Attachment

  A. fore weak

B. rear narrow

C. suspensory ligament weak

Teats, veins, quality

A. teats not even in size or placement

B. teats bulbous or too long

C. udder meaty

D. lack of veining

Type classification calls the owner's attention to the desirable and weak points in conformation of individual animals and to the transmitting ability of the herd sires and dams.

PRODUCTION RECORDS

Average production by milk cows in the United States in 1861 was less than 3,870 pounds. In contrast, one of the four Ayrshire cows imported in 1837 by John P. Cushing yielded 9,680 pounds of milk in a lactation. Native cows in the Sturtevant herd over 8 years averaged 4,605 pounds of milk (67 cow-years), and Ayrshire cows averaged 5,550 pounds of milk (67.9 cow-years). Eleven owners submitted to Secretary C. M. Winslow 614 records of Ayrshires from 1874 to 1890 which averaged 6,109 pounds of milk. Winslow considered this the average capacity for the breed under farm conditions.

The Home Dairy Test was adopted in 1884, with prizes offered for production of milk, butter, and cheese by individual cows, and for herds of six cows or more. Several cows produced 10,000 to 12,617 pounds of milk in a year. In 1894 the Committee on Tests stated: "Resolved, That this Association approves the plan of offering prizes to herds of Ayrshire cows on the basis of solids in milk and a certain percent of butterfat, as determined by the Babcock test.

"Resolved, That the Executive Committee be authorized to carry out this plan under such rules and regulations as it may prescribe."

Supervision of the tests was placed with the respective state College of Agriculture in 1895. Tests then were of 2 days' duration, but were extended to 365 days in 1896. Rena Myrtle 9530 completed the first record at the Vermont Agricultural Experiment Station, producing 12,175 pounds of milk, 468 pounds of butterfat. The Preamble of the Advanced Registry rules adopted in 1902, stated: "For the purpose of encouraging a better system of keeping milk and butterfat records, that we may obtain more and reliable records of the dairy yield of Ayrshire cows, we hereby adopt the following rules and regulations for the establishment of a system of Advanced Registry for Ayrshire cattle."

The requirements to qualify were:

	7 days		Year's record		
	Milk	"Butter"	Milk	"Butter"	(Fat)
2-year-old	200	8	5,500	325	192 6/7 lbs.
Mature form (5 years)	350	14	8,500	375	321 3/7 lbs.

Bulls entered the Advanced Registry when they scored 80 points and had two daughters qualify. They entered without scoring when four daughters had qualified. Butterfat tests were supervised for 2 days monthly by the Agricultural College representative. Owners reported daily milk production to the breed association monthly, from which the butterfat yield was computed. Home Dairy Tests were continued until 1911. The 2,598 Advanced Registry records completed before July 1917 averaged 9,555 pounds of milk, 3.95 percent and 377 pounds of butterfat. A higher requirement was set for the Advanced Registry in a 300-day Roll of Honor class established March 1, 1918. Those cows also must carry a living calf for 180 days during the test. Roll of Honor requirements were 6,000 pounds of milk and 250.5 pounds of fat as 2-year olds, up to 9,000 pounds of milk and 360 pounds of fat at mature age of 5 years.

Roll of Honor tests were lengthened to 305 days. Owners tended to delay breeding cows while on test, for higher production records. Generally only the selected cows were placed on test. The Association published only those records that exceeded the requirements.

HERD TEST

The Ayrshire Breeders' Association led by adopting a Herd Test in July 1925 which included *all cows in the herd, rather than selected individuals.* This called attention to good management of all cows and regular breeding. It measured a bull's transmitting ability rather than publicized a few higher producing daughters. Records of poor producers were omitted from the computed herd average after 1930 if their registration certificates were cancelled. Lifetime records became emphasized more than individual lactations. The Herd Test rules were replaced by Dairy Herd Improvement Registry (DHIR) in 1966 with all dairy breeds.

Average production of Ayrshires in the Herd Test is shown in

TABLE 6.1
AVERAGE PRODUCTION OF AYRSHIRE COWS IN THE HERD TEST

| Year | Cows | Actual production | | |
		Milk (lbs.)	Test (%)	Fat (lbs.)
1926	1,420	7,698	4.01	309
1930	2,712	7,992	4.02	321
1937	3,555	8,575	4.03	345
1947	8,830	9,052	4.05	377
1957	10,892	9,695	4.10	398
1968[a]	8,087	12,705	4.00	508

a. Computed to a mature equivalent 4.00 percent fat corrected milk.

Table 6.1. This was computed to a 305-day mature equivalent basis in 1957, as 10,478 pounds of milk, 4.10 percent and 432 pounds of butterfat. Records shorter than 240 days were excluded. Certificates of meritorious production were issued to animals exceeding a standard according to age. Eighty-eight cows exceeded lifetime production of 100,000 pounds of milk before 1969.

The 2,617 cows also tested for solids-not-fat averaged 10,705 pounds of milk, 8.7 percent and 933 pounds of solids-not-fat. The 254 cows tested for protein averaged 11,022 pounds of milk, 3.4 percent and 374 pounds of protein in 305 days.

A "production base" was computed from HIR records up to 305 days in length during the previous 3 years, beginning in 1959, for use in calculating Approvals, transmitting indexes, and other improvement programs.

The Advanced Registry was discontinued in 1926 but reinstated June 14, 1933, with higher requirements (10,000 pounds of milk, 400 pounds of butterfat for 2-year olds; 12,000 pounds of milk, 480 pounds of fat at 5 years old). Before the new requirements, Advanced Registry records averaged above 10,000 pounds of milk and 430 pounds of fat. Roll of Honor records averaged a little lower. Production tests are supervised under the "Uniform Rules for Official Testing" approved by the American Dairy Science Association, Purebred Dairy Cattle Association, and the breed associations.

Methods of supervision of DHIA records were modified in 1959–60 to be acceptable by all dairy breed associations. Supervision is by the superintendent of official testing of the respective state agricultural college.

The Ayrshire Breeders' Association maintains a file of lactation records for use with registration certificates, pedigrees, and sale catalogues. Preferred and Selected registration certificates are based on production records of ancestry. Analyses of production records and type classification scores determine an Approved Sire and an Approved Dam. The 100,000-pound Club certificate and an animal's lifetime production are determined from assembled records. Average yearly production of a herd, a type classification score of 0.835, and other qualifications are required for an annual Constructive Breeder Award.

Leonard Tufts began studies of desirable transmitting bulls through their daughters' records. His work was followed by a Research Committee of Mr. Tufts, James W. Linn, and Gus Bowling. E. C. Deubler, Henry B. Mosle, and C. T. Conklin served later. The West Virginia Agricultural Experiment Station assisted with analysis of records provided by the Association.

Secretary C. T. Conklin wrote: ". . . the late Leonard Tufts of Pinehurst, N.C. . . . patiently labored for a method of rating sires that would require a full and complete record of all daughters regardless of level of production."

APPROVED AYRSHIRE SIRES

The plan for Approved Ayrshire Sires adopted in 1942 resulted from these studies, requirements for qualifying being: (a) a bull must have at least ten daughter-dam pairs, using the first available lactation records; (b) all tested daughters must average at least 8,500 pounds of milk or 340 pounds of butterfat, with not under 3.9 percent fat; (c) the sire must have an equal parent index of at least

8,500 pounds of milk or 340 pounds of butterfat; (d) not less than 90 percent of tested daughters must each meet production requirements.

Some 101 bulls were approved before 1943. The standard was increased in 1945 to 9,000 pounds of milk, 3.9 percent and 360 pounds of fat; 60 percent of daughters to exceed this standard. The standard was raised in 1958 to 9,450 pounds of milk, 3.9 percent and 390 pounds of fat in 305 days on a 2× mature equivalent basis. Requirements were 10,000 pounds of milk and 400 pounds of fat, if test was disregarded. The base in 1961 was 9,750 pounds of milk and 400 pounds of fat. In 1963 this was increased to 10,000 pounds of milk, 3.9 percent and 410 pounds of fat in 305 days on a 2× mature equivalent basis. If test was disregarded, 11,000 pounds of milk and 440 pounds of butterfat were required.

Since 1946, bulls approved for production also were eligible for Approval on type transmission when ten or more unselected daughters had been classified with an average score of 0.825 or above. At least 50 percent of registered tested daughters 3 years old or older must have been classified. A bull was called Double Approved when qualified for production and type transmission. The average type score was increased to 0.835, and in 1963 to 0.850. Six bulls qualified as Approved Sires in 1967, bringing the total number to 840.

Bulls in artificial service were rated on achievements of the first 20 daughters from tested dams. Average production of the herd and of dams was published to indicate the environment when the records were made.

USDA Sire Performance Summaries for production, to which the Association's type ratings are added, superceded the Approved Sire program in 1968. This plan utilized DHIR and DHIA records, herdmate comparisons, adjustment for number of lactations and herds involved, and of estimated repeatability weighted by computer methods.

Approved Dams

A plan for Approved Dams in November 1942 required (a) the dam must have three or more daughters with first lactations of at least 8,500 pounds of milk, 3.9 percent and 340 pounds of butterfat; (b) 60 percent of her tested daughters must each have met production requirements; (c) 60 percent of female progeny above 3 years old must have production records; (d) 50 percent of daughters over 4 years old must exceed the requirements on a 305-day 2× mature equivalent basis. The index of a proven son was regarded equivalent to a daughter.

The production requirements for Approved Dams were increased gradually as the average production of Ayrshires on DHIR test increased. In 1970 the mature equivalent requirements for three daughters were 11,600 pounds of milk, 3.9 percent and 470 pounds of butterfat by not less than 60 percent of her registered progeny over 4 years old. With only two daughters, each must yield at least 12,180 pounds of milk, 3.9 percent and 487 pounds of fat. If fat test was disregarded, 12,760 pounds of milk or 517 pounds of butterfat were needed.

Few living cows would qualify as Approved Dams, yet the rating popularized their progeny for breeding purposes. Some 5,394 Ayrshire cows became Approved Dams through 1968, with 68 cows added in 1969.

100,000-POUND CLUB

Since 1933 the Association issued certificates to cows qualifying for the 100,000-pound Club in lifetime milk production. One such cow was Auchenbrain White Beauty 2d 21687, Imp.; calved in 1902 and imported by Penshurst Farm in Pennsylvania (Fig. 6.2). She produced 125,123 pounds of milk, 5,161 pounds of butterfat, as well as seven bulls (four in Advanced Registry) and five Advanced Registry daughters. By 1944 about 90 percent of American-bred Ayrshires were descended from her.

Imp. Garclaugh May Mischief 27944 (Fig. 6.3) produced over 125,000 pounds of milk in nine Advanced Registry records. Seven daughters were tested, and four sons served in famous herds, including Penshurst Man O'War 25200 (Fig. 6.4) at Penshurst Farm, and Penshurst Mischief Maker 18719 at Strathglass Farm. She exerted as great an influence on the breed as did Auchenbrain White Beauty 2d of an earlier generation.

The Association secretary wrote that:

It is the lifetime record that counts. Good dairy cows must wear
well and perform efficiently if profits over and above original
costs and maintenance charges are to be secured. Since the average cow does not come into production until 30 to 36 months
of age, she must begin her milking career with quite a charge
against her account. Very rarely will a first-calf heifer "pay
out" during her first lactation for all expenses from birth.
But with each succeeding lactation, an added credit . . . soon
more than balances accounts, putting such cows on a dividend-
paying basis.

Strathglass Lucky Puff had a cumulating record of 189,843 pounds
of milk, 4.0 percent and 7,598 pounds of butterfat in 5,235 days.
She was an Approved Dam with eight progeny averaging 9,041
pounds of milk, 4.22 percent and 381 pounds of butterfat per lactation. She and two daughters classified Excellent. Delchester Au-
dacious Betty 2nd had a lifetime production of 219,891 pounds of
milk and 8,676 pounds of butterfat in 5,439 days in milk.

FIG. 6.2. Auchenbrain White Beauty 2d 21687, Imp., born in 1902, was one
of the great transmitting dams of the breed.

FIG. 6.3. Imp. Garclaugh May Mischief 27944 was a great lifetime producer and transmitting dam. Her progeny included Penshurst Man O'War 25200.

FIG. 6.4. Penshurst Man O'War 25200 was one of the great improvers of the breed. Of 147 living Approved Ayrshire Sires in 1955, over 72 percent were descended from him. His double grandson Penshurst Man O'War 30th was in artificial service until 16.7 years of age. His 167 daughters had 748 records averaging 9095 pounds of milk, 4.09 percent and 372 pounds of fat. He had 19 Approved sons, 33 Approved daughters and 34 progeny with over 100,000 pounds of milk.

BREEDING AYRSHIRES

A 4.0 percent butterfat test is among the characteristics emphasized for Ayrshire milk. An example of this influence was seen at Sycamore Farm, once owned by Mrs. E. R. Fritsche, former president of the Association. An increase in fat test was attained largely through three transmitting sires whose immediate ancestors averaged mainly over 4.0 percent butterfat in their milk. The change accomplished by this herd is shown in Table 6.2.

TABLE 6.2
CHANGE IN AVERAGE BUTTERFAT PERCENTAGE IN MILK OF AYRSHIRES
AT SYCAMORE FARM THROUGH INHERITANCE.

Year	Number of cows	Days in milk	Average production			Proportion of herd testing under 4.00 percent
			Milk (lbs.)	Test (%)	Fat (lbs.)	
1926	42	293	7,192	3.85	277	57.1
1931	39	322	10,808	4.14	448	30.8
1935	52	317	10,312	4.27	441	13.3

T. V. Armstrong summarized published analyses of Ayrshire milk in 1959. Some 208 milk samples from 14 cows contained an average of 4.15 percent fat, 8.96 of solids-not-fat, 3.58 of total protein, 4.70 of lactose, and 0.68 percent of ash. He pointed out a trend, which DHIR records confirm, toward an increasing percentage of fat in Ayrshire milk. The Association encouraged securing records of protein and solids-not-fat contents of Ayrshire milk.

W. J. Tyler and George Hyatt, Jr., calculated a heritability coefficient of 0.3 for type among 1,601 Ayrshire daughter-dam pairs and 3,738 paternal sisters. The correlation coefficient between butterfat yield and type rating of 5,177 cows was 0.19.

J. C. McDowell of the USDA Bureau of Dairy Industry analyzed many DHIA records and found that the higher producing cows made the greater returns over feed cost on the average. Secretary David Gibson, Jr., reported: "The man that can get 200,000 lbs. of milk from 20 Ayrshire cows in a year is certainly doing a better job for himself and the breed than one who has 30 and getting the same yearly production. Largeness of operation is not the complete answer or a sound substitute for efficiency."

Slightly more than 100 male calves are born per 100 females.

Fewer bulls are registered although many are used in grade herds. Introduction of artificial breeding since 1938 and increasing numbers of cows per herd reduced the demand for bulls in natural service. This is reflected in the proportion of male registrations, as shown in Table 6.3.

POLLED AYRSHIRES

The polled character was not infrequent among native cows in Ayr, Scotland. Some of them entered into the foundation of Galloway cattle—a black polled beef breed. Polled Ayrshires persisted through the generations. They were present early in parts of Vermont, including Rutland County.

TABLE 6.3
INFLUENCE OF ARTIFICIAL BREEDING AND INCREASING SIZE OF HERDS
ON THE RATIO OF MALES TO FEMALES REGISTERED

Year	Ratio of males to females registered	Year	Ratio of males to females registered
1900	1:2.9	1940	1:4.8
1910	1:2.9	1950	1:6.1
1920	1:3.3	1960	1:11.4
1930	1:4.0	1970	1:14.6

Jacob S. Horst of Lancaster County, Pennsylvania, purchased a polled bull, Barberry Heights Peter Pan 4th, from Vermont in 1929. He owned a daughter of the famous Penshurst Charming Princess (horned). Some good polled Ayrshires descend from her, and several noted polled bulls trace to this herd. The polled bull Prince Perhaps of Bart, from the cow Dot's Polled Polly 129757, went through the Lancaster County Ayrshire Association sale in 1940. Three-fourths of his progeny from horned cows were said to be polled. His first daughters outyielded their dams. This bull was advertised under the caption "Why not polled Ayrshires?" in 1941 by Aaron H. Harnish of Lancaster, Pennsylvania.

The Ayrshire Breeders' Association authorized the symbol "X" before the registration number of polled Ayrshires. Breeders were invited in 1944 to report past or present Ayrshires born without horns. It was found that 175 had been recorded mostly in recent years. Later applications for registration bear this information.

Burt Froberg, La Fayette, Rhode Island, reported birth of a polled heifer calf in 1948. No polled blood was in her ancestry so far as he knew. A polled bull calf Grey Banks' Golden Mulley Boy was born to horned parents and grandparents. Harold B. Cobb, Bluffton, Illinois, reported this bull's first calf from a horned cow to be polled.

An editorial in *The Ayrshire Digest* (in December 1946, page 194) stated "For years it was believed that no Ayrshire was complete without horns. But in the meantime the percentage of dehorned Ayrshires has increased. In recent years they have won championships at representative shows, and they have sold well at auction sales. It is estimated that more than half of the breed is now dehorned. If it is more economical to keep dehorned Ayrshires, they will be dehorned Ayrshires. Fads and fashion cannot compete long against practical results."

A note from Scotland in *The Ayrshire Digest* (June 15, 1948, page 416) stated: "The dehorning of Ayrshires in both Scotland and England is receiving a great deal of attention. Principal reason is the increasing use of paddocks or courts in which cattle are confined during the winter months rather than maintained in the stanchions. English breeders seem to be most enthusiastic about dehorning."

The *genetic* character "polled" is dominant, hence their numbers will increase, especially since breeders formed the Polled Ayrshire Breeders' Association in 1945. Sixty-eight breeders then were in Pennsylvania, New York, North Carolina, Vermont, West Virginia, and Indiana.

Some polled Ayrshires trace from the polled bull Lucky Boy of Willow Springs, out of the polled cow Maggie Mason of Ira, owned by Clayton Fish of Ira, Vermont. The cow Dot's Polled Polly possibly inherited this character from her sire—Armour's Asa of Sand Hill. His short scurlike horns were attached loosely. *The Ayrshire Digest* stated that several of this bull's progeny had similar horns. This line was believed to be a mutation. From 1944 to 1957, 4,711 polled Ayrshires were registered. During 1969, 33 males and 352 females registered were polled. Three polled Ayrshires were exported to Scotland in 1948.

BLACK-AND-WHITE AYRSHIRES

Melrose Good Gift 14612 at the Kansas State Agricultural College herd was regarded by E. N. Wentworth as typical "black-and-white" color for the breed. His head and spots were a dark mahogany. When mated with unrelated red-and-white cows, his male calves turned to a blackish tinge when 2 to 4 months old, and were distinctly "black-and-white" within 4 months. Heifer calves were red-and-white from this cross. The cow Bangora was the original black-and-white cow in the herd. Wentworth concluded:

> If the factor for black-and-white color is represented by B, the hereditary constituents are as follows: BB is always black-and-white; bb is always red-and-white; Bb is always black-and-white in the male and red-and-white in the female.
>
> Black-and-white is a simple allelomorph of red-and-white in Ayrshire cattle.
>
> In the male the black-and-white character is dominant and in the female the red-and-white is dominant.
>
> Males heterozygous for the two characters are black-and-white, while females heterozygous for the two characters are red-and-white.

HEREDITARY DEFECTS

Crampy. A neuromuscular condition called crampy, progressive posterior paralysis, spasticity, or stretches is a recessive hereditary character in mature bulls and cows. Heterozygous (carrier) animals do not exhibit the condition. Symptoms in the homozygous condition may include periodic spasticity of the back muscles and/or one or both rear legs when under stress. They favor affected parts when rising, standing, or walking. Intensity increases with age and may result in death.

Defective skin. Defective skin occurs at birth in calves of several dairy breeds. Skin fails to develop on parts of the lower limbs, mucous linings of the mouth, or nostrils. Such areas soon become infected, and the animal dies of septicemia.

Dropsy or edema. Congenital dropsy, edema, or anasarca develops late during fetal growth. The uterus becomes greatly filled with fluids. Fetal tissues are swollen; rear legs may be enlarged, and the ears droop. The condition is believed to be due to a single recessive gene, expressed in homozygous condition.

Ear notch. An ear notch character occurs in varying degrees in some Ayrshires and Jerseys. Some 53 out of 6,358 classified Ayrshires had ear notch.

Wry-face. Wry-face—a bow in midline of the face—was seen in six Ayrshires among 6,358 animals examined. The character occurs in other breeds.

Wry-tail. Sacral vertebrae are longer on one side than on the other, causing wry-tail. Ayrshire fieldman M. H. Benson saw wry tail in a three-generation group of Ayrshires. About 1.4 percent of Ayrshires classified had wry-tail. Presence of wry-tail may be genetic evidence of some Jerseys among foundation stock in early development of the breed.

RECESSIVE LETHALS

A committee studied lethal and undesirable characters to develop a program. When a bull in artificial use transmitted lethal or serious functional abnormalities, a letter requested that his semen not be distributed. A recessive character crops out only when *both parents* possess the controlling gene. The owner of a cow that drops such a calf should analyze his own herd for half-sisters and their progeny, one-half of which may carry the same gene from a common ancestor.

ARTIFICIAL BREEDING

Fillmore Inca 56336 was used by the Litchfield County Artificial Breeding Association in October 1941. Ayrshire breeders placed desirably proved bulls with the Massachusetts Selective Breeding Association in August 1946 for use throughout New England. Other organizations maintained Ayrshire bulls or bought semen for their members, Penshurst Man O'War 30th being used widely. Some 58.8 percent of Ayrshires registered in 1969 were conceived from artificial service. Forty-six Ayrshire bulls were in active use or being sampled in 1969, and their frozen semen transferred to other organizations. To extend the search for capable bulls, an Ayrshire Genetic Breed Improvement Program was adopted by mail ballot in 1970. A rotating 5-year committee accepted at least five bulls annually that met eligibility requirements of ancestry. These bulls were placed under contract with the owner for A.I. sampling. The objective was for 30 to 40 daughters to meet production and type classification standards on which the committee and the owner decided future disposition. The cooperating semen-producing business participated. "As an incentive to secure tested and classified daughters of the young sires being sampled, a payment by the Ayrshire Breeders' Association of $5.00 be made to the owner of each daughter having completed a Standard DHIA or DHIR 305-day 2× record, and which has been Officially Classified." The Association's Records Division analyzed all records of ancestry and progeny.

Six Canadian Ayrshire bulls went to Scotland, including two sons of Selwood Betty's Commander, a Canadian-bred bull long in A.I. service at Eastern A.I. Cooperative, Ithaca, New York. His tombstone on the Ayrshire Breeders' Association ground proclaims him "The milkiest bull in the world" for transmission of high milk production.

Constructive Breeder Award

A Constructive Breeder Award was initiated in 1942 for breeders owning at least eight cows. Fifty percent of the cows that have calved must have been bred by the owner, and 65 percent owned at least 4 years. The herd must be on a testing program and have produced at least 9,500 pounds of milk, 3.9 percent and 390 pounds of butterfat per cow in 305 days that year. Seventy percent of females in milk must have been type classified, and scores average 0.835 points. Owners apply annually for this award. Forty-eight breeders qualified in 1969.

Approved Ayrshire Milk

In December 1937 a program was adopted to promote Ayrshire milk under a registered trademark—"Approved Ayrshire Milk." Milk from herds with at least 75 percent of registered cows was eligible. The milk must contain at least 4 percent fat and conform to Grade A standards. The herd must be free from brucellosis and tuberculosis. The herd, equipment, and processing methods were inspected before granting a license. Seven distributors were licensed in 1967. The trademark "Scotty Milk" also has been used since 1968 for milk with at least 2 percent fat.

EXTENSION SERVICE

J. G. Watson was the first field agent in 1916 for direct service to breeders. The program is educational, promotional, and service-to-aid with breeders' problems, state and local organizations, junior work, sales, and similar activities. A representative serves also as a type classifier, supplemented by the Association's office staff.

BREED PUBLICITY

Breed associations bring the good qualities of a breed before the public with breed pamphlets. News releases and annual Year Book, Advanced Registry, Herd Test, Approved Ayrshire Sires and Dams volumes are used. The modest annual Year Book was followed by *The Ayrshire Quarterly* on April 1, 1915, and *The Ayrshire Digest* (monthly) as breed publications. "The Ayrshire Cow—A Handbook for Breeders" appeared as an elaborate brochure in 1955. Exhibits at shows attract attention of the public and other breeders.

DISTINGUISHED SERVICE AWARD

The Board of Directors established a Distinguished Service Award in 1948, restricted "to recognize distinguished service for the advancement and improvement of the Ayrshire breed." The first ten awards were conferred on breeders in nine states, including one to Secretary Clifford T. Conklin. He contributed to developing association programs and policies for 25 years.

The office of Executive Secretary David Gibson, Jr., is The Ayrshire Breeders' Association, Brandon, Vermont 05733.

PUREBRED DAIRY CATTLE ASSOCIATION

The Purebred Dairy Cattle Association was organized on July 5, 1940, at Peterborough, New Hampshire, for problems of mutual interest among the dairy breeds. A "Court of Dairy Queens" at the National Dairy Show in 1940 and 1941, an essay contest on "The

Value of a Purebred Dairy Sire," unified dairy cow and bull score-cards, unified rules for official production testing, a code of sales ethics, health regulations for dairy cattle, regulations for artificial breeding and use of frozen semen, and other mutual problems have been considered by committees organized by the association. Recommendations of this group are passed for consideration to the separate breed associations. Membership in the organization consists of three representatives of each breed registry association, usually the president, secretary, and a member.

REFERENCES

Anonymous. 1843. Notice of annual meeting. *The Cultivator* 10:93.

Bowling, G. A. 1951. Ayrshire cattle in America—Efforts at breed improvement. *Trans. Highland Agr. Soc. Scotland* 62(ser. 5):1–16. Reprinted in *Ayrshire Dig.* 39(2):11, 14; 39(3):146–47, 238; 39(4):252–53. 1953.

———. 1964. The tools with which we work. *Ayrshire Dig.* 50(1):36–37; 50(2):74, 77; 50(3):120, 198; 50(4):220, 237.

Conklin, C. T. 1942. Ayrshires in America. In E. P. Prentice, *American Dairy Cattle*. Harper, New York.

Cushing, George. 1962. Bold new plan of action. *Ayrshire Dig* 48(6):313–14, 358–59.

Gibson, David, Jr. 1958. Executive Secretary's annual meeting report. *Ayrshire Dig.* 44(6):309–12.

———. 1968. New Ayrshire Performance Summaries. *Ayrshire Dig.* 54(12):438–39.

Gilmore, L. O., and N. S. Fechheimer. 1961. Abnormalities in cattle. *Ayrshire Dig.* 47(7):414–15.

Howard, Sanford. Characteristics of Ayrshire cattle. *U.S. Patent Off. Agr. Rept. 1863*, pp. 193–98.

Pratt, Zadock. 1862. The farming region of Greene and Orange Counties, New York, with some account of the farm of the writer. *Rept. of Comm. of Patents for the year 1861. Agriculture*, pp. 411–27.

Rice, V. A. 1943. Sparks from observer's observations. *Ayrshire Dig.* 29(2):5–7, 25, 37, 43–44, 50–51.

———. 1943. Chips off the old block. *Ayrshire Dig.* 29(12):14–16, 79–84.

———. 1944. Wanted: An improved sire index. *Ayrshire Dig.* 30(2):8–9, 36–37.

———. 1954. Herd averages and bull provings. *Ayrshire Dig.* 40(12):766, 960.

———. 1956. The evolution of approval. *Ayrshire Dig.* 42(1):5–6, 34.

Rotch, Francis M. 1862. Select breeds of cattle and their adaptation to the United States. *Rept. of Comm. of Patents for the year 1861. Agriculture*, pp. 427–69.

Strohmeyer, H. A., Jr. 1962. Photography of dairy cattle. *Ayrshire Dig.* 48(5):273–77.

Sturtevant, E. L. and J. N. 1875. *The dairy cow. A monograph of the Ayrshire breed of cattle.* A. Williams & Co., Boston.

Tufts, Leonard. 1931. Notes on breeding. *Ayrshire Dig.* 17(1):5–6, 24–26, 28.

———. 1938. Random samples—The basis for proven sire studies. *Ayrshire Dig.* 26(3):3–4, 38.

———. 1938. Ayrshire families in the making. *Ayrshire Dig.* 27(2):5–6, 38.

Tyler, W. J., and George Hyatt, Jr. 1948. The heritability of official type ratings and the correlation between type ratings and butterfat production of Ayrshire cows. *J. Dairy Sci.* 31:63–70.

Watson, J. G. 1918–21. A history of the Ayrshire cow. Published serially in *Ayrshire Quart.* 4(3):3–4 to *Ayrshire Dig.* 7(8):14–15.

Wentworth, E. N. 1916. A sex-limited color in Ayrshire cattle. *J. Agr. Res.* 6:141–47.

Ayrshire Digest

1925. Ayrshire herd test 11(5):7.

1926. Auchenbrain White Beauty 2nd, Ayrshire mother supreme. 12(12): 5–9.

1931. Superior Sires. 17(8):6–7.

1933. Ayrshires at the Columbian Exposition. 19(12):18–19, 55–57.

1934. Dairy Herd Improvement records given recognition. 20(2):6, 17.

1937. What's back of the Ayrshire breed? 23(12):5–6, 36–38.

1938. Rules for Approved Ayrshire Milk licenses. 24(4):5.

1939. How the Ayrshire came to Massachusetts. 25(5):13, 34–35.

1940. Tuft's Approved Sire program adopted. 26(12):10–12.

1941. How big should Ayrshires be? 27(9):5, 29–30.

1941. Type classification plan adopted. 27(12):8–9, 136.

1942. A plan for selecting Approved Dams. 28(12):16, 150.

1943. First Constructive Breeder Award announced. 29(7):7, 50.

1946. The Double Approved Sire is here. 32(7):5–6.

1946. Code of ethics adopted. 32(7):8, 66.

1947. Research—Key to the Ayrshire treasure chest. 33(12):12–13.

1955. A salute to Penshurst Man O'War. 41(3):105–07, 132.

1956. How the Approved Dam program works. 42(2):45–46, 87.

1957. The Ayrshire udder. 42(8):398–99, 432.

1957. P.D.C.A. works for all dairymen. 43(1):5, 38.

1958. All American Ayrshire award. 44(2):49–50.

1960. The Ayrshire score card useful as breed improvement tool. 46(12): 836–38.

1963. The story of Penshurst. 49(5):288–93.

1963. The Ayrshire cow. A handbook for breeders. 3d ed.

1967. Classification adds descriptive terms, provides more meaningful information. 53(2):56–57.

1968. The Ayrshire Approved Dam plan. 54(12):510.

1970. The Ayrshire genetic breed improvement program. 56(1):10, 13.

1970. Identity Enrollment Program proposed. 56(1):12–13.

1970. Ayrshire pedigree evaluation plan for young bulls. 56(1):15.

1970. New rules for Approval of Ayrshire sires. 56(1):15.

CHAPTER 7

BROWN SWISS IN SWITZERLAND

THE OLDEST breed of dairy cattle was developed in the mountain-
ous area of Switzerland, a country that occupies 15,940 square miles
with the plateau, Jura, and Alps regions. It is on the same latitude
as northern Maine, upper Michigan, and Washington state. The elevation is between 646 and 15,217 feet above sea level. About 77.5
percent of the area is considered productive land. The country comprises 25 cantons and half-cantons, inhabited by some 4 million
people.

Rainfall and temperatures vary considerably over the country.
Most of the young stock and some cows are taken to mountain
pastures for about 3 months in the summer (Fig. 7.1). Sudden
snowfalls may cover the mountain meadows for 3 or 4 days, neces-
sitating storage and use of hay. The cattle are herded; milk is made
into cheese, which is carried to the valleys. Many cows remain

Headpiece: Vignette of Brown Swiss cow.

stabled in the valleys; they are fed soiling crops largely of mowed pasture forage during the summer. All other grasses are cured into hay for the winter. Little grain is fed.

When Emperor Rudolph I of Hapsburg (Austria) died, the cantons of Schwyz, Unterwalden, and Uri formed the League of Three Communities on August 1, 1291. Freedom from feudal obligations to Austrian nobles was recognized by Emperor Henry VII in 1309

FIG. 7.1. Many heifers and some cows are taken to the high mountain meadows for about 90 days each summer.

and Frederick II in 1340. Other cantons joined between 1332 and 1513. The country was called Schweiz first in 1320, and commonly after 1336.

The Hundred Years War in the Middle Ages revolved around alpine possessions, cattle, pasture rights and privileges among cloister dependencies of five monasteries, and free farmers in Appenzell, Schwyz, Uri, and other areas. Cattle raising and dairy products were of great importance.

The treaty of Westphalia in 1648 liberated Holland and set the Swiss confederation apart from the German empire. The peace of Vienna after Napoleon's empire broke down brought the French, German, Italian, and Romanic areas of Switzerland to the present borders.

Four breeds of cattle are native to this country—Simmentaler (fawn and white), Brown Swiss, Freiburg (black and white) and Eringer—distributed as in Figure 7.2. Some 46 percent of the 1,646,-229 cattle in 1956 were Brown Swiss. They were concentrated in the eastern part of the country, with 10 percent or more of this breed in most of the cantons.

EARLY CATTLE

Archaeological studies were made of some 200 lake dwellings in Europe, mainly in Switzerland and Germany. Wild animals common in the region were *B. primigenius* Bojanus, bear, stag, wolf, and other large animals. A few specimens of *B. taurus trochoceros* Rutimeyer were found, but this species apparently became extinct. When Neolithic tribes migrated westward up the Danube valley into Europe, they brought a smaller species of domesticated ox— *B. longifrons* Owen, also known as *B. taurus brachyceros* Rutimeyer. Bones of this ox predominated in debris of the lake dwellings. This

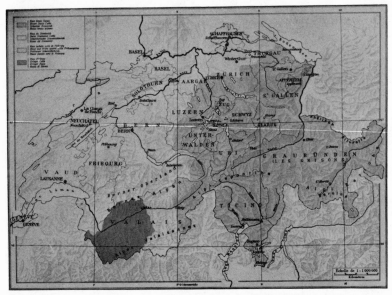

FIG. 7.2. Switzerland comprises 25 cantons and half-cantons. Brown Swiss cattle are bred in the shaded eastern area; Simmentaler or Fleckvieh cattle in the west; Eringer cattle in a part of Canton Wallis; and black-and-white Freiburg cattle in one canton. Up to 10 to 29 percent of the cattle in local areas of most western cantons also are Brown Swiss.

species is believed to have been the main ancestor of cattle in the plateau and mountain areas.

Helvetians (branch of the West Celts) lived in Switzerland between 500 B.C. and 58 B.C. and are ancestors of the present people. Romans invaded some parts between 58 B.C. and A.D. 200, in the late Bronze and early Iron Ages, as seen from relics and Roman coins in later lake dwelling sites. The size of some short-horned alpine cattle was increased by a larger strain brought by German immigrants about the first century after Christ. Servatius, field commander of the Goten King Theodore the Great, reported in A.D. 515 that the kind of cattle "appeared valuable on account of its size."

Kussnacht excavated bones typical of *B. brachyceros* (*B. longifrons*) at Gesslerburg on Mount Rigi. C. Keller identified numerous vertebrae, a few lower jaws, and other bones in 1919. J. Ulrich Duerst concluded from skeletal measurements that the large strains developed from a cross presumably in central Germany between *B. primigenius* and *B. brachyceros*.

Fossil skulls and Swiss folklore, according to Engeler, pointed toward the origin and descent of Brown Swiss cattle thus: (a) The race in its first domesticated form traced to the lake dwellers' peat cow, brought from the Near East under domestication by early Neolithic people. The peat cow and its descendants, the Celtic-Germanic cow, were maintained until the pre-Roman period in north, west, and central Switzerland, except in Graübunden. (b) The first mingling between *B. brachyceros* and *B. primigenius* cattle occurred in upper and middle Italy about the fourth century before Christ. These cattle were taken into Graübunden canton with migrations of pastoral people of Rhätur. This hybrid was restricted essentially to the Rhätischen valley of canton Graübunden.

The conquest and domination of the Helvetians by the Romans in 58 B.C. to A.D. 200 had not influenced the kind of cattle in central and northern Switzerland. Roman cattle did not enter this region. A hybrid between the small, short-horned Celtic cow (*B. longifrons*) and the wild urus (*B. primigenius*) was brought by German immigrants into northern and western Switzerland during feudal times.

EARLY MONASTERY CATTLE

The monk Meinrad cleared the forest at Einsiedeln in A.D. 861 and built a chapel for worship. Kaiser Otto III founded the cloister there on October 27, 947, and its abbot exhibited the dignity of a powerful sovereign, according to Duerst. Cattle breeding expanded. Duerst assumed that the hybrid German cattle were used there, since the district around Lake Zurich to the north had been colonized by Germans. He attributed the large size of the Rigi race of cattle to the German cattle influence in this mountain district.

Deaths of cattle in unusual numbers in 942 was mentioned in monastery records. Alps and meadows were owned by the monastery in 965. The Chronist of Muri (a branch of the monastery at Einsiedeln) mentioned a herdsman (Senn) taking cattle of 12 owners to a summit where he herded and milked them and made cheese for the owners from July to autumn. The owners were cloister retainers. A portrayal representing a raid of their animals on the night of January 6–7, 1314, pictured two cattle colored distinctly brown, one purple, one yellow ochre, and one seal red. A monastery inventory on March 28, 1544, listed 2 bulls, 34 cows, 30 oxen, and 17 other cattle. Rulers and influential people from Austria, Germany, and Hungary bought cattle in the canton three centuries ago.

Abbot August I wrote on August 12, 1607, to Count Johann von Recking concerning Swiss cows of a desired color: "We have obtained reply from our officials. They have directed that we have no such kinds with colors among our monastery cows." Nor could the abbot offer the buyer any bull with "curly hair." Engeler concluded that Einsiedeln and Swiss cows were more or less single colored then. Count Maximillian wrote to the Abbot on January 22, 1618, to buy 2 bulls of "beautiful large size and red color with large necks" and 10 or 12 milk cows of similar color, or if not available, then of "beautiful light weixelpraun."

In 1655 cows were being bred to calve in late winter and early spring at Einsiedeln to take advantage of summer pastures. Cattle at the monastery were described in 1786 as being black or brown, with no yellow, white, or red. This monastery and village are shown in Figure 7.3.

The *Sennengesellschaft* or herdsmen's society was founded in 1614. One objective during the next two centuries was to protect and defend their cattle against wild animals and "all other evil dangerous happenings." They reorganized in 1861 as the Farmers' Society of Einsiedeln.

LOCAL STRAINS OF SWISS CATTLE

At least 12 separate strains of local cattle were recognized by name before 1800 in six or more cantons. There were Schwyzer, Marsh, Toggenburger, Unterwalden, Urner, Haslithaler, and other local types in 1859. Some cattle had sway-backs, high shoulders, or crooked legs, as criticized at the first large exhibition of Swiss cattle in Langenthal in 1868. Wilhelm mentioned three chief groups of solid-colored Alpine cattle in 1872.

The Schwyzer or Rigi race (in the Mount Rigi district) varied from light gray-brown to dark brown with a light stripe along the back, about the eyes, and muzzle. The cattle were ruggedly built, excellent for milk, work, and meat. The Rigi race then was found in cantons Schwyz, Lucerne, Zug, and southern Zurich. Smaller

FIG. 7.3. The monastery and village of Einsiedeln as they appear today.

brown cattle were in northeastern Switzerland. Wery mentioned wider distribution of the large cattle in 1883; he said that the cows had good udder development and yielded 2,400 to 3,000 liters (5,433 to 6,816 pounds) of milk yearly.

Differences between strains of brown cattle were decreasing from valley to valley, as observed at the Neuenberg exposition in 1887. The Gotthard tunnel connecting Flüclon on Lake Lucerne with the Ticine valley above Milan, Italy, in 1880 increased traffic in cattle. Health of cattle was under supervision of district inspectors when exhibited at the cattle markets. Health certificates were required for rail shipments. Lenhert mentioned greater uniformity of type in 1896, and that the most beautiful cows were found south of Lucerne on Mount Rigi.

Heinrich Abt, early herdbook leader, described Brown Swiss cattle as dark brown to silver gray, seldom with white spots under the belly. White spots elsewhere on the body debarred winning a premium. White on the forehead, muzzle, flank, elbow, or in the switch was frowned upon.

Advancement in cattle breeding on the high mountains accompanied improvement of the alps and meadows and an adequate plane of feeding. Experience with cattle was behind the proverb— "Hunger im Stall, Hunger überall."

CATTLE SHOWS

Competitive showing of Brown Swiss cattle began with a cantonal show in 1805. In Canton Schwyz, prize competitions began in 1857, following international expositions in Paris in 1855 and 1856. Cattle under 12 months old did not compete. United States Consul Tanner commented in 1887 that "the system of having shows offers additional stimulus and incentive to have fine cattle that has caused the cattle of Europe to be *pushed* . . . to the high point which they have attained. . . . It has caused common stock to disappear entirely from Europe."

Dr. Lytin invented a measuring instrument in Baden-Baden, recognized by the German Agricultural Society for use when judging animals. Hugo Lenhert recommended its use in Switzerland in 1873.

The Society of Swiss Farmers sponsored a scale of points for judging Brown Swiss cattle before 1882. Shows were regarded as necessary in demonstrating desirable breeding types. The show system differed between cantons but periodic judges' conferences brought more uniformity. A conference called by the Swiss Department of Industry and Agriculture at Wadenswil in July 1893, incorporated body measurements into the scorecard, based on measurements of prize animals at previous national expositions. Cattle from the higher altitudes were smaller than those from lower zones "due to feed supply."

Scoring and Classification

Animals are inspected, scored, and measured for type classification, as in the shows. The bull and female scorecards have ten anatomical divisions valued at ten points each for perfection. Body measurements have been established for different ages; those for mature bulls and cows are given in Table 7.1. Consideration when scoring an animal is given to age, condition, disposition, pregnancy, whether stabled or "alped" with relation to activity, stability, environment, and climatic conditions.

Breed Characteristics

Brown Swiss cattle are a triple-purpose breed: for milk, work, and meat. Milk production is of prime importance, with meat and working ability secondary. A mature cow should weigh 600 kilograms (1,320 pounds). Since the cows are required to graze on the mountains at times, they need rugged feet and legs. Good health, regular reproduction, and longevity are characteristics. Breeding herds are free from tuberculosis and brucellosis, and mastitis seldom occurs. The cattle adapt themselves readily in almost all climates. They are found widely in all ordinary high altitudes.

Good cows produce 10,000 pounds of milk, 4.0 percent and 400 pounds of butterfat in 300 days after the third calf. Production is reduced significantly by summer grazing on the mountains or when held all year at high altitudes. All animals are used for meat ultimately. It was held that females calving when 2 3/4 to 3 years old should live longer than those calving when 2 years old.

Approved Breeding Bulls

Cantonal authorities in 1883 approved bulls for breeding which had been inspected by a commission at annual competitions. The animals were judged on strong constitution, development, capacity for milk, live weight, estimated slaughter yield, and transmission of desirable qualities to their progeny. The premium moneys were paid after an approved bull had served a year, or a "commended" bull served 6 months in an area. Less premium money went to herdbook

TABLE 7.1
RANGE OF MEASUREMENTS (IN CENTIMETERS) FOR BROWN SWISS BULLS AND COWS OF VARIOUS AGES AS CLASSIFICATION STANDARDS

	Bull		Cow	
	Range	Average	Range	Average
Height at withers	136–152	144	124–138	131
Height at hips	136–152	144	127–141	134
Shoulder to pinbone	159–189	174	145–165	155
Depth of chest	75–87	81	65–75	70
Width of chest	52–66	59	36–52	44
Chest circumference	216–246	231	175–205	190
Length of rump	53–65	59	47–55	51
Width of hips	52–64	58	50–60	55
Width of thurls	52–62	57	46–54	50
Width of pins			33–41	37
Circumference of cannon	23–27	25	18–22	20
Weight, pounds	1,765–2,380	2,075	1,100–1,440	1,270

bulls, the remainder to cows and breeding heifers, breeding society syndicates, and progeny groups.

Cattle Improvement

The Swiss Federation and the cantons passed laws and appropriated moneys for improvement of dairy cattle at several times from 1848 to 1958. These laws were enacted for leadership rather than paternalism, illustrated in the motto "Through state-help to self-help."

Appropriations were for specific purposes "to be used for the improvement of agriculture." The Cantons matched federal appropriations. Federation laws were flexible to comply with the respective cantonal rules. The laws of 1893 and 1958 provided that premium moneys be withheld from a prize bull until he had been used in the canton for at least 6 to 9 months, and until a prize cow dropped a calf by a prize bull. Some subsidies contributed toward organization of local breeding syndicates. Premiums were awarded for bull families and groups of breeding cattle "of which the progeny shall be entered in a regular permanent breeding register." Furthermore, "the subsidized associations are required to present annually their animals, of show age and entered in the breeding register, in the show of families or of groups. The amount of the federal subsidy toward the cost of the organization is set according to the number and quality of animals awarded prizes during these shows."

Local show authorities reported the details to the Department of Industry and Agriculture.

HERDBOOKS

The monastery at Einsiedeln kept a private herdbook from 1775 to 1782. The Swiss Agricultural Society began a *Schweizerisches Herdsbuch* for Brown Swiss cattle in 1879 but discontinued it after 2 years because of opposition to a "bureaucratic" endeavor. The present herdbook resulted from combined efforts of the agricultural society, governmental influences, and laws (1891 to 1896), based on developments among the cattle breeding syndicates. The Central Herd Book for Brown Swiss was founded in 1893 with its office now at Zug.

CATTLE BREEDING SYNDICATES

The first cattle breeding syndicate to maintain a breeding or local herdbook was organized in Zurich in 1887. The government appropriated a small sum toward cost of organizing additional syndicates, and granted premiums based on accuracy of the secretary's record books. By 1969, 844 syndicates were active in 22 cantons with 26,830 members owning 221,835 herdbook Brown Swiss cattle. The syndicates owned most of the bulls used by their members. Herds averaged 8.4 females per herd.

The Federation of Brown Swiss Cattle Breeding Syndicates was formed in 1897. Their purposes remain the same: to establish a clearly defined breeding objective; unify judging practices; provide milk and butterfat controls; facilitate procurement of breeding bulls at central bull markets; test bulls; hold cow family shows; publicize the breed; and influence legislation and rules affecting cattle. They assemble the central herdbook from the local bull syndicate records. Hans Eugster is director of the Swiss Herd Book, with its office in Zug.

MILK CONTROL

Selected cows and herds are tested in about 1,500 circuits under the herdbook society. The local official in each circuit supervised milking one day each month and submitted milk samples to the central laboratory for butterfat determinations. Records are entered on punch cards for machine calculation. In 1961, 99,056 Swiss cows averaged 8,034 pounds of milk, 3.88 percent and 312 pounds of fat in 305 days. The 38,891 cows on the plains averaged 8,839 pounds of milk, 3.89 percent and 342 pounds of butterfat.

AUTUMN BULL SHOWS

The Federation has sponsored an autumn bull show and market at Zug since 1897 and at Saipans since 1940. The central office was at Bunen (1897–1910), moved to Lucerne (1910–38), and now occupies permanent quarters at the showgrounds at Zug. A group of 73 syndicates withdrew in 1911 to form the East Switzerland Brown Cattle Society but reunited with the federation in 1935. The fall show at Raperswil continued but has been held at Zug since 1956.

Each local syndicate maintained one or more bulls, their members owning 50 to 1,500 herdbook cows. Each local syndicate secretary (a) kept a list of herdbook (registered) cattle, and branded the insignia on their horns; (b) kept a local herdbook record of breeding; (c) entered birth reports of calves from herdbook cows; (d) kept a register of young stock, with earmarks; and (e) kept the local show and milk production records. Each breeding syndicate was entitled to one delegate at the federation assembly for each 150 herdbook cattle.

Model rules concerning care, health tests, and management of bulls were suggested by the herdbook office as the basis for a contract between a bull syndicate and the caretaker. These rules suggested frequency of use, interval between services, health of cows brought for service, provisions for feeding and exercise, and veterinary care. Permanent records were kept of services and the identification of each female bred.

RECOGNITION OF HERDBOOK BULLS

A plan begun in 1897 recognized a bull for 1 year or for life. For the 1-year recognition, the bull was 9 months old and scored at least 80 points. The sire and dam must have been of recognized ancestry for two generations. The dam and both granddams must have milk production records. This recognition was awarded at cantonal shows or at the federation's bull market at Zug.

Lifetime recognition was given to bulls 18 months or older, with a higher standard for milk production of ancestors. The plan instituted in 1941 recognized potential breeding quality of bulls based on the following:

a. Bulls under 18 months must score 85 points; older bulls, 87 points or more.

b. At least four parents and grandparents scored 87 points or more.

c. The dam or both granddams earned the lactation insignia, with a minimum of 3.9 percent butterfat.

d. At least three of the parents and grandparents rated as members of a breeding family with fertility recognition. At least one insignia must be on the sire's side and one on the dam's side of the pedigree.

Class I. A bull qualified by four requirements, a to d.

Class Ib. A bull met three of the four requirements.

Class II. A bull met two of the requirements.

Class III. A bull met one of the requirements.

A canton might temporarily recognize a bull that was underage at the regular show. Such recognition held only until the next show. Recognition at the Federation's bull market applied for the entire country.

Premiums for Superior Bulls

The Department of Agriculture granted small premiums to encourage retaining superior bulls in service. The bull must have served at least three seasons, and his type score must have rated among the highest one-third in the canton that year in order to receive a prize. He must have been free from tuberculosis and brucellosis. Sixty percent of the females served (at least 30 in the year) must have been with calf. The parents and both grandparents must have been registered, and the dams and granddams must have qualified in the production register. He must have at least 20 "marked" daughters or sons of good quality. A breeding family show was required after the fourth season. Favorable results would qualify him for life.

No prizes were awarded to bulls stabled exclusively. Young bulls were required to be "alped" during the local mountain grazing season, or until 10 days before the Federation's bull market. Certificates were issued and the bulls recognized at the Federation's bull market. Some bulls were pastured in valley meadows, or exercised at least 2 months before exhibition at the market.

Artificial Breeding

Since June 5, 1941, the Brown Swiss Federation required that all calves conceived artificially be identified by a birth report and records confirming parentage. A law in 1958 required that one or two official stations be established to replace two private studs with about eight bulls. Some 23,634 cows were inseminated in 1964–65. Bull stations were regarded as helping small farmers and establishing early proof of the bulls' transmitting ability with daughters under milk control.

Three stations used 46 bulls for artificial breeding during 1967–68. Semen was imported from the United States and used experimentally to produce about 400 calves, which were tattooed for identification. The Institute for Animal Breeding at the University of Berne made 357 blood typings with Brown Swiss in 1968 for parentage determinations.

SPECIAL TYPE AND UDDER CLASSIFICATION

The Brown Swiss Cattle Breeders' Federation instituted a special type and udder classification for cows in 1957 which was conducted especially for "bull mothers." Such classifications were conducted in the spring and summer by experts from the herdbook office. Cows at these classifications must have scored at least 85 points; they generally were under milk recording and were from 1 to 6 months in lactation. They were judged according to the following score-card, and ranked into five groups corresponding somewhat to the ratings of Excellent to Fair in the United States. Descriptions of the body conformation and udder were entered in the herdbook and on the pedigrees.

BROWN SWISS SCORECARD FOR
CONFORMATION AND UDDER CLASSIFICATION

Body conformation score:

   Size—good, large, small, oversize

   Depth and Width—good, little depth, shallow

   Fleshing—good, heavy, poor

   Feet and legs—good, coarse, fine, defective

   Color—good, light, dark, belly spots

      Total score (10 to 5 points)

Udder score:

   Form—square, goat-shaped, bulbous, hanging, divided

   Size—large, medium, small, long, broad, short, shallow

   Texture—glandular, medium, fleshy

   Fore udder—good, too small

   Rear udder—good, too small, broad, scanty attachment

   Udder skin—pliable, medium, thick

      Udder score (5 to 2½ points)

Teats, veins:

   Distance—symmetrical, front wide and rear narrow, front narrow and rear wide; sideways wide, medium, narrow

   Length—good, long, short

   Thickness—good, fine, thick, full of cracks

   Tip—round, flat, crater-form, sloping

   Between teats—none, distance___, too close

Udder veins—marked, average, lacking
  Total for teats and veins (5 to 2½ points)
Total score, udder, teats and veins (10 to 5 points)
Ease of milking—easy, difficult
Notes (for entries by the judge)

The judge underscored the descriptive word either once, or twice if the word applied markedly. The gradings or ratings were as follows:

	Body conformation	Total udder
Without faults	10	5
Very good	9	4½
Good	8	4
Fairly good	7	3½
Satisfactory	6	3
Unsatisfactory	5	2½

FAMILY SHOWS

Breeding family shows were established in 1901 to recognize animals with desirable transmitting ability. A bull was at least 3 years old and a cow 5 years old before a breeding family show could be held for them. Three kinds of family shows were held:

1. Bull breeding families consisted of a bull with at least ten sons. Owners of a Brown Swiss bull applied to the Brown Swiss Federation for a show at the time of the Federation's bull market. Two Federation representatives judged these sons, considering their numbers, quality, and arrangement, which included all the sons exhibited at the market in Zug.

2. Bull breeding families by old bulls could be judged at a local association show. These included at least 30 sons and daughters over 6 months old, or 20 if this included over half of his marked progeny. Half the progeny must be cows that had dropped one or more calves. The bull was not required to be present.

3. Cow families descended from one cow included her sons and daughters, and the next generation through the female line. Sons that had passed through the Federation's bull market could be considered. The foundation cow did not need to be present. Cows must have had production records. Those that had calved two or three times must have rated an Ⓒ; with four to five calvings, at least two

Ⓒ , and six or more calvings rated three Ⓒ brands for milk production. The butterfat test must have averaged at least 3.7 percent. The owner of the foundation cow must have been a member of the local association or syndicate. A Federation representative judged the animals according to quality as very good, good, or satisfactory and prepared a short description of the group. They received a diploma, and the breeding family insignia Ⓩ was recorded.

Breeding families classified with at least 60 percent of First Class sons were rated Silver Medal; with at least 70 percent of First Class sons they rated the Vermeil Medal; and those with at least 80 percent of First Class sons rated a Gold Medal. The same medal was awarded only once to the same parent. Breeding families that were judged First Class at the Federation's bull market or in the local syndicates were awarded the breeding family insignia Ⓩ , and properly recorded. This insignia was for life. The exhibitor bore the cost of conducting breeding family shows.

FERTILITY STAR

Regularity of reproduction was recognized with cows by awarding a six-pointed fertility star ✱ when they had dropped six normal calves within 7 years. Such recognition also emphasized longevity. It made necessary the continued maintenance of permanent records by the local syndicate secretaries. Such cows frequently bore several brands for milk production as well as the fertility star for regular reproduction.

Private records of production were kept early by some owners. A Brown Swiss cow at Grignon School yielded 65 kilograms (143 pounds) of churned butter in 68 days in 1865. Forty cows on an estate near Scham averaged 8,126 pounds of milk, 3.36 percent and 273 pounds of butterfat in lactations averaging 286 days. United States Consul S. H. M. Byers cited an average yield per cow of 5,315 pounds of milk, supplied to the Anglo-Swiss Condensed Milk Company in 1881, from cows fed largely on grass and hay. Brown Swiss cows at the Einsiedeln monastery averaged 6,356 pounds of milk in 1,667 lactations between 1872 and 1903 while part of them were on the Sihltal alp from June to mid-September. Secretary Heinrich Abt wrote in 1905 that "the associations have the facts about proving for production. It is especially valuable that milk production of each cow and the progeny of each breeding bull be proved, since bulls descended from poor milking cows can transmit this quality to their offspring. Therefore associations should acquire no sire that does not come from a good milking cow."

BULL MOTHERS

The Federation of Brown Swiss Cattle Breeders' Syndicates began supervising milk production of cows whose owners wished them declared "bull mothers." These herdbook cows scored well for conformation. Their milk was weighed daily, verified by a supervisor, and fat tests determined twice monthly at a central laboratory by Gerber test. Seven cows averaged 9,367 pounds of milk, 365 pounds of butterfat in 323 days, with tests between 3.67 and 4.17 percent fat.

HERD TEST

Supervision of production in eight herds began in November 1902. The 28 cows averaged 9,179 pounds of milk, 3.88 percent and 357 pounds of butterfat in 357 days and were dry 58 days. The Federation proposed that members record all first-prize cows for milk production. Records were up to 365 days in length from 1923 to 1931, then for 305 days. Some 174,693 lactations completed during 1969 averaged 9,203 pounds of milk, 3.86 percent and 355 pounds of butterfat in 338 days. They received largely grass and hays. Cows qualified for the lactation brand insignia ⊙ with the following production:

Lactation	Calving within 14 months		Not calving within 14 months	
	Milk	Butterfat	Milk	Butterfat
	(pounds)		(pounds)	
First	6,600	245	7,260	268.4
Second	7,480	288	8,250	305.8
Third	8,360	323	9,240	341.0

If the cows were on alpine pastures for 2 months, the requirements were reduced by 800 pounds of milk and 33 pounds butterfat if at altitudes of 1,200 to 1,600 meters; and 1,320 pounds of milk and 46.4 pounds butterfat if at altitudes above 1,600 meters.

Cows on the plateau (plains) averaged 9,843 pounds of milk, 3.90 percent fat, compared with 8,025 pounds of milk in the mountain region during 1962–63.

The milking rate of 144 cows in eight sire groups was reported in 1964. They averaged 4 months after calving and 3.28 previous lactations. Maximum milk flow of the groups was 4.09 to 6.37 pounds (1.8 to 2.8 liters) per minute at the peak, and took 7.7 to 11.3 minutes to complete milking out. Forequarters yielded 44.2 and the rear quarters 55.8 percent of the daily production. In the autumn of 1968, two mobile four-quarter milking machines found an average of 4.88±1.28 pounds of milk per minute and 1.01 pounds of strippings. A Milko-Tester replaced the Gerber butterfat test in the Central Laboratory in 1969.

Transmitting Bulls

A program initiated in 1942 recognized bulls with the Ⓢ brand on the horns for good transmission of milk and butterfat to their daughters. The herdbook office designated 22 bulls in 1957. At least 60 percent of the daughters of producing age must have records, and at least half of them with two lactations. The requirements were (a) a comparison of the daughters' production with the average of the milk control society, and (b) at least ten daughter-dam comparisons for actual production with relation to the breed average and altitude zone in the year the records were made. Milk records were for 270 to 290 days and were not adjusted for summer pasturing on the mountains ("alping"). If the average milk yield was slightly below the standard for recognition, there could be higher butterfat percentages and vice versa. Records were compared on a spot graph. If daughters increased in milk and butterfat over the dams, an Ⓢ insignia had been earned.

The following averages were computed by the herdbook office for 48 proved bulls investigated in 1957:

	Bulls earning ⓒ	Bulls without ⓒ
Number of bulls proved	23	25
Herdbook daughters per bull	50	64
Daughters having production records	35(69%)	42(65%)
Average milk yield, pounds	9,347	9,058
Butterfat yield, pounds	372	356
Butterfat test, percent	3.98	3.93
Daughter-dam comparison		
Milk, pounds	+248	− 46
Butterfat test, percent	.08	+.03
Compared with milk control association		
Milk, pounds	+102	−211
Butterfat test, actual percent	+.10	+.05
Production index		
Milk, pounds	9,413	8,879
Butterfat, pounds	383	349
Butterfat, percent	4.07	3.93
Production, corrected for "Alping"		
Milk, pounds	9,969	9,316
Butterfat, pounds	406	366
Butterfat test, percent	4.07	3.93

The conference of experts described the aims of raising cattle: "Brown Swiss cattle are bred for a combined yield of milk and beef. Milk yield and high butterfat percentage are chiefly stressed. Vigorous constitution, health, prolificacy, steadiness for alpine pasturing and rough grazing, efficient assimilation of forages and high adaptability."

Cattle in Switzerland have been free from tuberculosis since 1958. Bang's disease (brucellosis) also has been eradicated. H. Glättli is director of the Brown Swiss Cattle Breeders' Federation. The headquarters building houses the herdbook activities and is adjacent to the showgrounds of their annual bull market. Publications of the Federation are in French, German, and Italian.

REFERENCES

Anonymous. 1970. The breeding value factor, *Mitteilungen* 6:669–73.
Abt, Heinrich. 1905. *Das Schweizerisches Braunvieh.* (Rev. ed., 1911.) Huber & Co., Frauenfeld.
Adam, B., and M. Anker. 1859. *Abbildungen der Rindviehrassen und Schlage in der Schweiz.*
Anker, M. 1858. *Bericht uber die erste schweizerische Viehaustellung in Bern.*

Baumgartner, R. 1872. *Die schweizerische Rindviehrassen.* Solothurn.

Beauchamp, E. R. 1887. Swiss cattle and dairy products. In cattle and dairy farming. *U.S. Consular Repts.* Part 1, pp. 303–19.

Byers, S. H. M. 1887. Statistics of Brown Schwytzer cattle. In Cattle and dairy farming. *U.S. Consular Repts.* Part 1, pp. 303–19.

Duerst, J. U. 1928. Kulturhistorische Studien den schweizerischen Rindviehzucht. *Schweiz. Landwirtsch. Monatsch.*

———. 1931. *Grundlagen der Rinderzucht.* Springer, Berlin.

———. 1942. Historische Forschungen uber den Ursprung und die Urformen der Hoherviehschlage der Schweiz. *Z. Zuchtungskunde* 17(5):149–58.

Engeler, W. 1943. *Die Haltung der Zuchtstiere.* Bern.

———. 1943. Geschichte, Methoden und zukunftige Organisation von Zuchtfamilianschauen und Nachzuchtsuchungen. *Schweiz. Landwirtsch. Monatsch.* 10(7).

———. 1947. *Das Schweizerische Braunvieh.* Frauenfeld.

———. 1955. *Anleitung zur Beurteilung der schweizerischen Braunvieh.* Frauenfeld.

———. 1956. Brown Swiss cattle. Comm. Swiss Cattle Breeding Assoc. Bern.

———. 1959. *Ergebnisse der Haltepramien und der Zuchtfamilienschauen im Jahre 1958.* Zug.

———. 1959. *Leistungsprufte Stiere des schweizerischen Braunvieh im Jahre 1958.* Zug.

Engeler, W., and J. Decking. 1958. *Die Auswertung der Milchkontrollergebnisse beim schweizerischen Braunvieh Kontrolljahr 1957/58.* Zug.

Engeler. W., K. Keller, and H. Meli. 1964. Results of ease of milking tests in Brown Swiss cows. *Mitt. Schweiz. Braunvieh Verb.* 1964(4):187–91.

Hehn, V. 1876. *Kulturpflanzen und Haustiere in ihren Ubergang aus Asien.* 3rd ed. Berlin.

Henne am Rhyn. 1878. *Geschichte des Schweizervolkes und seiner Kultur von den altesten Zeiten bis zur Gegenwart.* Vol. 1. 3rd ed. Leipzig.

Hobson, A. 1929. Agricultural survey of Europe. Switzerland. *USDA Tech. Bull. 101,* pp. 1–64.

Kaltenegger, J. 1883. *Die historische Entwicklung des Bundner Viehs.* Aarau.

Keller, C. 1919. *Geschichte der schweizerischen Haustierwelt.* Frauenfeld.

Keller, Ferdinand. 1878. *The lake dwellings of Switzerland and other parts of Europe.* Trans. by J. E. Lee. 2 vols. J. J. Little & Ives Co., London.

Kick, W. 1878. *Lehrbuch der Rindviehzucht nebst Berechnongen nach den neuesten Stande der Wissenschaft und Erfahrung.*

Kramer, H. 1912. *Das schonste Rind.* 3rd. ed. Berlin.

Lardner, D. 1832. *The history of Switzerland from* B.C. *110 to* A.D. *1830.* Philadelphia.

Lenhert, Hugo. 1896. *Rasse under Leistung unser Rinder.* 3rd ed. Berlin.

Mason, F. H. 1887. Swiss cattle. In Cattle and dairy farming. *U.S. Consular Repts.* Part I, pp. 287–297.

Muirhead, Findlay. 1923. *The blue guides. Switzerland.* Muirhead & Rossiter, London.

Muller, F. 1896. *Das schweizerische Braun- und Fleckvieh.* Bern.

Ringholz, P. O. 1908. Geschichte der Rindviehzucht im Stifte Einsiedeln. *Landwirtsch. Jahrb. Schweiz* 22:413–508.

Rutimeyer, L. 1861. *Die Fauna der Pfahlbauten in der Schweiz.* Basel.

Schuppli. 1891. *Monographie des schweizerischen Braunvieh.* Aarau.

Tanner, G. C. 1887. Cattle breeding in Europe and in the United States. In Cattle and dairy farming. *U.S. Consular Repts.* Part 1, pp. 41–56.

Weckherlin, August. 1827. *Abbildungen der Rindvieh und anderer Haustier-rassen auf den Privatgutern seiner Majestat des Konigs von Wurtemberg nach dem Leben gezeichnet und lithographiert von Lorenz Ekeman Alesson.*

Werner, H. 1888. Die Rindviehschlage der Schweiz. *Landwirtsch. Jahrb.* 5:177.

———. 1892. *Die Rinderzucht.* (2nd. ed., 1902.) Berlin.

Wery, M. G. 1883–84. *Ann. Inst. Natl. Agron.* Ser. A(9):141–212.

Wilckens, Martin. 1876. *Die Rinderrassen Mittel-Europas.* Wien.

Wilhelm, Gustav. 1872. *Die Rindviehrassen der Alpenlander.* Wien.

CHAPTER 8

BROWN SWISS IN AMERICA

THE GOOD reputation of Brown Swiss cattle preceded them to the United States. Francis Rotch wrote in 1861 that "in France, especially, they [Brown Swiss cattle] are much esteemed, and at the agricultural show at Grignon we saw numbers of them in milk, and their performance at the pail compared favorable with those of the common short-horns which were considered the best milkers in the establishment."

Henry M. Clark of Belmont, Massachusetts, visited Switzerland in 1869. He bought a bull and seven heifers through Gottfried Burgi about Mount Rigi and in Canton Schwyz. One heifer dropped a bull calf. These cattle received prizes from the Worcester Agricultural Society in 1873, 1874, and 1875. Christine 6 and Geneva 7 were reported by the Society to have produced 415 and 376 pounds of milk, respectively, in 7 days in 1875, and "it took 7 to 8 quarts of milk to make a pound of butter on the average."

Ten other importations from Switzerland included 15 bulls and 111 females. Pregnant females dropped 15 bulls and 16 heifer calves that were registered. A foot-and-mouth disease embargo prevented further importations. However, six more bulls and a heifer came in four shipments via Mexico between 1908 and 1931. The foundation stock totalled 38 bulls and 135 females registered, from which the breed developed in the United States. A bull imported to Canada in April 1969 was registered so that his semen could be used for artificial breeding.

BROWN SWISS ASSOCIATION OF AMERICA

Six breeders of Brown Swiss cattle attended the New England Fair at Worcester, Massachusetts, and organized the Brown Swiss Association of America on September 8, 1880. Their interest was in maintaining authentic pedigrees and promoting the breed. The Association name has been changed several times. It was incorporated under Wisconsin laws in 1925. The Association business was conducted in the secretary's home while the Association was small. Secretaries were located in Massachusetts, Connecticut, New York, and Wisconsin, in turn. The Association now is located at 800 Pleasant Street, Beloit, Wisconsin 53511. Officers are president, vice-president, and seven directors. Marvin L. Kruse succeeded Fred S. Idtse as secretary-treasurer in 1963; his address is P. O. Box 1019, Beloit, Wisconsin 53511.

The officers and directors divided the United States into seven districts in 1941, and eight in 1968, with about equal membership, subject to revision each 5 years. A director is elected in each district to serve a 3-year term. The secretary and two field representatives were active in the extension program.

There were 1,266 active, 2,746 inactive, 747 junior, and 27 honorary members in 1966. Life membership was obtained on recommendation of a member, approval of the directors, and payment of a $25 membership fee. Corporation membership in 1968 did not exceed 10 years and was renewable. Partnership membership terminated on any change in their organization, or death of a partner. Membership and herdbook privileges could be withdrawn by the

Board of Directors for due cause, after a hearing. Junior membership privileges have been granted since November 1965.

Annual meetings are held in November. Members may vote by proxy on an approved form filed with the secretary before the meeting. Proxies are voted according to the member's instructions. by a committee, or another member.

SWISS RECORD

A list of all imported Brown Swiss cattle and their descendants was published by the Association in 1880. The policy was: "Only such animals as are proved to have been imported from Switzerland, or to have descended from such imported animals will be considered thoroughbred." The pamphlet was named the *Swiss Record*. Volumes published in 1889, 1891, 1895, 1899, and 1901 were consolidated into Volume 1 in 1908. Twenty-five volumes were published before that form was discontinued.

Compulsory ear tattooing to identify registered animals was adopted in 1940. Over 2,000 Brown Swiss have been exported to Cuba, Mexico, Central and South America, Canada, Iraq, Afghanistan, and other countries.

The Board of Directors approved the uniform rules of the American Dairy Science Association and Purebred Dairy Cattle Association concerning artificial breeding, frozen semen, and blood antigen identification of animals conceived by artificial insemination. Colonel Harry of J. B., sons of Jane's Royal of Vernon, and other bulls exerted a wide influence on the breed through artificial services.

PROVISIONAL REGISTRATIONS

The Purebred Dairy Cattle Association authorized study of an open herdbook system similar to those operating where the dairy breeds originated. Such a plan would allow owners of herds with lapsed registrations to re-establish purebred status, and some high grade animals to merit registration. The plan was adopted in 1953. Requirements for provisional registration involved four generations of females. A *Foundation Cow* of predominately Swiss characteristics and size was tattooed, and produced 10,000 pounds of milk, 400 pounds of butterfat in 305 days on a 2× mature equivalent basis. Her type was at least Good Plus in major breakdown ratings. Matings through two generations required a registered bull of equal producing transmission. The *Third Generation Cow* met these requirements, plus production of 11,500 pounds of milk, 450 pounds of butterfat, and a Very Good mammary system. This cow qualified for full registration. Eight animals qualified as Foundation Cows in 1954. Their type ratings included one Excellent, six Very Good, and one Good Plus animal. Seven heifer calves were reported as First Generation females, and one as a Second Generation heifer calf. One cow became registered.

Opposition and solicitation of proxy votes gave a majority vote at the 1958 annual meeting to discontinue provisional registration. The Board of Directors closed this provision on November 13, 1958, without affecting previous entries.

PROPOSED IDENTITY ENROLLMENT

In April 1968 the Board of Directors of the Purebred Cattle Association recommended a plan for provisional registration to the members whereby the sires of unregistered Brown Swiss cows might become known. Nearly half the official DHIA production records were unusable for the breed because the sires were not identified. This loss reduced numbers of records usable for USDA Sire Summaries that contribute to breed progress. The Association president appointed nine persons prominent in related dairy fields: Ray Denkenbring, Richard Kellogg, Thomas Lyons, Norman E. Magnussen, Miles McCarry, C. E. Meadows, Eugene Meyer, Robert Schroeder, and Donald E. Voelker—with Directors Bernard Monson, Howard Voegeli, and former secretary Fred S. Idtse—to consider the problem. They concluded that a reliable system of identifying unregistered Swiss cows was needed urgently and that it must be simple and expeditious. At least 10,000 unregistered Brown Swiss cows are on DHIA test, with many owners interested in advancement.

The Board proposed a program of three steps:

Step 1. A cow of Brown Swiss characteristics would be inspected by an Association representative for color and conformation. An ear tag with official identification number would be inserted, recorded in the Association office, and a certificate issued at a fee. A female fetus or daughters then in the herd also would be subject to Step 1. (In November, 1970, the Board of Managers ruled that female fetuses with correct breeding identification were eligible for Step 2.)

Step 2. Subsequent female offspring by a registered Brown Swiss bull would be eligible for entry under present rules in the Identity Enrollment Herd Book with the prefix C.I.E. (Certified Identity Enrollment) preceding the name, and she must exceed breed average in production.

Step 3. A female in the Identity Enrollment Herd Book, meeting production and at least Good Plus in classification would have "Certified" stamped on her Identity Enrollment certificate. Production and type classification requirements will be based on Cow Indexes as soon as these indexes are available for registered cows on DHIA-DHIR test and official type classification. Progeny of C.I.E. Dams, sired by a registered Brown Swiss bull would be eligible for entry in the official herdbook by a transfer fee. The Directors reserved the right to cancel certificates of animals that in their sole judgment fell below desired qualifications.

Active members voted in favor of Identity Enrollment, which the Board of Directors then approved unanimously. Secretary Marvin L. Kruse inspected the first Swiss cow during the Florida Dairy Conference at Gainesville on May 14, 1969. Two 4-H Club heifers also were inspected as typical of Brown Swiss characteristics.

The Show

Brown Swiss cattle have appeared at shows in the United States since 1873. They were displayed as "dual-purpose" cattle because of size. The Association voted in 1906 to promote Swiss as a strictly *dairy breed* to compete at the National Dairy Show. They have been shown as a dairy breed since 1907.

The first scale of points was published in an enlarged Volume 1 of the *Swiss Record* in 1889, based on 100 points for perfection. The scale of points was revised in 1912, 1930, and 1939 and then replaced by the unified Scale of Points approved by the Purebred Dairy Cattle Association in 1942. The unified Scale of Points was revised in 1957 by redistribution of the points and modified descriptions.

H. E. Goode described influence of the shows thus:

> [In type] a general change has been noted. . . . The coarse head has given way to one of more refinement, which is indicative of the dairy quality. The heavy shoulder has been refined, the topline straightened, the rump elevated, the thighs thinned, the udder has become attached high behind and extended farther forward, and the teats have been spread and reduced to a more convenient size. These changes have been added to the general dairy appearance of the breed and fortunately neither size, constitution, nor capacity has been sacrificed.
>
> . . . the show yard started the Brown Swiss on the rapid road toward improvement in dairy characteristics and the official test has been the factor which has demonstrated the productivity of the breed.

Canton shows in the United States with Brown Swiss cattle were initiated by Fred S. Idtse in Wisconsin on May 21, 1938, patterned after Jersey "parish" shows and the earlier black-and-white shows in Utah. Educational judging contests at the canton shows have increased interest in desirable dairy type.

A Bell Ringer program was begun in 1957 to recognize Brown Swiss animals of desirable type even if displayed at only a single canton or state show. The goal was to support canton shows and recognize desirable individuals. Owners submitted glossy photographs of winning animals, which were judged by a committee of canton show judges in each state. Photographs of winning Bell Ringer animals have been the subjects of an annual national Bell Ringer judging contest. Winners are announced at the annual meeting of the Association in November.

MODEL SWISS COW

In 1928 R. C. Keister of Chicago was paid $500 to paint a model Brown Swiss cow. He studied photographs of noted animals, guided by officers and directors and by his visits to good Swiss herds. A preliminary painting was displayed before a group of breeders at the National Dairy Show in 1930. The model was approved after minor changes at the annual meeting in 1930. A four-color picture appeared in *The Brown Swiss Bulletin* of January 1931.

The model cow was patterned after no particular animal, yet it bore close resemblance to the noted producer and show winner prominent at the time—Hawthorn Dairy Maid 6753. The model stressed size of udder, dairy temperament, and style in advance of many cows of the breed. The purpose of the model was an attempt to unify judging standards, which had been criticized previously. The first judging conference sponsored by the Association was held in 1942 at two canton shows in Wisconsin.

A new model cow planned by Vernon Hull, Paul Bennetch, and Fred Gauntt and prepared by Displaymasters, Inc., of New Jersey, was approved in March 1955. This scale model embodied improvements in dairy character and mammary development patterned after Jane of Vernon and her descendants.

Herd Classification

Herd classification was adopted by the Brown Swiss Association in November 1942 after it was tried unofficially in Illinois. Herd classification was described: "as valuable as the show ring has been and will continue to be in focusing the attention . . . on desirable type. . . . Only a small percentage of animals are shown, and in the show ring judging is done on a comparative basis and no attempt is made to classify the individual animals in terms of the breed score card."

Type classification measured the degree of excellence in type as compared with an ideal. It furnished a permanent record of desirable characters and faults of individual animals. Registration certificates of animals that classified Poor were cancelled, and only female progeny were registered from Fair animals.

Four leading judges were appointed as official classifiers: F. W. Atkeson, Elmer Hanson, C. S. Rhode, and James Hilton. They worked together in four herds in Illinois and Wisconsin to develop uniform methods. A "breakdown" classification rated anatomical parts of each animal separately, improving over the system used earlier with Ayrshire, Holstein-Friesian, and Jersey cattle. The classification plan follows.

Rating		Score of animal	Revised score
		(points)	(points)
Excellent	(E)	90 or more	
Very Good	(VG)	85–89	
Good Plus	(GP)	80–84	
Good	(G)	70–79	75–79
Fair	(F)	60–69	65–74
Poor	(P)	Fewer than 60	Fewer than 65

Experience soon showed need to increase the scores of Good, Fair, and Poor classes. Detailed ratings were published, based on the four main divisions of the unified dairy scorecard adopted in 1942. These were:

General appearance	30 points
Legs and feet	
Rump	
Dairy character	20 points
Body capacity	20 points
Mammary system	30 points
Fore udder	
Rear udder	

All females in the herd past 3 years old are classified, but classifications for bulls are optional with the owner. A cow classified Excellent may be re-inspected after an interval of 12 months or more for reclassification. Animals rating Excellent may be rerated Excellent up to five times (three times under 10 years old).

Criticism of mammary development may be indicated in remarks for loose attachment, L.A.; undesirable teat placement, T.P.; or objectionable teat, O.T. Congenital defects (present at birth) are indicated, such as crampy legs, C.L.; blindness, B.; wry-tail, W.; wry-face, W.F.; cross-eyes, C.E.; parrot mouth, P.M.; wing shoulders, W.S.; or screw tail, S.T. An animal found to be a "weaver" is classified Poor. Undersized animals are marked "small" in remarks, and reduced one point in the final rating.

An increasing proportion of animals are rating Excellent and Very Good, with smaller proportions among the Good Plus, Good, Fair, and Poor ratings. The breakdown classifications were published in *The Brown Swiss Bulletin* in groups under the sire, and in the *Records of Production and Type Classification of Brown Swiss Cattle*. This practice allows breeders to study type characters of different families for use in selecting breeding animals.

Secretary Fred S. Idtse prophesied in *The Brown Swiss Bulletin* of February 1943 that "just as a cow's production is used and of value in pedigree long after she is gone, her official type rating will also be used and of value." The breakdown rating and remarks of an animal's weakness and desirable characteristics allowed analysis of the conformation. It was an advance over previous methods. The strong characters, revealed from the averages of classification scores are dairy character, body capacity, legs, and feet. All three characters have been improved in recent years. Udder attachment also has

TABLE 8.1
PERCENTAGE OF BROWN SWISS WITH BREAKDOWN RATINGS OF POOR TO EXCELLENT IN TYPE CLASSIFICATION

	1942–57	1942–65			1966–67[a]			
	E, VG	E, VG	E	VG	G+	G	F	P
Legs and feet	35.66	44.28	9.34	45.23	35.11	7.75	2.27	0.29
Rump	27.60	34.93	7.11	34.32	38.48	14.97	4.78	0.34
Dairy character	51.80	64.91	24.96	53.62	19.25	1.97	0.19	
Body capacity	42.97	55.09	18.26	51.20	27.34	3.01	0.19	
Fore udder	24.52	31.88	7.32	32.61	37.53	15.47	6.24	0.83
Rear udder	27.51	35.59	8.87	37.49	36.38	13.33	3.76	0.17

Key: E, Excellent; VG, Very Good; G+, Good Plus; G, Good; F, Fair; P, Poor.
a. 4,711 Brown Swiss were rated for the first time, excluding 2E, 3E, etc.

been improved. Part of these improvements can be attributed to transmitting ability of Jane of Vernon and her descendants. These improvements are shown in Table 8.1.

Effective in April 1967 the Excellent rating may be conferred up to five times, as follows: under 5 years, Excellent; 5 or more years, 2E; 9 or more years, 3E; 12 or more years, 4E; 15 or more years, 5E. This plan recognized retention of desirable type to advanced age.

The Brown Swiss Descriptive Type Classification was modified in April 1968 with 12 subdivisions:

Stature—upstanding, intermediate, low set.

Front end—shoulders smoothly blended, chest strong and wide; desirable strength and width; coarse shoulders and neck; narrow and weak.

Back and loin—straight, full crops, strong wide loin; straight, weak crops; low front end; weak loin and/or back.

Rump—long and wide, nearly level; medium width, length, or levelness; high and/or coarse tail head; narrow, especially at pins; sloping.

Hind legs—strong, clean, flat bone, squarely placed, clean flat thighs acceptable; sickled and/or close at hock; bone too light or refined; coarse or blemished hock.

Feet—strong, well-formed; acceptable, with no serious faults; front feet toe out; shallow heel; weak pastern.

Fore udder—moderate length and firmly attached; moderate length, acceptable attachment; short; bulgy or loose; broken and/or very faulty.

Rear udder—firmly attached, high and wide; intermediate in height and width; low attached but firm; narrow and pinched; loosely attached and/or broken.

Udder support and floor—strong suspensory ligament and clearly defined halving; acceptable; floor too low, tilted and/or uneven floor; broken suspensory ligament and/or weak floor.

Udder quality—soft and pliable; intermediate; could not determine; meaty, persistent edema.

Teat size and shape—plumb, desirable length and size and squarely placed; acceptable, no serious fault; wide front teats; other undesirable teat placement; objectionable teats size and shape.

Miscellaneous terms—dry; small, frail; OCS; blindness; cross-eyed; parrot jaw; wry-face; wing shoulders; post legged (too straight); crampy legs; spread toes; wry-tail; screw tail; weaver; udder quartered; other notes.

The description was entered on the classification report by coded numbers.

PRODUCTION RECORDS

The earliest production records of Brown Swiss cows in the United States (for the cows Christine 6 and Geneva 7) were published by the Worcester Agricultural Society in Massachusetts in 1875. Several breeders kept private records. Bessie 11 produced 82,274 pounds of milk in 10 years (1878–87). During 3 of these years she was credited with 1,683 3/4 pounds of butter churned from her milk.

Brienz 168, imported by Scott & Harris, yielded 245 pounds of milk, 9.32 pounds of churned butter in a 3-day milking trial supervised by the University of Illinois at the American Dairy Show in Chicago in November 1891. She was 11 years old, weighed 1,410 pounds, and was milked three times daily. This was the highest daily butter yield known to Professor W. A. Henry of the University of Wisconsin to that date. She won a first prize at the World's Columbian Exposition in Chicago in 1893 when 13 years old.

REGISTER OF PRODUCTION

The Brown Swiss Association established a Register of Production on May 10, 1911. Merney 2859 completed two records previously in the University of Wisconsin dairy competition: 13,643.7 pounds of milk and 554 pounds of butterfat as a 5-year old, and a later record of 14,674.8 pounds of milk with 596.9 pounds of butterfat. Requirements to qualify for the Register of Production were 220.5 pounds of butterfat in 365 days beginning at 2 years 6 months of age, up to 330 pounds of butterfat at 6 years or older. This standard was increased in 1922 to 250.5 and 360 pounds of butterfat at the respective ages. A bull was admitted to the Register of Production when three daughters out of different dams had qualified.

A 305-day test was established in 1922 which required that the cow also drop a living calf within 14 months of previous calving. A Farmer's Class was established later with the same requirements in 305 days. This required a living calf also, and that the cow be milked twice daily after the first 15 days.

HERD IMPROVEMENT REGISTER

The Association established a Herd Improvement Register in March 1932. Production of every cow in the herd was supervised and *all records published*. The herd test increased in popularity. The average production of Brown Swiss cows in the several classifications is listed in Table 8.2.

A tally was made of 6,116 Register of Production records completed prior to April 1940. The average butterfat percentages ranged between 2.92 and 5.74 percent, the mode being between 4.0 and 4.1 percent (see Table 8.3). The average butterfat test in

TABLE 8.2
AVERAGE PRODUCTION OF BROWN SWISS COWS

Number of records	Length (days)	Times milked	Milk (lbs.)	Test (%)	Fat (lbs.)	Class
2,018	365	3	15,618	4.01	626	ROP
2,348	365	2	12,723	4.13	525	ROP
1,009	305	3	13,145	4.07	536	ROP
2,371	305	2	12,563	4.05	509	HIR
2,540	305	2	10,740	4.13	444	ROP
48,287	305	2	9,675	4.04	391	HIR
Actual production of mature cows 1968–69						
	305	2	13,357	4.10	548	DHIR

records of the breed appears to have increased less than 0.1 percent since the Register of Production was established in 1911. This slight increase is due largely to avoiding bulls from dams whose milk had a low butterfat test.

Methods of supervising DHIA records were modified on a trial basis in 1958. They were accepted by the several breed associations for the breed programs.

SIRE AND COW RECOGNITION

The Association established a Sire Recognition program in 1968, based on USDA Sire Summaries and average type classification scores of daughters. If fewer than 30 tested daughters were considered in a Sire Summary, 50 percent of them must have been registered or positively identified by blood typing.

A *Superior Sire* was rated on a sliding scale with predicted differences in milk production of daughters over herdmates as follows:

Predicted difference in milk (pounds)	Repeatability (percent)
+400	50
+435	40
+470	30
+500	20

Repeatability is based on numbers of daughters, lactations, and herds represented. At least ten daughters must have been classified and their scores average 83 points for mammary system and 83 points overall.

Daughters of a *Qualified Sire* must rate similarly in average type

TABLE 8.3
AVERAGE BUTTERFAT TESTS IN REGISTER OF PRODUCTION AND HERD
IMPROVEMENT REGISTRY RECORDS COMPLETED BY BROWN
SWISS COWS BEFORE APRIL 1940.

Fat Percentage	Number	Range
5.7	1	5.74
5.6	0	
5.5	1	
5.4	0	
5.3	2	
5.2	4	
5.1	8	
5.0	15	
4.9	30	
4.8	54	
4.7	93	
4.6	161	
4.5	224	
4.4	357	
4.3	473	
4.2	572	
4.1	702	
4.0	715	Mode and Median
3.9	710	
3.8	637	
3.7	512	
3.6	374	
3.5	233	
3.4	128	
3.3	56	
3.2	33	
3.1	14	
3.0	5	
2.9	2	2.92

classification score. The predicted difference in milk production must at least equal +200 pounds, with a 17 percent repeatability. The Superior and Qualified ratings are not permanent but must be earned each time a USDA Sire Summary is computed or type classification calculated by the Association. Those failing to maintain the status are referred to as Previously Superior or Previously Qualified Sires, respectively. The summaries are available three times each year.

When published sire recognition data are presented, they include date of summary, registration name and number of bull, owner, percent of incomplete records, number of daughters with records, their average milk and butterfat, predicted difference, repeatability, number of classified daughters, and the average type score in the various breakdowns. The *Registered Brown Swiss Sire Performance Summaries* are available thrice yearly on a subscription basis, and replace the former volumes of *Records of Production and Type Classification of Brown Swiss Cattle*.

The Board approved use of Cow Indexes for all registered Brown Swiss cows, based on DHIA-DHIR records and type scores, when the indexes become available.

BREEDING BROWN SWISS

Some hereditary characters are distinctive among Brown Swiss. The brown color was common in Switzerland during the feudal period. It may vary from light to dark brown. White spots are disfavored but sometimes appear back of the navel. A fillet of lighter colored hair encircles the muzzle. Albinism occurs rarely.

The average gestation period of 291 days is longer than in other breeds. Calves are large, attaining mature size and weight later than in other breeds.

The change in policy from dual purpose to strictly dairy type in 1906 resulted gradually in improving dairy character (light natural fleshing) and refinement of head, neck, and dewlap. Size and shape of the teats have been improved through selection of breeding animals. Jane of Vernon's descendants contributed greatly to this character as well as to refinement without loss of size. Rugged legs and feet were emphasized in Switzerland, where cattle are "alped" for about 90 days on mountain meadows during the summer.

The relation between butterfat production and type was investigated with records of 3,161 daughter-dam pairs representing 284 Brown Swiss sires between 1950 and 1953. Sires with five or more daughter-dam pairs were included. The genetic correlation was positive between production and the components used in type classification. K. R. Johnson and D. L. Fourt concluded that selection on type should bring slow genetic improvement in production.

Some 428 samples of mixed Brown Swiss milk from 39 herds between Massachusetts and the Pacific coast were analyzed at the

University of Illinois (Bulletin 457). They ranged between 2.92 and 6.74 percent butterfat, and 7.99 to 11.79 solids-not-fat, averaging 4.02 percent fat and 9.3 solids-not-fat. Protein, lactose, and ash averaged 3.61, 5.05 and 0.73 percent, respectively.

Increased numbers of cows in herds and artificial breeding have reduced needs for bulls, favoring close selection of those reared for breeding purposes. This change is seen in the ratio of males to females registered.

Swiss Record	Year	Male:female ratio
Volume 1	1869–1908	1:1.47
Volume 10	1925	1:1.55
Volume 20	1942	1:2.15
Volume 25	1946	1:2.16
	1960	1:7.85
	1969	1:7.54

There were 66 Brown Swiss bulls in 16 studs among bulls of the dairy breeds in 30 artificial breeding organizations in the United States during 1969. Frozen semen transfers between studs enabled wide use of selected animals. Some 48.8 percent of Brown Swiss registered in 1969 resulted from artificial inseminations. During 1969, 52.5 percent of dairy cows in the United States were bred artificially to those selected bulls. Bulls from which semen is frozen must be blood typed. If semen is sold, the living sire and dam also must be blood typed.

Wry-tail is not an uncommon recessive minor defect in some Swiss cows. A few albinos have occurred. A few calves have shown a recessive character resembling epilepsy, which they outgrow. Other recessives include crampy or progressive posterior paralysis, and lack of resistance against the two types of lump jaw—the fungus or bone type, and bacterial type affecting soft tissues. When recessive genes are present in a herd, the occurrence may be reduced gradually by using sires known not to carry the gene.

BREED PROMOTION

Brown Swiss cattle have been exhibited at shows in the United States since 1873. Production records were published first in pamphlets, and in a separate book in 1923. Brown Swiss were advertised as "The Farmer's Cow," esteemed for size, ruggedness, quiet disposition, longevity, and good production.

Booklets containing pictures, facts, and production records of Brown Swiss cattle have been published since 1918. Volume 5 of *Records of Production and Type Classification of Brown Swiss Cattle* appeared as a supplement to *The Brown Swiss Bulletin*. It supplanted the *Swiss Record* for tracing pedigrees.

The Brown Swiss Bulletin first appeared in July 1922. The breed magazine contained reports of breed activities, production records, show and sales reports, and type classifications. It was also a medium for advertising the breed.

Breeders of Brown Swiss cattle and the Association secretary first contacted the public at farmers' meetings and shows. The Association advertised moderately in farm magazines. A full-time fieldman began to serve in 1938. The secretary and two fieldmen served in 1969.

OUTSTANDING ANIMALS

A few famous Brown Swiss cattle brought prominence and improvement to the breed. The bull Imp. Bonepart 141 was imported in 1884 as part of a large shipment, and was used widely by John E. Eldridge. This bull appears among the ancestry of many cattle.

Imp. Junker 2365 was brought from Switzerland by E. M. Barton to Illinois in 1906 and used a full natural lifetime. The bull was Grand Champion at the National Dairy Show in 1907 to 1910, and sired many good producing daughters.

Swiss Valley Girl 2150 was shipped from Iowa as a yearling to the herd of F. W. Hull at Painesville, Ohio. She produced 13,113 pounds of milk and 495 pounds of butterfat on two milkings daily as a 13-year old. She dropped three bulls and nine female calves. Fourteen records of her daughters average 13,221 pounds of milk, 4.01 percent and 530 pounds of butterfat on yearly test. Two daughters won five Grand Championships at the National Dairy Show. Her daughter Swiss Valley Girl 10th 7887 was a prize winner at the National, and produced 27,513 pounds of milk, 1,106 pounds of butterfat in 365 days as a 12-year old.

Jane of Vernon 29496 (Fig. 8.1) was a great cow among all dairy breeds. She was bred by Orbec D. Sherry of Viroqua, Wisconsin; she won Grand Championship at the Waterloo Dairy Cattle Congress in 1932–36 and at the National Dairy Show in 1936. She was sold to Judd's Bridge Farm, where the bull Colonel Harry of J. B. 48672 and the daughters Jane of Judds Bridge 78476 and Jane's Chloe J. B. 109895 were produced. Her older son, Jane's Royal of Vernon 28594, and his four full sisters—Jane of Vernon 2nd 43893, Jane of Vernon 3rd 55496, Jane of Vernon 4th 65385, and Jane of Vernon 5th 65386—went to Lee's Hill Farm in Morristown, New Jersey. They won Grand Championships at the National Dairy Show in 1938, 1940, and 1941; they were first-prize Get of Sire and first and second prize Produce of Dam groups in 1938.

Jane of Vernon produced 23,569 pounds of milk, 1,075.58 pounds of butterfat as a 4-year old; 21,880 pounds of milk, 1,039 pounds of butterfat at mature age. Six daughters completed 15 Register of Production records which averaged 16,700 pounds of milk, 4.5 percent and 754 pounds of butterfat. Her older son, Jane's Royal of

FIG. 8.1. Jane of Vernon 29496 classified Excellent in all categories except Very Good rump. She was an excellent producer. She transmitted good production and type to all of her progeny.

Vernon 28594 (Fig. 8.2), was a popular sire. Her second son, Colonel Harry of J. B. 48672, sired over 1,000 registered progeny, largely by artificial service. Grandsons and later descendants have been used widely in artificial breeding.

FIG. 8.2. Jane's Royal of Vernon 28594 sired some of the highest producing Brown Swiss cows. His daughter Royal's Rapture of Lee's Hill 115541 produced 150,216 pounds of milk, 8,438 pounds of butterfat in 6 Register of Production records; classified Excellent, and won the Grand Champion female at Waterloo in 1949.

RESEARCH PROJECTS

The Association has sponsored a study of composition of Brown Swiss milk at the University of Illinois, and is interested in methods of protein analysis of milk as a possible breed program. It sponsored projects to determine normal growth standards, length of gestation in Brown Swiss cattle, and other projects.

The breed has spread widely since 1869, considering that only 22 bulls and 119 females were imported and an additional 16 bulls and 16 females were imported in dam. An appreciable export trade has developed, especially to Latin American countries.

REFERENCES

Alvord, H. E. 1899. Breeds of dairy cattle. *USDA Farmers' Bull. 106.*

Atkeson, F. W. 1943. Effects of war-time conditions on breed association activities. *Brown Swiss Bull.* 21(7):7–20.

Atkeson, F. W., H. I. Ibsen, and F. Aldridge. 1944. Inheritance of an epileptic type character in Brown Swiss cattle. *J. Hered.* 35:45–48.

Becker, R. B. 1953. American contributions to better dairy cattle. *Hoard's Dairyman* 98:736–39.

Becker, R. B., C. J. Wilcox, and W. R. Pritchard. 1961. Crampy or progressive posterior paralysis in mature cattle. *J. Dairy Sci.* 44:542–47.

Burlingham, L. 1942. Brown Swiss in America. In E. P. Prentice, *American dairy cattle.* Harper, New York. Pp. 412–19.

Davis, H. P. 1917. Breeds of dairy cattle. *USDA Farmers' Bull. 893.*

Goode, H. E. 1918. Fifty champions of the dairy breeds. *Kimball's Dairy Farmer.* Waterloo, Iowa.

Harris, George M. 1946. Some early history of the Brown Swiss Association. *Brown Swiss Bull.* 25(1):13–15.

Henry, W. A. 1891. The chemical test for butterfat. *Breeders Gaz.* 20:469–70.

Holt, Frank L. 1926. Brown Swiss breed of dairy cattle. In T. R. Pirtle, *History of the dairy industry.* Mojonnier, Chicago.

Idtse, Fred S. 1953. The open herd book. *Brown Swiss Bull.* 32(3):11–13.

———. 1958. Annual meeting. Minutes. *Brown Swiss Bull.* 37(6):67, 74–79.

Inman, Ira E. 1940. *Brown Swiss cattle. Facts and figures.* Beloit, Wis.

Johnson, K. R., and D. L. Fourt. 1960. Heritability, genetics and phenotypic correlations of type, certain components of type, and production of Brown Swiss cattle. *J. Dairy Sci.* 43:975–87.

Overman, O. R., et al. 1939. Composition of milk of Brown Swiss cows. *Illinois Agr. Exp. Sta. Bull. 457.*

Overman, O. R., R. J. Keirs, and E. M. Craine. 1953. Composition of herd milk of Brown Swiss breed. *Illinois Agr. Expt. Sta. Bull. 567.*

Plumb, Charles S. 1920. The Brown Swiss. In *Types and breeds of farm animals.* Rev. ed. Ginn, New York. Ch. 12.

Rotch, Francis M. 1862. Article on select breeds of cattle and their adaptation to the United States. *Rept. Comm. Patents for the year 1861. Agriculture,* pp. 427–69.

Wilson, Charles C. 1952. So proudly we hail the queen—Jane of Vernon. *Brown Swiss Bull.* 30(8):11–12, 67–68.

Brown Swiss Bull.
    1922. Vol. 1.
    1931. Model Swiss cow. 9(7).
    1943. Herd classification. 21(8):7, 12.
    1951. Why classify? 29(8):11–12, 66.
    1960. Type classification, like production testing, should be made to pay—not cost. 39(1):11–14.
    1964. Jane of Vernon—Her influence continues. 43(1):11–18.
    1965. Brown Swiss "Tomorrow's dairy cow today." Origin, history and programs for advancement. 44(1):11–13, 15, 83, 86–87.
    1968. Directors take big steps for breed improvement. 46(11):13–15.
    1968. Registered Brown Swiss Sire Summaries. First issue appears; lists Superior and Qualified Sires. 47(1):11–18.

1968. Proposed Identity Enrollment program revisions . . . adoption recommended by the Board of Directors. 47(8):22–23.

1969. Brown Swiss descriptive type classification. 47(8):47.

1969. Centennial year Brown Swiss U.S.A. 48(1):11–21.

Brown Swiss Cattle Breeders' Association

Annual reports.

Registered Brown Swiss Sire Performance. 1968.

Rules and regulations governing the Register of Production. Adopted May 10, 1911.

Rules and requirements of the Register of Production of Brown Swiss cattle. Adopted Oct. 12, 1922.

Hoard's Dairyman

1948. Brown Swiss gestation period. 93:904.

1951. Changes in character of the Brown Swiss cow. 96(18):723, 750.

Swiss Record. 1889–1925. Vols. 1–25.

CHAPTER 9

DUTCH BELTED

L ITTLE EARLY history of Dutch Belted cattle has been recorded. Gurtenvieh, "canvassed," or belted cattle were regarded well for milking, ability to fatten, and beautiful conformation. August Weckherlin described them in 1827 in Canton Appenzell in northeastern Switzerland and the Tyrol mountain valleys in Austria. W. Kick mentioned them in 1878, but they were not described in the reports on cattle of Switzerland or the Netherlands by United States consuls in 1885. Hugo Lehnert stated in 1896 that top crosses with Brown Swiss cattle caused disappearance of the Gurtenvieh from the mountain area, leaving occasional white spots among descendants.

Headpiece: Vignette of Dutch Belted cow.

Breed Comparisons in Wurtemberg

Wurtemberg once was a kingdom of 7,530 square miles, north of Switzerland. The king conducted detailed trials on his estates with 15 breeds of cattle: Alderneys, five breeds from England, three from Switzerland, Friesians, Hungarian, small zebus from India, and three strains from his kingdom. He obtained Gurtenvieh or "canvassed" cattle twice from Canton Appenzell across Lake Constance in northeastern Switzerland, the first group in 1810. The trials were reported by two German writers and by John H. Klippart of Ohio and Secretary F. L. Houghton of the Holstein-Friesian Association of America.

Since color descriptions of Gurtenvieh in Appenzell and Wurtemberg were identical with those of Lakenfeld or Dutch Belted cattle, it is believed they were the same breed and had a common origin.

On the average, the king's cows weighed 950 pounds and calves 73 pounds at birth. The Gurtenvieh cows averaged 5,056 pounds of milk yearly in the trials, while Friesians averaged 6,548 pounds and Brown Swiss 5,764 pounds of milk. Twelve other breeds gave lower yields. C. F. Schmidt, Royal Court Domain Counsellor, recorded results of the trials.

No written evidence has been traced, yet it is believed from genetic evidence that cattle with the hereditary white belt were moved by nobility during or soon after the feudal period. Farmers owned them in the local mountain area, but only nobility and wealthy landowners possessed them in the Netherlands. There was published evidence of movements of Friesian and Brown Swiss cattle by prominent people in these areas.

Early History in the Netherlands

William Aiton toured Holland about 1830 and reported: "I saw at Haite Lust, near Haarlem, the seat of M. Van Dervolet of Amsterdam, a stock of ten cows that were all black in the head, neck, fore and hind quarters, and the trunk of the body white. One cow in that herd was of a brown colour where the others were black, but, like them, had a white trunk."

Frank R. Sanders, who bred Dutch Belted cattle near Bristol,

New Hampshire, toured the Netherlands in 1907 and visited with breeders. He wrote:

> from the records obtainable, and from conversation with several of the older breeders of Holland, it seems that these cattle began to flourish about 1750, and no doubt the system of selection by which this marvelous color breeding was attained dates back into the seventeenth century. One breeder says his father informed him that there were gentlemen of wealth and leisure near what is now Haarlem, North Holland, who conceived the idea of breeding animals to a certain color, chiefly with the broad white band in the center of the body, with black ends. These noblemen had large estates, and it is claimed that for more than 100 years they and their descendants worked upon the perfection of the peculiar color markings until they produced Dutch Belted cattle, pigs, and poultry.

Herds of belted cattle were few, and owners seldom sold animals of these color markings to commercial herds. No breed organization was formed. Registrations were cared for by the Netherland General Stamboek at the Hague.

G. J. Hengeveld, head inspector of the Dutch Herd Book of North Holland wrote: "The Dutch Belted cattle stand equal to the best cattle in our country for milking, breeding and fattening. They are owned by gentlemen farmers, but they cannot be bought. They are highly appreciated on account of the rarity of their color."

NAME OF THE BREED IN HOLLAND

The name Lakenfeld, Lakenvelder, or Veldlarker was applied to the breed in the Netherlands, the name meaning literally a field or blanket of white, conveying the idea of a white body with black ends. Sanders observed that many Dutch Belted cattle in the Netherlands bore white markings on one or more legs. A photograph of 19 Dutch Belted cattle in the Purmerand, North Holland, in 1906, showed white markings on 18 of 36 legs visible, and white extended above the pasterns in five instances. A picture of 14 cows in the herd of William Jochems, The Hague, Holland, appeared in Volume 13 of the *Dutch Belted Herd Book of America*. Of 22 pasterns and feet visible, two had white markings. A typical Dutch Belted cow was photographed in the herd of W. Jochems, Duindigt, near The Hague, by the author in 1938 (Fig. 9.1).

Sanders estimated in 1907 that Lakenvelder cattle in the Netherlands numbered about 1,000 head, largely in Utrecht and North Holland provinces.

DUTCH BELTED CATTLE IN AMERICA

The first Dutch Belted cattle were brought to his estate near Goshen in Orange County, New York, in 1838 by D. M. Haight, United States Consul to Holland. He made two importations later. The

FIG. 9.1. A good type Dutch Belted cow owned by W. Jochems, Duindigt, in the Netherlands.

great showman P. T. Barnum purchased several cattle from a Dutch nobleman in 1840, and exhibited them with his circus as "a rare and auristocratic breed." The cows proved to be good milkers, and he placed them on his Orange County farm. H. W. Coleman made a small importation in 1848 to his estate near Cornwall, Pennsylvania. In 1906 a bull and two cows were registered as imported by H. W. Lance, of Peapack, New Jersey. T. R. Pirtle stated that 69 Dutch Belted animals were imported before 1920.

EARLY DEVELOPMENT IN THE UNITED STATES

In 1884 Adrian Holbert wrote of buying Dutch Belted calves earlier. His son Jessie Holbert managed the herd, but had lost 42 animals by fire in 1883. Ten young cattle survived in an adjoining building. D. C. and Jeremiah Knight bought a bull and a heifer from D. M. Haight in 1852. Ten years later the son inherited the herd, some of which descended from one noted milker.

Dutch Belted Cattle Association of America

Resolved, That recognizing the importance of a trustworthy Herd Book of Dutch Belted Cattle, which shall be accepted as final authority in all questions of pedigree and for the preservation of the purity of the breed, we do hereby form ourselves into an Association for keeping and publishing a Herd Book to advance our interests as breeders of these cattle.

Resolved, That Dutch Belted Cattle entitled to registry must be pure bred, black and white, with a continuous belt around the body. White feet or small black spot in belt somewhat objectionable, but not a disqualification.

Resolved, That an animal must be of good form, finely proportioned, preserving all the qualifications of the milk form.

Resolved, That the Herd Book be published as soon as convenient after 100 Certificates of Registry be issued.

N. W. Howell was elected President. H. R. Richards of Easton, Pennsylvania, served 25 years as secretary.

The long interval from importations and founding the *Herd Book* made it impossible to trace all pedigrees completely to animals imported from 1838 to 1848. The executive committee inspected all animals for dairy conformation and color markings for entry in Volume 1 of the *Herd Book*. The committee claimed "that in our system of selection we have a more uniform class of cattle than has ever been registered in any herd book in existence." The organization incorporated under New Jersey laws on July 12, 1909, noting: "The objects of the Dutch Belted Cattle Association of America throughout America, is to issue trustworthy Herd Books of record for the maintenance and preservation of purity of the breed, and as final authority on all questions of pedigree and to regulate standards of excellence from time to time as the interests of the members may occasion." Membership increased gradually, limited by the few animals imported and by exportations to other countries.

Volume 1 of the *Herd Book* contained entries from 31 herds. One bull bred in the herd of D. M. Haight was registered by his estate.

Eight bulls and 17 cows from this herd were registered by later owners. Of 46 bulls and 177 cows in Volume 1, 104 traced in part to the Haight herd. Others probably were so descended, but records had not been kept that established this point. Two-thirds of the entries in Volume 1 were in Orange County. Even in 1891 Orange County claimed half the Registered Dutch Belted cattle in this country.

Registered herds were established in 32 states, largely in the Cornbelt. Dr. and Mrs. J. G. DuPuis owned the largest herd—White Belt Dairy—near Miami, Florida. Active members in 1960 were located in California, Florida, Illinois, Iowa, Minnesota, Missouri, New York, Oklahoma, Oregon, Tennessee, Texas, Vermont, and Wisconsin. Some 2,143 males and 5,233 females were registered in the first 14 volumes of the *Herd Book*. From autumn 1957 to November 1959 some 45 bulls and 152 females were registered. Consideration was given by the association to acceptance of unregistered heifers as purebreds where it could be certified that all grandparents were registered. This plan was submitted for a vote of the members. If adopted, entries could be received from six other states that had not kept up registrations during the inactivity and death of a former secretary.

The Show Ring

D. E. Howatt, H. B. Richards, and S. B. Heaton were appointed "to adopt a scale of points to be published as part of the second volume of the *Herd Book*." This scale of points follows:

Dutch Belted Cows

Points   *Counts*

1. Body color black, with a clearly defined continuous white belt. The belt to be of medium width, beginning behind the shoulder and extending nearly to the hips.   8
2. Head comparatively long and somewhat dishing; broad between the eyes. Poll prominent; muzzle fine; dark tongue.   6
3. Eyes black, full and mild. Horns long compared with their diameter.   4

4. Neck fine and moderately thin and should harmonize in
   symmetry with the head and shoulders.   6
5. Shoulders fine at the top, becoming deep and broad as
   they extend backward and downward, with a low chest.   4
6. Barrel large and deep with well-developed abdomen;
   ribs well rounded and free from fat.   10
7. Hips broad and chine level, with full loin.   10
8. Rump high, long and broad.   6
9. Hindquarters long and deep, rear line incurving. Tail
   long, slim, tapering to a full switch.   8
10. Legs short, clean, standing well apart.   3
11. Udder large, well developed front and rear. Teats of
    convenient size and wide apart; mammary veins large,
    long and crooked, entering large orifices.   20
12. Escutcheon.   2
13. Hair fine and soft; skin of moderate thickness of a
    rich dark or yellow color.   3
14. Quiet disposition and free from excessive fat.   4
15. General condition and apparent constitution.   6
                                                              ____
Perfection   100

Dutch Belted Bulls

The scale of points for males shall be the same as those given for females, except that No. 11 shall be omitted and the bull credited 10 points for size and wide spread, placing of rudimentary teats, 5 points additional for development of shoulder, and 5 additional points for perfection of belt.

Sixteen Dutch Belted cattle were exhibited at the World's Columbian Exposition in Chicago in 1893 by H. B. Richards, of which 13 were sold to Mexico City after the show. From one to three herds of Dutch Belted cattle were exhibited at the National Dairy Show in 1907 to 1913, with 16 to 41 animals competing.

PRODUCTION RECORDS

The first production record mentioned was with the entry of "No. 33 Idell—Bred by and property of J. E. Wells, Chester, N.Y.; calved January, 1885; sired by Fritz (11); dam, Betsy, purebred, milk record of 10,599 pounds in 11 months; belted irregular. General condition good." Betsy, unregistered dam of foundation cow No. 33 Idell, had the first production record published in the *Herd Book*.

Dutch Belted cows were entered in the milking contest at the Pan-American Exposition at Buffalo, New York, in 1901. Their average production was 4,978 pounds of milk and 169.4 pounds of butterfat in 120 days.

Advanced Registry rules were approved at the annual meeting on May 9, 1912. Records were started a year later. Seven cows owned by Dr. H. W. Lance, Peapack, New Jersey, averaged 10,224 pounds of milk, 3.53 percent and 361.2 pounds of butterfat. Two cows were mature age. Herds owned by Dr. J. G. DuPuis of Lemon City, Florida, and J. A. Wilson, Brunswick, Maine, were entered in turn and others later. The leading producers in succession were:

	Milk (pounds)	Test (percent)	Butterfat (pounds)
Peapack Duchess 1390, Imp.	13,065.0	3.43	447.64
Peapack Anna 1487	13,159.0	3.68	484.31
H. W. Lance, Peapack, N. J.			
Ferndell 1961	13,477.9	3.73	501.10
Dr. J. G. DuPuis,			
Lemon City, Florida			
Glenbeulah's Beauty 2172	13,295.8	3.99	531.19
A. Quackenboss,			
St. Cloud, Wisconsin			
Gem of Columbia 2038	17,268.2	3.67	633.86
Dr. J. G. DuPuis			
Angelina 2641	16,022.6	4.17	668.07
O. A. Leonard,			
Tolland, Connecticut			
Gloria 3231	16,545.9	4.71	780.09
J. A. Wilson, Brunswick, Maine			
Loraine of Brunswick 3020	18,211.2	4.48	816.53
J. A. Wilson			

A 305-day Roll of Honor class was established in 1928. White Belt Honey 2719 owned by Dr. J. G. DuPuis, was the first cow to qualify in this class. She produced 10,713.7 pounds of milk, 3.81 percent and 408.43 pounds of butterfat and carried a living calf 272 days of the 305-day record. Ninety Advanced Registry records were published in Volume 13 of the *Herd Book*. The average production in different age classes is listed in Table 9.1. Breeders discontinued testing in the Advanced Registry several years ago.

Interest in production records has revived, with thought of establishing a form of herd test. Some 99 cows of all ages and stages of lactation were milked twice daily in one herd. Their actual average production during May 1960 was 895 pounds of milk, 3.6 percent and 32 pounds of butterfat. This is creditable production under commercial farm conditions with two milkings daily.

MODEL DUTCH BELTED COW

A model Dutch Belted cow was designed in 1947. It accentuated length of body, well-sprung ribs, chest, and barrel capacity. The

TABLE 9.1
AVERAGE PRODUCTION AT DIFFERENT AGES BY DUTCH BELTED COWS[a]

Age classes	Number of records	Average production		
		Milk (lbs.)	Test (%)	Butterfat (lbs.)
Mature	39	11,587.8		439.54
Senior 4-year-old	10	10,783.5		434.72
Junior 4-year-old	7	10,494.5		405.35
Senior 3-year-old	11	9,829.3		405.38
Junior 3-year-old	10	8,320.8		307.37
Senior 2-year-old	6	10,241.9		436.85
Junior 2-year-old	5	7,575.5		292.87
Yearling	2	6,908.8		290.01
Average	90	10,418.8	3.89	405.84

a. All but a few records were made on 2 milkings daily.

FIG. 9.2. A registered Dutch Belted cow at Roselawn Farm in Florida.

withers were fine and the shoulders blended smoothly into the body. The triple wedge shape was typical of dairy cattle. The top-line was straight, hips broad, and rump level. Legs were short, wide apart, and placed squarely. The udder was large, well developed in front and rear, and attached strongly to the belly wall. The teats were of convenient size and placed squarely at well-spaced inter-vals on the floor of the udder. The head and neck showed breed character and refinement. Horns were short and curved inward. The color was black with a wide white belt around the middle. The feet, legs, and switch were black. The model typified the Scale of Points of the breed. Twenty three breeders registered 72 animals, and 25 Dutch Belted animals were transferred to new owners in 1970.

BREEDING DUTCH BELTED CATTLE

Dutch Belted cows weigh 900 to 1,400 pounds and mature bulls from 1,600 to 2,000 pounds. The calves average around 75 pounds at birth, males being slightly heavier than females. Light natural fleshing is desirable.

When the wide white belt extends too far, the fore udder and lower part of the rear legs may be white from the hoofs upward. As with all predominately black breeds, some animals in the Neth-erlands and in the United States carry the gene for recessive red hair coat.

The secretary of the Dutch Belted Cattle Association of America is James H. Hendrie, P. O. Box 358, Venus, Florida 33960. The president is John G. DuPuis, Jr., 6000 N.W. 32nd Avenue, Miami, Florida 33142.

REFERENCES

Aiton, William 1830–31. On the dairy husbandry of Holland. *Quart. J. Agr.* 2:328–45.
Becker, R. B. 1933. Recessive coloration in Dutch Belted cattle. *J. Hered.* 24: 283–86.
Hendrie, James. 1957. History of Dutch Belted cattle. *Dutch Belted Cattle Assoc. of America.* Miami.
Houghton, F. L. 1897. *Holstein-Friesian Cattle.* Brattleboro, Vt. P. 39.
Kick, W. 1878. *Lehrbuch der Rindviehzucht nebst Berichtingen nach dem neuesten Stande der Wissenschaft und Erfahrung.* Berlin.

Klippart, John W. 1865. Report of an agricultural tour in Europe. *Ohio Agr. Rept.* Ser. 2:17–280.

Kuitert, K. 1921. Color inheritance in cattle. *J. Hered.* 12:102–09.

Lehnert, Hugo. 1896. *Rasse und Leistung unser Rinder.* Berlin.

Pirtle, T. R. 1926. *History of dairying.* Mojonnier, Chicago. P. 44.

Plumb, Charles S. 1920. *Types and breeds of farm animals.* Rev. ed. Ginn, Chicago.

Weckherlin, August. 1827. *Abbildungen der Rindvieh und anderer Haustierrassen auf den Privatgutern seiner Magistat des Konig von Wurtemberg nach dem Lieben gezeichnet und lithographiert von Lorenz Ekenean Alessen mit beigefugten Text von August Weckherlin.*

Miscellaneous Breed Publications

Dutch Belted Herd Book, 1886–1929. Vols. 1–13.

Dutch Belted Cattle Bull. and Live Stock J.

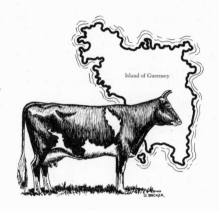

Island of Guernsey

CHAPTER 10

GUERNSEYS IN THE CHANNEL ISLANDS

THE CHANNEL ISLANDS are located about 33 miles north of St. Malo, Brittany, and about 75 miles south of Weymouth, England, between latitude 49° and 50° North. This is the same latitude as northern Labrador, but the warm Gulf Stream moderates the climate. Average rainfall is 38 inches annually; mean temperature is between 49.5° F. in January and 59.7° F. in August. The Islands include Jersey (45 square miles); Guernsey (24.5 square miles); Alderney (3.06 square miles); Sark (2 square miles); Herm (400 English acres); Jethou (100 English acres); and Breqhou (smaller).

EARLY HISTORY

Cave dwellers of the Mousterian civilization and later the makers of "bell beaker" pottery inhabited the Channel Islands before the arrival of the Celts, who brought knowledge of working metals. At

Headpiece: Vignette of Guernsey cow and Island of Guernsey.

least 27 cromlechs have been discovered on four islands. Bones of ox, pig, and sheep and parts of eight human skeletons were found in La Hogue Bie cromlech on Jersey along with "beaker" type pottery.

Jersey was occupied by the Romans for some 400 years, and was named Caesarea. They knew Guernsey Island as Sarnia. The Channel Islands came under French rule in the sixth century A.D.

Norsemen invaded France (Gaul) in A.D. 286 and several times thereafter, landing on the Channel Islands in 856. The French king, Charles the Simple, made peace with Jarl Rollo, who led a Norse invasion in 912. Charles gave his daughter Sisele in marriage and granted the province of Neustria (Normandy) to Rollo in return for Rollo's conversion to and baptism in Christianity and his promise to serve the French king. So Norse adventurers became "Normans," with Rollo as the Duke of Normandy. The Channel Islands were part of his dukedom. James Wilson of Ireland believed the Norsemen brought some polled dun or silver-gray cattle to countries in which they settled.

John of Ditton (1309) mentioned establishment of an abbey in the Vale, taking possession of the land, requiring the inhabitants to pay every tenth sheaf of grain as tithe, the eleventh as campart, and a ponnage tax on each house. Castles were fortified in 1327 and 1377 "to contain all the people of the island, and their cattle and other effects" against invasions. Charles Kitts, in *The Guernsey Cow, Her History and Her Records*, wrote that monks from St. Michel in Brittany founded the abbey St. Michel du Vale about A.D. 960. William the Conqueror added the Vale, Castel, St. Saviours, and St. Peter-in-the-Wood to their lands providing they taught the natives agriculture. This involved introduction of cattle from Breton (Froment de Leon) which were colored solid red, red-and-white, or fawn-and-white, were of small size, and yielded high quality milk. A few years later, monks from Cherbourg founded an abbey on Alderney and one in St. Martin's parish, bringing Norman brindle cattle by barge (the Isigny cattle). The latter cattle were larger than those from Brittany and were good milkers. Crosses of these cattle furnished the main foundation of present Guernseys.

The Feudal Period

William the Conqueror, seventh Duke of Normandy, invaded England in 1066, won the battle of Hastings, and introduced feudalism. The Channel Islands, already divided into parishes, were ruled by him as their duke. When King Phillip of France annexed the Normandy mainland to the French crown in 1204, the Channel Islands remained loyal to the Duke of Normandy (English king). The islanders' main income in the sixteenth century came from agriculture, fisheries, and home-knit woolen garments ("Jerseys").

Two fiefs were originally on Guernsey during feudal times. Land was divided into large farms (bordages). Houses of the tenants (bordiers) were surrounded by the plowlands (corvées). The latter were divided into long strips called ox-gangs or bouvées. Oxen furnished by each bouvée pulled the common plow for the whole field. A tenant was first allowed to enclose his field in the seventeenth century. The bordiers finally were abolished in 1857. The feudal ruler (Duke of Normandy) received several taxes: fouage or hearth tax, ponnage for mast eaten by pigs in the manor's woods, verp for "ownerless" beasts straying onto his pasture, and other taxes.

Cattle were mentioned on the Channel Islands in 1581, 1613, and 1617 in witchcraft trials. James I levied taxes on the export of cattle hides in 1607. Doctor Heylin, chaplain of the Island garrison in 1629, mentioned in his survey of Guernsey and Jersey that the islanders possessed "enough of cattell both for themselves and for their ships."

Laws of the Bailiwick of Guernsey over four centuries dealt with cattle in several ways. Exports of beef tallow were prohibited in 1535. In 1537 people could not keep more cattle than they had land to provide feed. In 1542 animals that appeared starving were to be buried well off the road, and cattle could not be tethered or left unguarded on the roadway. In 1580 stray cattle were to be impounded in the crop season—a period defined by the Royal Court. In 1581 persons found milking the cows of others were subjected to a penalty. A number of ordinances regulated the people who could place their animals on the commons (communal lands of feudal times).

THE ENCLOSURE ACT

The Enclosure Act of the Island of Guernsey in 1717 restrained cattle at large, thus permitting owners to plan matings to improve their herds. A local common was retained along the coast in St. Martin's parish. Grazing of public commons by cattle and sheep was discontinued by order of His Majesty in Council on August 4. 1830.

Laws regulating the harvest of seaweed (vraic), essential for fuel in winter and fertilizer, were first passed in 1611, modifications of which stand even today. The seasons of harvest were established. and regulations as to method of harvest provided that the boats would not tear the roots of seaweed from the rocks.

LAWS PROHIBIT IMPORTATION OF CATTLE

In 1814, the Royal Court forbade landing of any bulls or lean cows into the Bailiwick of Guernsey, subject to a fine and confiscation of the animals. This law was requested by constables of nine parishes and was opposed from St. Peter Port. The laws were made more stringent 5 years later—allowing importation only for slaughter—to protect the Island cattle against misrepresentation of cattle from France, trans-shipped and sold in England as being *from the Channel Islands*. Any cattle imported for slaughter had to be killed within 4 months. The law of February 17, 1824, forbade importation of *any heifer* even for slaughter. A register of imported animals was kept to assure compliance with the law. The law of April 26, 1824, added heifer calves to the prohibited list. Calves born to imported cows had to be vealed within 8 weeks.

A new stringent law was passed on October 3, 1842, revoking the two laws of 1824 but adding to their provisions. Specifically, all imported animals (no bulls allowed) must be branded with an "F" at least 4 inches high and large in proportion on the left cheek and right thigh before leaving the dock.

These laws dealt with animals to be slaughtered as meat for the inhabitants. Other laws gave protection against infectious diseases (as foot-and-mouth disease) which might be brought in straw or bedding. Cattle from Jersey were not "foreign" and sometimes were brought as a bride's dowry.

EARLY ENGLISH WRITINGS

Eighteenth- and nineteenth-century writers mentioned "Alderney" cattle and the richness and color of their milk and butter. Thomas Hale (1758) described Alderney cattle as short-horned, squarely built, not tender, requiring rich feeding, and yielding a quantity of good milk. George Culley (1786) said these cattle were kept on the estates of the nobility and gentry because their exceedingly rich milk supported the luxury of the tea table. The Alderney cattle were fine-boned and light red or yellow in color; the beef was generally yellow, "very fine in the grain, and well-flavoured." Thomas Dicey (1798) attributed "smallness of their cattle . . . to the shortness of their grass." Guernsey and Jersey cattle were remarkable "for the peculiar goodness of their butter."

William Berry (1814), regarded as the most accurate Guernsey historian, mentioned gathering vraic twice yearly to manure the land, and described a crop rotation of wheat, wheat, barley or oats, clover, and parsnips. The latter were considered "the best winter feed for milch cows." Whole milk was churned for butter "of a fine yellow or golden color."

Thomas Quayle made the *Survey of Agriculture of the Channel Islands* in 1815 for the Board of Agriculture (England). He wrote: "Guernsey cattle are larger-boned, taller, in every respect more stout and coarsely made than those of Jersey." The Guernsey cow's carcass was bulkier, its haircoat less fine, and its color usually darker. Guernsey cows usually brought a higher price for export than Jerseys.

Jonathan Duncan wrote in 1841:

> The tethering of all cattle, the use of the spade, and the general culture of clover, lucerne (alfalfa), parsnips, turnips, and mangel-wurzels, add wonderfully to the means of sustenance of all animals.
>
> It is an invariable practise, throughout Guernsey to tether the cattle, staking them by the horns, by means of an iron or wooden peg, attached to a halter about 12 feet in length. This is shifted about four or five times a day, allowing a fresh range of 2 to 5 feet each time. By this system, the most is made of the grass, for none can be trodden down or wasted. . . .

The supposed general average that a cow can yield throughout the year is one pound of butter or eight quarts of milk during the 24 hours. It is observed that the fattest cows are not the best milchers, and the best milchers will not always produce the largest quantity of butter.

Guernsey cows tethered on pasture are shown in Figure 10.1.

FIG. 10.1. Guernsey cows tethered on pasture. They are advanced 2 or 3 feet at a time to graze but not to trample the fresh forage.

The average annual export of cattle from the bailiwick from 1822 to 1827 was 420 from Guernsey and 75 from Alderney. In 1832 the export total was 632, and in 1833, 553 cows, heifers, and calves were exported.

In 1901 J. de Garis of St. Saviours stated to Miss Edith Carey, author of *Guernsey Folk Lore,* that farm parents often gave the bride a cow to help set up housekeeping—a practice which accounted for cows being moved between Jersey and Guernsey before the herdbooks were established.

An Agricultural Society

La Gazette de Guernesey of January 17, 1817, published a notice (in French) by D. de Lisle Brock and Pierre Bredthafft to the effect that:

> His Excellency Major General Bayly . . . having met many citizens who agree with him, the following resolutions have been agreed upon,
> 1. That there should be formed a Society of Agriculture with the object of acquiring and disseminating the most useful knowledge on matters concerning the country in general, but that the principal object shall be to conserve and improve the breed of our cows, so greatly appreciated by everyone.

People were invited to become members and to pay a fee which would be used to buy farm implements as models, and to award prizes for the best specimens of cows, bulls, and other animals.

Two resolutions were adopted:

> That the conformation, beauty and good qualities of cows and bulls and their improvement shall receive the attention of this Society. . . .
> That it shall be recommended to each parochial committee to present their views to the central committee on the means to be taken to insure that a sufficient number of bulls of the best type and of the best strains in the different parts of the Island be kept . . . asking [the States] that a reasonable sum be granted annually; . . . and to occupy themselves generally with the improvements which may be introduced for the cultivation of the land and for the improvement of sheep, pigs, and horses.

Thirty bulls (1½ to 3 years old) were recommended. Five judges and the Society officers visited the parishes and made the selections. The first parish show of cattle was held June 4, 1817, to select the bulls to be used in Vale parish.

Shows for Cattle

The first Island show was held on June 25, 1817, with five judges officiating. The scale of points for bulls at the first Agricultural Society show allowed points for eight items, as follows:

Points

1. Type of sire and dam. Dam considered to be a good cow producing yellow butter — 5
2. Conformation and general appearance — 2
3. Colour — 3
4. Distinctive qualities indicating that the strain produces yellow butter — 2
5. Beauty of head which is not all white, beauty of horn, the eyes alive and bright — 3
6. Neck fleshy, shoulders and chest wide — 2
7. Hind quarters not pointed, straight back — 2
8. Good legs which do not cross behind in walking — 1

Total points — 20

Animals that received prizes during the first four years were colored pale red, dark red, red-and-white, brindle, and black-and-white.

John Jacob commented (1830) that because of the Society's activities ". . . great improvements have been made in their breed of cattle." The States (legislature) granted £60 per annum at that time for premiums devoted to improvement of the cattle.

The "Société d'Agriculture de Guernesey" was chartered on September 16, 1842, upon petition of 20 persons representing the ten parishes; it took over a show organized by an earlier society. "Royale" was added to the name after Queen Victoria (1838–1901) granted right of charter in 1842. That same year a new scale of points (in French) allowed 28 points for bulls and 30 for cows. A cow scoring less than 21 points was not awarded a prize. Shows increased in importance and influence. For example, the show in 1857 entered 31 bulls, 89 cows, and 89 heifers. Bulls were required to remain on the Island for a period for breeding purposes before their owners received the prize money. Cows and heifers also must have calved or have become pregnant before their prizes were paid. The scale of points was revised to 100 points for perfection in 1883.

The Herd Book Council of seven members arranged for local shows to be held at 2-month intervals to qualify bulls and cows for the *Herd Book*. Cows must have been at least 42 days after calving. Bulls—with the exception of those in the artificial insemination service—were qualified at 12 months for use in the owner's herd only, but at 15 months they became available for public service.

The Scale of Points for judging bulls in 1958 follows:

Points

1. Size and General Appearance. Bulls four years old and over, about 1500 lb.; masculine throughout, frame well furnished with muscle, characteristic of a male animal, but not beefy; alert and attractive expression, general poise indicating balance and symmetry of movement. Colour, shade of fawn with or without white markings.   16

2. Head. Masculine but not coarse, clean cut lean face, broad between bright eyes, gentle expression, strong sinewy jaw, muzzle broad and buff coloured; horns medium length, not coarse.   12

3. Neck and Shoulders. Neck long with well developed crest, clean throat; shoulders powerful; withers fine, backbone rising well between shoulder blades.   5

4. Chest. Deep and wide between and behind forelegs, body deep; ribs long, deep, well sprung and wide apart.   10

5. Back. Long and straight from shoulder to junction of tail, broad and level across loins and hips.   10

6. Hindquarters. Hook bones wide apart; thighs long, flat and muscular but not beefy. Tail long and fine, neatly set between pelvic bones. Rump long and wide, and level between hooks and pin bones.   10

7. Legs. Fine. Forelegs straight, wide apart and squarely set. Hindlegs well apart, nearly perpendicular from hock to fetlock with point of hock directly in line with pelvis bone; wide curve from flank to hock joint. Legs free in motion, but no tendency to sweep or turn.   12

8. Teats. Of fair size, wide apart and squarely placed; escutcheon well developed.   5

9. Quality. Rich golden pigment throughout as indicated in ears, end of tail bone, around eyes and body generally, yellow at base of horns, hoofs amber colored. Hide thin and loose to touch.   20

———

100

Bulls were accepted only as Pedigree Stock. Cows were accepted as Pedigree Stock or as Foundation Stock (that is, the sire and dam were not necessarily on record, or the birth of a calf was not reported on time).

The dam of a bull must have exceeded the requirements for Advanced Registry by 10 percent in Scheme A (pregnant within 5 months of previous calving) or by 20 percent in Scheme B (not pregnant within the time set). The dam must appear for inspection with her young bull, or have been scored previously. She must score 84 points, of which not fewer than 19 points were for udder.

The Scale of Points for cows was as follows:

Points

1. Size large but not coarse. Alert and attractive expression; general poise indicating balance and symmetry of movement. Color, shade of fawn with or without white markings.   5
2. Head. Fine and long, lean face broad between the eyes. Horns fine and slightly curved. Eyes large, and bright with gentle expression. Muzzle wide and buff colored.   8
3. Neck and Shoulders. Neck long and thin, sloping gradually into the shoulders; clean throat, withers fine and wedge shaped.   6
4. Body. Abdomen large and deep; loin wide and thin, chest deep and wide between and behind forelegs. Ribs well sprung, deep and wide apart.   10
5. Back. Long and straight from shoulder to juncture of tail, broad and level across loins and hips.   8
6. Hindquarters. Pelvis wide. Tail long and fine, neatly set between pelvis bones. Rump long and wide, and level between hooks and pin bones.   6
7. Legs. Fine. Forelegs straight, wide apart and squarely set. Hindlegs well apart, nearly perpendicular from hock to fetlock, with point of hock directly in line with pelvis bone; wide curve from flank to hock joint. Legs free in motion with no tendency to sweep or turn.   8
8. Udder. Of good size and extended well in front; firmly attached to the body, well up behind, elastic but not fleshy, quarters evenly balanced, symmetrical and free from division. Teats of medium size, well apart, evenly placed and hanging perpendicularly.   25
9. Milk veins prominent, long and tortuous and branching, with deep fountains. Veins on udder clearly defined. Escutcheon wide, high and broad with thigh ovals.   8
10. Quality. Rich golden pigment throughout as indicated in ears, end of tail bone, around eyes and body generally, yellow at base of horns, hoofs amber coloured. Hide thin and loose to touch.   16

100

THE QUEEN'S CUP SHOW

Queen Victoria initiated the annual Queen's (King's) Cup Show by contributing ten guineas each for the two best bulls and five guineas for the best herd of four cows. At a later show the best cow received the latter trophy. The Queen's Cup Show is the apex in quality on the separate islands of Guernsey, Alderney, and Sark (see Fig. 10.2).

A Cup was awarded to the best bull in the Artificial Insemination Center in 1957, and a Cup was open to competition among other bulls. Under the latest rules, owners of bulls exhibited their animals along with the dam or her scorecard, the sire, and both granddams or their scorecards. Bulls of sufficient age were represented by six or more progeny over 4 months old. All were considered by the judges in making the awards.

Bulls winning the Cups were available for public service for 1 year at a service fee approved by the Society. After this period, the Society's brand, "E.R.," was placed on the horn, and the bull's name was engraved on the Cup.

FIG. 10.2. Guernsey cows lined up before the judges at the King's Cup Show.

A Queen's Cup was awarded to the best cow, already in the Advanced Registry. Winners could not compete a second time.

GUERNSEY HERD BOOK

Island breeders differed on the necessity for keeping a formal pedigree herdbook. Some people contended there was no need for one since all Island cattle were purebred. Others thought only the better animals should be entered as breeding stock to improve island cattle. In 1878 James James initiated a private herdbook, which included entries based on show winnings and selection. James's *Guernsey Herd Book* entered 178 animals. Sixty-three animals were added to an edition in 1879. The society finally accepted the herdbook of James James as a foundation, and added to it as follows: "That all the animals receiving prizes during the three previous years shall be admitted . . . also all animals which shall receive prizes subsequently, together with their progeny. Good attention being taken that the animals admitted shall possess the superior quality in regard to the milk and butter qualities."

Volume 1 of *The Royal Guernsey Agricultural Society Herd Book* was published in 1882. After 1885, progeny of registered cattle were eligible for entry in the *Herd Book* without inspection at the shows. Unregistered females that qualified at a show or passed inspection by the herdbook examiners were registered as Foundation Stock, with the initials F.S. following their registration number.

A rival herdbook, the *General Herd Book of the Island of Guernsey*, was begun in 1881 because of differences of opinions as to animals eligible for registration. The American Guernsey Cattle Club sent Secretary William H. Caldwell to the Island in 1901 to study entry requirements of the two herdbooks. Recognition of the *General Herd Book* was withdrawn in 1902, and it was discontinued. An "Alderney Branch" of the Agricultural Society's herdbook was recognized; Volume I was published in 1907.

After July 1, 1918, only a bull whose dam and both granddams were in the Advanced Registry might be registered. Such a bull must be inspected with his dam, at a local show. Fat production of the dam and the sire's dam must exceed Advanced Registry requirements since 1922. Now the requirements must be exceeded by 10 percent under Scheme A, or by 20 percent under Scheme B involving pregnancy requirements.

All registered bulls 15 months old in the bailiwick of Guernsey were required to stand for public service. Service records were kept in duplicate. About 1,200 heifers and 150 bulls were registered annually before World War II. Registrations were continued during the war even under German occupation, but no shows were held. Some decline in registrations occurred after the war. The 1965 Census showed 48 bulls more than 12 months old, 25 younger bulls, 2,623 cows, 348 heifers in calf, and 834 other heifers, exclusive of bulls in the A.I. stud. Artificial breeding decreased the demand for bulls.

PRODUCTION RECORDS

Few production records were cited in early Guernsey literature, except statements that a cow would average a pound of butter daily for the year and some produced twice as much for short periods.

An average yield of 627 imperial gallons per year was attained by J. Boyd Kinnear's herd on the Island. The society provided in 1885 that an Official Milk and Butter Test Book be maintained for sworn private and public production tests conducted under certain rules. Records of milk and butter production of four cows for 7-day periods were published in 1888. Records were begun on a yearly basis in 1911, applying monthly butterfat tests to the owner's milk weights, as was practiced then in the United States. This was the Advanced Registry of the Society.

The founding of the Advanced Registry, in the judgment of Charles Kitts (official test supervisor), was a landmark with the Guernsey breed. Records during the first year showed an average of 9,516 pounds of milk, 5.05 percent and 480 pounds of butterfat, mainly from mature cows. Requirements to qualify for the Advanced Registry were 250.5 pounds of butterfat for animals 2 years old and up to 360 pounds in 365 days for those 5 years or older. The standards were increased in May 1938, and now are 6,000 pounds of milk, 293.5 pounds of butterfat for animals 2 years old, up to 9,000 pounds of milk, 425.0 pounds of butterfat for animals 5 years and older. The average test must equal or exceed 4.25 percent butterfat. Both the Babcock and Gerber butterfat tests were recognized.

HERD TEST

A Herd Test was established on Guernsey in 1931. It was discontinued in 1937 when the English Guernsey Cattle Club did not recognize the records for importation of animals to England. The objectives originally stated for this system of mass testing were: "To find by simple testing and record keeping under ordinary conditions those cows which were low producers as to be unprofitable to the owner. To find out the bulls which are prepotent for high production. To encourage the use of such bulls as fully as possible and to discourage the use of those which transmit low production."

In 1958 the States of Guernsey and the State Committee for Agriculture took the first steps toward compulsory production testing of all cows. Limited facilities for keeping records of so many cows made it necessary to proceed gradually. The State Committee proposed that a sole Compulsory Recording Scheme for entire herds apply to all recognized production records on the Island. The groups agreed to conduct a Voluntary Herd Recording Scheme for six years under the Society's supervision. The States of Guernsey passed a resolution on May 25, 1954: "To authorize the States Committee for Agriculture and Fisheries to make an annual grant in respect of each milk producer whose entire herd of cows is continuously milk and butter-fat recorded under the scheme for the time being administered by the Royal Guernsey Agricultural and Horticultural Society, such grants being payable to the said Society. . . ."

The rules barred individual selective testing while the Society was under the grant. The Society supervised the Herd Test, and Advanced Registry requirements remained unchanged. About 80 percent of the cows were under Herd Test in 1959. The recording staff was transferred in July 1961 to a laboratory at Beau Sejours under a subcommittee of the States Committee for Agriculture and Fisheries.

DIPLOMA OF MERIT

Beginning in April 1951 bulls were awarded a Diploma of Merit when ten or more daughters qualified for the Advanced Registry. Two daughters by artificial insemination were equivalent to one daughter by natural service.

Cows that produced 50,000 pounds of milk before they were 8 years old were recognized with a Certificate of Merit. An Award of Merit was earned by a cow with three Advanced Registry records.

ARTIFICIAL BREEDING

A Guernsey Artificial Breeding Center was established on the Island in 1956, partly to control any reproductive diseases and to make promising bulls more widely accessible. Every registered bull must have been available for public service previously, with few restrictions.

The States Committee for Agriculture and Fisheries and the Herd Book Council drafted rules for operation of the stud and appealed to breeders for bulls. Of 20 offered, four virgin bulls were selected to begin the program in 1956. The demand for artificial service was so great that two more bulls were added in 1957. A veterinary officer was appointed to manage the service in 1958. When restrictions resulting from two cases of foot-and-mouth disease were removed, semen was sent to Alderney, Sark, and Brecqhou. Between 70 and 80 percent of the cows on Guernsey were artificially inseminated in 1958.

The isolated location of the Channel Islands contributed to relative freedom from some diseases. Tuberculosis and brucellosis were not serious factors. Foot-and-mouth disease outbreaks were eradicated by slaughter of the herd and quarantine until danger was past.

Lethal recessive characters were discussed by the Society in 1958. A *Herd Book* rule was adopted in 1957 for detection of and protection against some conditions: "All calves born dead or dying within 12 hours of birth must be notified to the Secretary of the Royal Agricultural and Horticultural Society before being disposed of. (This includes premature births of 5 months onwards). Owners failing to do so will render themselves liable to a fine not exceeding £10 for any one infraction."

Bulls that produced lethal recessive characters since February 1958 were to be withdrawn from public service after April 1960. A bull calf could be qualified by the Herd Book Council for the owner's private use and test matings. Such test matings against lethal recessives were not considered public service. They involved mating with at least eight known carriers and 25 daughters of known carrier cows or bulls. The test was completed at calving of the last such mating.

A recognized defect of the spermatozoal acrosome cap caused the Society to rule on April 1, 1960, that "all bulls aged 12 months or over at the date of sale which have not proved themselves [fertile], must be examined by a Veterinary Surgeon before being sold."

OUTSTANDING GUERNSEYS

May Rose 2d 3251 P.S. (Fig. 10.3) was a great transmitter. Highly Commended in 1893, she won first prizes in 1896 and 1897. As Claremont May Rose 3648 E.G.H.B., she won first prize at the

Royal (Winchester) show in 1897. She gave 48 and 55 pounds of milk in a day when 16 years old. This cow had marked refinement, dairy temperament, level topline, ruggedness, and capacity, but she had weak hind legs in advanced age. Her udder was capacious and strongly attached. Her son Imp. King of the May, two daughters, and granddaughter Imp. Itchen Daisy 3d came to the United States.

FIG. 10.3. May Rose 2d 3251 P.S. or Claremont May Rose 3648 E.G.H.B. was recognized as one of the greatest transmitting cows of the Guernsey breed.

The number of milking cows in 1963–65 averaged over 2,600, with a slight compensating increase in heifers raised for replacements. The area of land per cow in dairy farms decreased 10.6 percent from 1954 to 1965. Cow numbers were low in 1954, yet in 1964, 20.4 percent more cows yielded 38.6 percent more milk. This represented an increased yield of 15 percent per cow. In 1953, 349 holdings averaged 14.95 acres, compared with 227 holdings averaging 20.2 acres in 1964 with 11.45 cows per farm. Fewer acres produced root crops. The larger herds justified more acres for silage production. Lucerne (alfalfa) increased in the new seedings. Fewer acres produced wheat or some other crop. The 1964 agricultural census showed 1.75 acres (4.37 vergees) per cow on dairy farms on Guernsey.

Increased practice of artificial breeding in many countries reduced the demand for bulls. The ratio of male to female registrations decreased from 1 male to 7.66 females in 1944–53, to 1 male per 18.85 females in 1956–62. Exports of bulls decreased from an average of 50.3, to about 4 bulls per year.

REFERENCES

Anonymous. Jan. 17, 1817. [Formation of an Agricultural Society.] In *Gazette de Guernesey*.

Anonymous. Before 1938. *The Channel Islands and parts of Brittany and Normandy*. Ward, Lock & Co., London.

Anstad, D. T., and R. G. Latham. 1865. *The Channel Islands*. 2nd ed. London.

Berry, William. 1815. *The history of the Island of Guernsey, part of the ancient Duchy of Normandy, from the remotest period of antiquity to the year 1814*. London.

Brock, D. de Lisle, and P. Bredhafft. Jan. 17, 1817. [Formation of an Agricultural Society.] In *Gazette de Guernesey*.

Caldwell, William H. 1925. *Langwater Guernseys*. Peterborough, N.H.

———. 1941. *The Guernsey*. Peterborough, N.H.

Carey, Edith F. 1903. *Guernsey folk lore*. London.

———. 1904. *The Channel Islands*. London.

Coleman, J. 1875. *The cattle of Great Britain*. C. Cox, London.

Collas, Harold A. 1937. Farming in Guernsey. *Guernsey Breeders' J.* 51:265–67.

Culley, George. 1786. *Observations on live stock*. G. G. J. & J. Robinson, London.

de Guerin, Basil C. 1945. Guernsey herd safe on Island. *Guernsey Breeders' J.* 68:105–6, 119.

———. 1947. *History of the Royal Guernsey Agricultural and Horticultural Society*. Guernsey.

Dicey, Thomas. 1798. *An historical account of the Island of Guernsey from the first settlement*. London.

Duncan, Jonathan. 1836–37. *Guernsey and Jersey Magazine*. Vols. 1–4. London.

———. 1841. *The history of Guernsey, with occasional notices of Jersey, Alderney and Sark, and biographical sketches*. London.

French, T. G. 1944. The Alderney herd. *Guernsey Breeders' J.* 66:522, 528.

Hale, Thomas. 1758. *Complete body of husbandry*. 2 vols. London.

Hazard, Willis P. 1872. *The Jersey, Alderney and Guernsey cow*. 10th ed. Porter & Coates, Philadelphia.

Herman, H. A. 1956. The Island of Guernsey begins breeding artificially. *Hoard's Dairyman* 101:823.

Heylin, Doctor. 1656. *A survey of the two ilands, Guernsey and Jersey; with the isles appending, according to their politic, and forms of government, both ecclesticall and civill*.

Hills, Charles L. 1917. *The Guernsey breed*. Kimball, Waterloo, Iowa.

Inglis, H. D. 1834. *The Channel Islands of Jersey, Guernsey, Alderney, Sark, Herm and Jethou*. Whittaker & Co. London.

Jacob, John. 1830. *Annals of some of the British Norman Isles constituting the Bailiwick of Guernsey.* J. Smith, Paris.

James, James. 1878, 1879. *The Guernsey Herd Book.* Guernsey.

Jeremie, John. 1821–1822. *Account of Guernsey.*

Jones, W. A. 1962. Abnormal morphology of the spermatozoa in Guernsey bulls. *Brit. Vet. J.* 118:257–61.

Kitts, Charles. 1922. *The Guernsey cow, her history and her records.* Guernsey.

LeCornu, C. P. 1859. The agriculture of the Islands of Jersey, Guernsey, Alderney and Sark. *J. Roy. Agr. Soc. Engl.* 20:32–67.

LeCouteur, Col. J. 1845. The Jersey, misnamed Alderney, cow. *J. Roy. Agr. Soc. Engl.* 5:43–50.

Low, David. 1842. *On the domesticated animals of the British Isles.* Longmans, Green & Co., London.

MacCulloch, Robert. 1852, 1856, 1864. *Recueil d'Ordonnances de la Cour Royale de l'Isle de Guernesey.* Vol. 1–3. (Vol. 4 by Arthur W. Bell.) Guernsey.

Pitts, J. L. 1889. *Guernsey and its bailiwick.*

Prentice, E. Parmalee. 1940. *The history of Channel Island cattle. Guernseys and Jerseys.* Harper, New York.

Quayle, Thomas. 1815. *General view of the agriculture and present state of the Islands on the Coast of Normandy, subject to the Crown of Great Britain.* London.

Syvret, George S. 1832. *Chroniques des Isles de Jersey, Guernsey, Auregny et Sark.*

Tupper, F. B. 1854. *The history of Guernsey and its Bailiwick; with occasional notices of Jersey.* Stephen Barby, Guernsey.

Wallace, Robert. 1888. *Farm live stock of Great Britain.* 2nd. ed. Oliver & Boyd, London.

Warren, J. P. 1926. *Our own Island.* Guernsey.

Wilson, James. 1909. *The evolution of British cattle.* Vinton & Co., London.

Miscellaneous Breed Publications

General Herd Book of the Island of Guernsey. 1881–1902.

The Guernsey Herd Book. 1882–. Royal Guernsey Agricultural and Horticultural Society. Vols. 1–.

Royal Alderney Agricultural Society Herd Book. 1907–. Vols. 1–.

Royal Guernsey Agricultural and Horticultural Society annual reports. 1842–.

CHAPTER 11

GUERNSEYS IN THE UNITED STATES

THE FIRST Channel Island cattle in the United States were called "Alderneys" and hence may have been from Guernsey or Jersey. An Alderney cow was reported in the *Memoirs of the Philadelphia Society for the Promotion of Agriculture* (Volume IV, page 155) as yielding 8 pounds of butter in a week as a 3-year old. Imported in 1815, she was owned by Richard Morris in 1817. After the next calving she yielded 8 1/8 and 8 3/4 pounds of butter in 2 weeks. Her owner, Reuben Haines, commented that the butter was "so rich a yellow." The yellow color of Guernsey butter is due to carotene from the feed. It is a precursor of vitamin A.

Nicholas Biddle of Philadelphia imported three "Alderney" heifers in September 1840 which were registered with their progeny after the herdbook was established. However, Captain Prince imported three animals from Guernsey in 1833 and sent them to his brother on Cow Island, New Hampshire. Secretary William H.

Caldwell traced the pedigrees of 154 animals to this importation in 1896 and entered them in the Herd Register. Some excellent animals descended from this importation, including Glencoe's Bopeep 18602, Grand Champion at the National Dairy Show in 1910, 1911, and 1912.

Dr. L. H. Twaddell visited the Channel Islands in 1865 and described Guernseys and Jerseys as differing in conformation. The Fishers and Fowlers of Philadelphia were prominent early importers. The Massachusetts Society for Promotion of Agriculture made an importation in 1874 upon which three excellent herds were founded. A bull and ten females imported prior to 1865 were entered in the *Herd Register*. Six bulls and 54 females were imported during the next 10 years. S. C. Kent and associates of Pennsylvania imported and sold more than 1,000 animals at auction during 1881 to 1890. Interest among 16 owners led to organizing the Guernsey Breeders' Association in Philadelphia on February 5, 1884.

F. S. Peer of Ithaca, New York, imported over 3,000 Guernseys from the Channel Islands, England, and Canada in a 29-year period. W. W. Marsh, F. Lothrop Ames (Langwater Farm), and H. McKay Twombley brought some great transmitting cows. Joseph L. Hope, manager of Florham Farm, made their selections and advised with W. W. Marsh concerning the prominent Cherry family from Alderney. Between 1833 and 1937, 550 bulls and 12,362 females were imported and entered in the *Herd Register*. Imp. Itchen Daisy 3d 15630 was great among them (Fig. 11.1).

AMERICAN GUERNSEY CATTLE CLUB

Edward Norton of Farmington, Connecticut, brought "Alderney" cows to improve farming conditions and organized a joint-stock creamery to market the milk. He invited Mason C. Weld to discuss Guernsey cattle before 11 local farmers. They subscribed money to import a bull and 14 cows, which arrived in 1876. These men organized and elected Mr. Norton as secretary-treasurer. Breeders from Connecticut, Massachusetts, New Jersey, New York, and Pennsylvania met in New York City on February 7, 1877, and organized the American Guernsey Cattle Club, with Norton as secretary-treasurer; they also established the *Herd Register* for entry of animals that traced wholly to importation from the Island of Guernsey. After July 1882 imported cattle or the sire and dam of young animals were required to have been entered in the Island of Guernsey's *Herd Register*.

The *Herd Register* was published through Volume 45 in 1933. Since then a short pedigree has been included on each registration

FIG. 11.1. Imp. Itchen Daisy 3d 15630 was imported by Florham Farms and spent her later life at Langwater Farm. She was a class leader with two A. R. records; a consistent prize winner, dam of 6 females and 3 transmitting sons.

certificate. The *Performance Register,* begun in 1942, contained show winnings, production and progeny records, and type classifications. These volumes replaced the *Herd Register*.

Club officers are a president elected for a one-year term and often re-elected, first vice-president, second vice-president, and 12 directors. A director may be elected for a second term of 5 years. The Board appoints a secretary-treasurer responsible to the Board for Club operations. Each of ten districts is represented by a director, with five directors-at-large elected by secret ballot.

The districts, as established in 1969, are: District 1, New England states and New York; District 2, Pennsylvania; District 3, Delaware, Maryland, New Jersey, District of Columbia and Foreign, North Carolina, Virginia, West Virginia; District 4, Alabama, Florida, Georgia, South Carolina, Tennessee; District 5, Michigan, Ohio; District 6, Wisconsin; District 7, Illinois, Indiana, Kentucky; District 8, Arkansas, Colorado, Louisiana, Mississippi, New Mexico, Oklahoma, Texas; District 9, Iowa, Kansas, Minnesota, Missouri, Nebraska, North and South Dakota; District 10, Alaska, Arizona, California, Hawaii, Idaho, Montana, Nevada, Oregon, Utah, Washington, Wyoming. Elections were arranged to attain this number by 1974.

The Club office is organized as administration (secretary-treasurer and office manager), comptroller, data processing, *Guernsey Breeders' Journal*, herd register, pedigrees and classification, printing, production testing, promotion, and extension. The president, vice-president, secretary-treasurer, and comptroller fill the same positions with Golden Guernsey, Incorporated. Several Club directors also serve on the Golden Guernsey Committee.

To become a club member, a breeder's application must be endorsed by a member as reliable and desirable. A club representative investigates and reports on the applicant to the Board of Directors. Two negative votes debar membership. The membership fee is $50. Sustaining membership was attainable in 1966 for $10 down payment, plus $10 yearly for 7 more years for life membership. Joint or corporate memberships are for 20 years. There were 3,666 active members and 1,926 active Junior members in 1970.

Animals entered in Volume 1 of the *Herd Register* were colored fawn, orange, lemon, red, yellow, brown, cream, mulberry, brindle, very dark, and black. Fifty-eight bulls and 149 cows were solid colored. These colors mixed with white described 431 bulls and 1,198 females. No colors were listed for some animals. The tendency is toward more uniformity in color and other characteristics. A sketch of an animal appeared on the registration certificate since 1909; with a two-generation pedigree since 1933. Tattoo identification could replace sketches in 1966, but the majority still were sketched. Transfers of ownership are inscribed on the certificate of registration in the Club office. The Club established a pedigree service department after discontinuing the printed *Herd Register*.

The Shows and Scales of Points

When different breeds of cattle were separated at shows in the United States, there was a classification for Alderney or Channel Island cattle. Division into Guernseys and Jerseys in 1871 followed similar action by the Royal Agricultural Society in England in 1870.

Local and state agricultural societies sponsored shows and fairs, which were supported by membership fees, concessions, gate receipts, and by some state appropriations. The "Court of Last Re-

FIG. 11.2. Shuttlewick Levity 101850 combined type and production. She was Grand Champion at the National Dairy Show in 1927 and 1929, and her dam, Langwater Levity 70293, was Grand Champion in 1923.

sort" in type standards and show winnings was at regional shows, followed in 1906 by the National Dairy Show.

A great Guernsey cow at the national show was Shuttlewick Levity 101850. She and her dam each won Grand Champion awards at the National Dairy Show (Fig. 11.2).

The American Guernsey Cattle Club's first scale of points—in Volume 1 of the *Herd Register* in 1884—stressed yellow color of skin secretions, emphasized the escutcheon, and allowed 30 points for mammary development and 30 points for body conformation. The scale of points was revised in 1898, 1913, 1918, and 1935 based on careful studies and use by judges in the shows. A simplified scorecard in 1939 was divided into five major anatomical parts. The Purebred Dairy Cattle Association adopted a unified scorecard for cows in 1942 and revised it in 1957. Separate breed characteristics included size, color, and horns. A unified scorecard for bulls, published in 1944, also was revised then.

IDEAL GUERNSEY

After the "true type" Holstein-Friesian bull and cow had been developed, Chairman Charles L. Hill and members of the Guernsey type committee studied many photographs with artist Edward H. Miner. Miner's color painting of the "Ideal Guernsey Cow" was displayed at the annual meeting in 1926. The "Ideal Guernsey Bull" picture was completed in 1927, and the Guernsey ideal types were publicized in a brochure.

A new painting of the "Ideal Guernsey Cow" was completed in 1960 by the True Type Committee and artist Ralph Knowles, a Guernsey breeder in Maine. The model illustrated size, length of body, and strength of legs. The stance of the rear legs allowed attention to fore and rear udder attachments. The cow was dehorned in line with prevailing management practice.

GUERNSEY ALL-AMERICAN

Since 1954 nominations of leading show winners in each class have been made for the Guernsey All-American animals, and their photographs have been submitted to a Selection Committee. A panel of judges has selected the All-American Guernsey in each show-age class for the year. A Junior All-American competition was initiated in 1956 for 4-H Club and Future Farmers of America contestants.

TYPE CLASSIFICATION

The American Guernsey Cattle Club was the last dairy breed association to adopt type classification as an official program. Classification began unofficially in Huntington County, Indiana, in 1939. When the program was adopted in 1946, cows with Advanced Registry or Herd Improvement Test records were eligible to be classified, and heifers were eligible after dropping their first calf. Classification was permitted later, but the ratings were published after the animal completed a production record. Bulls were not classified. The ratings of daughters were grouped under their sires in the Performance Register. A full-time employee of the Club used the breakdown scorecard system and declared a total score for each

TABLE 11.1
The Proportion of Guernsey Cows in Different Type Ratings

| | | Cows classified | | | |
| | | 1947–60 | | 1970[b] | |
Ratings	Score	Total	Percent	Total	Percent
Excellent	90–100	3,334	02.4 [a]	57	00.5
Very Good	85–89	37,055	26.7 [a]	3,592	33.2
Desirable	80–84	60,110	43.3	5,244	48.5
Acceptable	75–79	30,753	22.2	1,771	16.4
Fair	70–74	6,886	05.0	156	01.4
Poor	Under 70	637	00.4	0	00.0

a. Cows rated Excellent or Very Good were required to be inspected at subsequent classifications.
b. Classified for first time.

TABLE 11.2
A Comparison of the Average of the First 15,000 Type Ratings of
Guernsey Cows with Those of 34,179 Classified in 1956–60[a]

Ratings	Overall ratings	General appearance	Dairy character	Body capacity	Mammary system
	Average of first 15,000 Guernseys classified				
	(percent)				
Excellent	1.5	2.3	7.4	12.0	2.0
Very Good	21.4	24.1	45.8	37.3	18.0
Desirable	47.4	43.7	38.8	37.6	43.5
Acceptable	25.1	25.9	6.9	11.1	29.8
Fair	4.1	3.7	1.0	1.3	5.7
Poor	0.5	0.3	0.1	0.7	1.0
	Average of 34,179 cows classified in 1956–60[a]				
	(percent)				
Excellent	2.1	1.5	10.0	8.8	1.7
Very Good	28.1	22.3	47.1	45.8	20.1
Desirable	44.7	47.4	31.0	33.1	44.4
Acceptable	20.5	24.7	9.9	10.6	28.4
Fair	3.9	4.0	1.9	1.6	5.1
Poor	0.1	0.1	0.1	0.1	0.3

a. A total of 65,546 cows were classified in this period. Of these, 34,179 cows had one or more official production records and were used in this analysis.

cow. The Club's leadership in appointing a full-time classifier afforded greater uniformity. The status of Guernsey type classification is shown in Table 11.1.

When a herd was submitted for classification, all cows under 8 years old that rated Excellent or Very Good previously were inspected for possible change. All other cows were examined for possible change of rating, and all ratings were published. No separate provision concerned cows that classified Fair or Poor.

When 15,000 Guernsey cows in 36 states had been classified, Earl N. Shultz tabulated their ratings into subdivisions of the breakdown scorecard on a percentage basis. The cows generally rated well in dairy character and body capacity. Improvements have been made in general appearance (shoulders, rump, feet, and legs), fore udders, and teat placements.

Recent analyses of 34,179 Guernsey cows having production records and rated for type in 1954–60 confirmed previous conclusions concerning average improvement in dairy character, body capacity, and overall rating. Further attention is needed with udders, shoulders, feet and legs, and general appearance. This computation was partly from analyses by Secretary R. D. Stewart. See Table 11.2.

COLOR SECRETION

When 1,044 Guernsey cows were classified in Waukesha County, Wisconsin, skin secretion inside the ears, in the tail, and elsewhere was rated light, medium, or dark yellow color (bright or dull), and small, medium, or large in amount. Observations were made in 32 herds producing Golden Guernsey milk during barn feeding in January, on pasture in June, with 922 cows again in November, and 26 selected cows in the next June. Overall ratings of Excellent, Medium, or Poor were given to each cow having the color secretion in the ears, tail, and escutcheon and on the nose, udder, and hoofs. Numerical ratings of 1, 2, and 3 were assigned for good, average, and poor color, respectively. Analysis of 10,423 color ratings and 11,459 butterfat tests of milk from these cows was summarized as follows:

Overall color rating	Number of cows	Butterfat tests Average (percent)	Milk color Average score
1—Good	26	4.78	4.86
2—Average	134	5.03	4.53
3—Poor	67	5.19	4.61

These observations indicated, subject to further verification, that cows with highly colored skin secretion likely would produce milk with good color. Based on these conclusions, the revised Unified Dairy Score Card for bulls and cows stated under Guernsey characteristics: "A bright golden yellow pigmentation on the nose, around the eyes, in the ears, in the escutcheon, around the udder and at the point of the tail is favored."

Subdivisions of the type classification standard in 1967, based on the official Dairy Cow Score Card, are:

General appearance
  Breed character
  Shoulders
  Feet and legs, fore and rear
  Rump
Dairy Character

Body capacity
Mammary system
  Fore udder
  Rear udder
  Teats

The classifier indicates in a "remarks" column any congenital defects, such as distinct wry-tail, wry-face, crossed eyes, parrot mouth, winged shoulders, broken down pasterns, excessively straight or crooked hocks, definite breaking of fore or rear udder attachment, or blindness apart from accidental.

Lawrence O. Colebank classified 228,508 Guernseys over 17 years as Official Guernsey Classifier. On retiring in 1971, he was succeeded by Merton B. Sowerby.

PRODUCTION RECORDS

The earliest published production records of Guernsey cows in the United States were private churned butter tests. The "Alderney" cow imported as a yearling in 1815 by a Mr. Wurts was credited with producing 8 pounds of churned butter in a week as a 3-year old. Private records by many owners helped to publicize Guernseys over native cows. Kathleen 38 was credited with 22¼ pounds of butter in a week. Eighteen churn test records exceeded 20 pounds, and 147 others were above 14 pounds of butter in 7 days. A yearly record of 12,856 pounds of milk was claimed for Imp. Lily Alexander 1059 in 1889.

Dr. S. M. Babcock developed a practical test for butterfat in whole milk at the University of Wisconsin in 1890. See Figure 13.5. It simplified the analysis for butterfat and gave a good method of measuring production—better than "butter" records. *Hoard's Dairyman* of February 28, 1891, reported a 7-day record of 283 3/4 pounds of milk, 13.59 pounds of butterfat by Imp. Regina 2691, a Guernsey owned by George Hill & Son, Rosendale, Wisconsin. The Babcock test was used publicly at the World's Columbian Exposition in 1893.

The Guernsey Breeder's Association, with members in Delaware, Maryland, New Jersey, and Pennsylvania, recommended in 1894: "that premiums be offered to breeders to induce them to breed for test. . . . The advancement of the breed demands more thorough and systematic testing." This district group held a competition based on yearly milk and butterfat production. Composite milk samples were analyzed by a chemist; the Association reserved the right to take samples if results appeared abnormal. King's Myra 5339 won first prize with 539.5 pounds of butterfat and Imp. Beauty des Domaines 3d 4933 won second place with 535.4 pounds of butterfat in 365 days.

The American Guernsey Cattle Club offered $300 in prizes in Home Butter Tests on a yearly basis for single cows and five-cow herds in a contest that began November 1, 1898. Five cows in the Sarnia Farm herd of George Hill & Son (Charles L. Hill) produced an average of 6,845 pounds of milk, 5.67 percent and 389 pounds of butterfat.

Advanced Registry

In 1896 the American Guernsey Cattle Club appointed William H. Caldwell and Charles L. Hill to draft rules for semiofficial testing. Testing began under the Club on November 1, 1898. Owners reported milk weights and submitted 1-day composite milk samples each month to the respective state experiment station for butterfat analysis. A station representative supervised three tests at the owner's farm in the year. The next year, monthly tests were supervised at the farm. Glenwood Girl 6th 9113 was the first cow admitted to the Advanced Registry with 12,184 pounds of milk, 572 pounds of butterfat in a year. The Club adopted an Advanced Registry in May 1901 based on such semiofficial yearly tests.

Entry requirements for 7 days or 365 days were set as follows:

Age	Milk	Butterfat
	(pounds)	
2-year-old	6,000	250.5
in 7 days		10.0
5-year-old	10,000	360.0
in 7 days		15.0

A bull entered the Advanced Registry when two daughters had qualified.

W. W. Marsh (Iowa Dairy Farm, Waterloo, Iowa) stated:

Advanced Registry work has placed the production value of cows on a scientific basis, and substituted accuracy for opinion and guesswork. Advanced Registry work will be the cornerstone of all breeding operations in all the dairy breeds from now on. But the structural beauty of the cow will always claim the attention of the constructive breeder. The reward of success both financially and otherwise, will go to that breeder who is able to combine great capacity with beauty of form. There is utility in beauty; there is beauty in utility.

The herd of W. W. and Charles Marsh was regarded as one of the four leading Guernsey herds in America, based on importations, breeding, production testing, and show ring achievements.

Recognizing reproduction to be essential in registered cattle, a double letter class of Advanced Registry Roll of Honor was established May 16, 1917, requiring that a cow carry a living calf for 265 days during the record. After 3 years, this requirement was recognized as too severe because owners tended to milk the cows almost to calving time for the largest possible record. A 305-day two-time milking or Farmer's Class was instituted in 1921. The cow then had to carry a living calf for 200 days during the record. The fat requirement was increased later, and the calving requirement was shortened. All production requirements were discontinued in 1946. Average production of Guernsey cows of different type classifications during 1956–64, is presented in Table 11.3.

Imp. King of the May 9001 exerted a wide influence on the Guernsey breed as a sire of production and desirable type (Fig. 11.3).

TABLE 11.3

AVERAGE PRODUCTION OF GUERNSEY COWS OF DIFFERENT TYPE CLASSIFICATION RATINGS, 1956–64[a]

| | Number of | | Numerical rating | 305-day 2X mature equivalent | | |
	Cows	Lactations		Milk (lbs.)	Test (%)	Fat (lbs.)
Excellent	1,200	4,982	91.4	11,749	5.0	582
Very Good	20,203	65,804	86.5	10,481	4.9	514
Desirable	29,387	80,919	82.3	9,603	4.9	470
Acceptable	11,634	28,266	77.3	8,800	4.9	431
Fair	1,926	4,100	72.3	8,197	4.9	399
Poor	41	70	65.6	7,563	4.9	369
Total	64,391	184,141				

a. A total of 23,678 DHIR records accepted during 1970 averaged 11,155 pounds of milk, 4.64 percent and 518 pounds fat in 305 days on a 2X mature equivalent basis.

HERD IMPROVEMENT REGISTRY

A Herd Improvement Registry division of the Advanced Registry was established in January 1930. All cows under 12 years old in milk in the herd were required to be tested each year. A cow's record could be omitted from the herd average if her registration certificate was cancelled before the eleventh month on test. Over 95 percent of the cows on test in 1958 were in the Herd Improvement Registry.

DAIRY HERD IMPROVEMENT REGISTRY

Some registered dairy cows were on Advanced Registry, HIR, and DHIA test simultaneously for their separate benefits, and some on DHIA test only. The Purebred Dairy Cattle Association studied methods and costs and concluded that more cows might be on breed programs if supervision were unified for acceptability. The

Executive Committee approved rules in 1958 for acceptance of combined HIR-DHIA records. The new rules required approval of DHIA supervisors by the state official in charge and required that cows be identified from registration certificates. Surprise check tests with preliminary dry milkings were required when a cow or herd produced at a high butterfat level. After a 1-year trial in Pennsylvania, the records were accepted during 1959–60 by all breed associations as Dairy Herd Improvement Registry (DHIR) records.

Over 100,000 records were used to determine the factors for conversion to a 305-day 2× mature equivalent basis. The average of 23,678 production records in 1970 was 11,155 pounds of milk, 4.64 percent and 518 pounds of butterfat in 305 days. The TeSa method of analyzing milk for butterfat was approved by the directors for use, along with the Babcock test that has been applied for so long.

COMPOSITION OF GUERNSEY MILK

Guernsey milk averaged above 4.8 percent fat for a number of years. Butter churned from Guernsey milk was noted for rich color for over two centuries. The Nebraska station observed that this

FIG. 11.3. Imp. King of the May 9001 was a leading transmitter. He was a son of Imp. Itchen Daisy 3d 15630, as well as grandsire of her sons Ne Plus Ultra 15265 and Itchen Daisy's May King of Langwater 17345.

color was due to the vitamin A potential value being stored as carotene rather than as colorless vitamin A. Carotene from the leafy forages also was stored in Guernsey body fat. Some 1,061 lactations by Guernsey cows in 14 herds analyzed in cooperative regional projects ranged from 4.21 to 5.10 percent of fat, 3.53 to 3.78 percent protein, 9.07 to 9.36 percent solids-not-fat, these averaging 5.01, 3.63, and 9.20 percent respectively. The New Jersey station analyzed milk from cows in one Brown Swiss, two Guernsey, two Holstein, and two Jersey herds. Although high fat and high protein contents tended to be associated, A. W. Hobler reported that "many cows may test three or four points more in fat that test less in protein than the cow with less fat," although this was not general.

Environmental factors, breed, and individual hereditary differences modify or limit composition of milk. Milk from 188 complete lactations of Guernsey cows in Arkansas averaged 4.5 percent fat, 3.2 of protein, and 9.2 percent of solids-not-fat, with ranges of 3.9 to 5.4 percent fat, 2.4 to 3.7 of protein, and 8.3 to 9.8 percent of solids-not-fat. Such wide ranges permit selective breeding over a long period.

GOLD STAR GUERNSEY SIRE AND DAM

The Gold Star Guernsey Sire and Gold Star Guernsey Dam awards recognized good transmitting ability, as adopted in August 1958. To become a Gold Star Guernsey Sire in 1963, ten or more of a bull's registered daughters must have production records at least 15 percent above the breed average, which then was 10,810 pounds of milk or 525 pounds of fat. Sixty percent of his daughters 4 years or older, or 30 daughters bred in four or more herds met requirements regardless of age. Classification scores for ten or more daughters (at least 60 percent of those 3 years or older and each eligible daughter classified) must average at least 82.5 points.

An artificial insemination Gold Star Sire program began in 1964. At least 50 A.I. daughters with herdmates in the USDA Sire Summary must have daughters in at least ten herds with a *predicted difference* of at least plus 400 pounds of milk. At least 15 classified daughters must have an average score of 81.5 percent. No type requirement applied for the A.I. Silver Star Production Sire award, but at least 51 percent of daughters must be registered. Summarized daughters must average 9,500 pounds of milk. No type requirement was needed of them. Summarized daughters must have a predicted difference of plus 350 pounds of milk, with at least 25 percent repeatability.

A registered Guernsey cow qualified as a Gold Star Guernsey Dam based on at least three progeny. All official records of the three highest progeny must average 20 percent above the breed average, which was 12,760 pounds of milk and 510 pounds of fat in 1970. Each daughter must have scored 82.5 or above. A son may be included if he has received a previous recognition award. Some 235 dams qualified in 1968.

GOLD STAR GUERNSEY BREEDER AWARD

Starting in 1959, Gold Star Guernsey Breeder Award certificates were granted on an annual basis, based on nine requirements. Applications were made before April. All purebred Guernseys over 8 months old owned by the applicant must be registered or their births reported. At least ten registered cows must have produced to meet requirements and the herd be on test currently. All cows in the herd must have completed an official record in the year, and the production must exceed breed average by 15 percent. The owner shall have bred 70 percent of the cows meeting production requirements. Eighty percent of cows with production records must have been classified, with the last rating averaging at least 82.5 percent. The herd must be free from tuberculosis with a clean test within 12 months. The herd must have been state certified as free from brucellosis, or have passed a clean test within a year. Milk-ring tests were not accepted. The applicant must be a member of the American Guernsey Cattle Club and an active member of a State Guernsey Breeders' Association. Sixty-seven Gold Star Breeder Awards were earned in 1968 in 20 states.

SIRE SUMMARIES

Sire Summaries of bulls with ten or more tested daughters were published in the *Guernsey Breeders' Journal* during 1954 and onward. The latest type score and all production records of daughters were computed on a 305-day $2\times$ mature equivalent basis. The Board of Directors adopted the USDA Sire Summary Program for 1968 with screened tape production records of all registered Guernseys on DHIR test.

Revision of sire summaries with improved analyses of DHIA-DHIR records resulted in new recognition requirements approved by the Board of Directors on April 30, 1968. A Guernsey bull became a Gold Star Sire when his progeny qualified simultaneously in production and type classification. At least 20 daughters made an average production of 9,500 pounds of milk on a 305-day $2\times$ mature equivalent basis. Their *predicted difference* over herdmates must equal at least plus 250 pounds of milk, with a repeatability of at least 60 percent. At least 25 percent of daughters in the production summary (minimum of ten) must be registered, and have type ratings in the last official classification that averaged 82.5 points.

To qualify as a Silver Star Guernsey Sire, at least ten daughters in the last production summary, 51 percent of which were registered, must average at least 9,500 pounds of milk on a mature equivalent basis, and have a *predicted difference* over herdmates of plus 350 pounds of milk in 305 days, with a 25 percent repeatability. His registered daughters, at least 60 percent of those in the production summary, must have type ratings averaging at least 83.5 points.

The same requirements applied for the Gold Star Guernsey Herd, except that the herd lactation average of all cows must exceed breed average by 25 percent irrespective of being home-bred. The production requirement for 1967 was either 12,850 pounds of milk or 620 pounds of fat on a 305-day $2\times$ mature equivalent basis.

Breeding Guernseys

F. Lothrop Ames improved the body conformation of Guernseys appreciably over 21 years with prepotent animals at Langwater Farm in Massachusetts. Production records and type classification are tools to continue spread of improvements through selection of breeding animals. Guernsey milk is valued for solids-not-fat and yellow color of the products—a trademark of the breed. Significant variations occur among individuals in butterfat and protein contents of milk. These are inherited independently and can be modified by selection of breeding animals.

Goals in breeding Guernseys are to retain good production, body capacity, dairy character, udder quality, and yellow color of the dairy products. Greater size, which is needed in many herds, may be attained in part by improved feeding during growth. Old faults needing continued attention are winged shoulders, sway backs, udder attachments, crooked hind legs, and weak pasterns. A few Guernseys possess the dominant gene for polled. Smoky nose and wry-tail are nondominant characters of little importance. Wry-face sometimes occurs, as does parrot jaw, which may be a recessive character. Progressive posterior paralysis or "crampy" is a recessive character that develops in adult animals within certain lines. Lack of resistance against lump jaw (actinomycosis and actinobacillosis) is an inherited nondominant character in some family lines.

Recessive characters controlled by a single gene are observed in one out of four cases when two "normal" parents that possess the gene are mated. Such characters can appear in one out of two progeny when a "normal" carrier is mated with an individual possessing the trait. When both parents exhibit the trait, it will be evident in all their progeny. This explains why recessive characters crop out more often in line-breeding or inbreeding, and less often when out-crossing is practiced. When a recessive character becomes evident in a herd, both the sire and dam were carriers. On the average, 50 percent of the dam's half-sisters by a common ancestor also may possess the same gene.

Secretary Max Dawdy tabulated 11,065 Guernsey sires with USDA Sire Summaries in 1970. Approximately one-third of them were Plus Proven for milk, similar to earlier analyses of DHIA sires in Table 22.4. Plus Proven sires tended to produce sons with similar ability. He gave some ways for early proving of young sires before extended use.

One Guernsey male was registered for each 1.7 females in Volumes 1 to 45 of the *Herd Register* up to 1933. Cooperative artificial breeding began in 1938. Greater numbers of cows per herd on fewer dairy farms resulted in registration of 1 male to 20.5 females in 1970. Some 56.3 percent of Guernseys registered in 1970 resulted from artificial inseminations.

Oklahoma investigators studied production and type classification records of 1,981 daughter-dam pairs of Guernsey cows representing 511 sires in 239 herds. There were 8,533 production records and 4,172 type classifications between 1947 and 1955. The daughters averaged 8,803 pounds of milk, 425 pounds of butterfat, and their dams had 8,688 pounds of milk, 426 pounds of butterfat. Type scores averaged 82.08 and 81.94 percent respectively. This study

TABLE 11.4
COMBINED HERITABILITY ESTIMATES

Traits	Number of studies	Heritability estimates	
		Average	Range
General appearance	4	0.22	0.14–0.33
Dairy character	4	0.18	0.07–0.30
Body capacity	4	0.23	0.13–0.28
Mammary system	4	0.18	0.09–0.24
Legs and feet	4	0.18	0.14–0.23
Milk yield	57	0.28	0.17–0.60
Milk fat, percent	31	0.55	0.32–0.90
Milk solids, percent	9	0.55	0.35–0.71
Protein, percent	9	0.55	0.35–0.76

was included among the heritability estimates of conformation and certain milk traits summarized by Howard C. Dickey of Maine. The combined heritability estimates published by Dickey in 1970 are shown in Table 11.4. Careful selection and culling of sires could achieve gradual improvements in each trait in the succeeding generations.

Robert D. Stewart reported that from 34,179 records by Guernsey cows "the cows that classified the highest in each of the major parts of the score card were also the highest producers per lactation." Only the latest classifications were used. All Excellent and Very Good cows under 8 years old must be rerated at each successive classification of the herd.

The strongest characteristics of Guernsey cows have been dairy character and body capacity. Mammary system and overall general appearance need attention when breeding for improvement of type.

Some 182 Guernsey bulls were among the dairy bulls of six breeds active in artificial breeding organizations in the United States during 1969. They included several Gold Star Guernsey bulls.

PROVISIONAL REGISTRATION

A plan for Provisional Registration of unregistered Guernseys was adopted in May 1970. Acceptable proof is required of three direct crosses to a registered Guernsey bull; or the females must have been in a purebred herd unregistered for at least 10 years (under signature of the owner or a Club member). The female must have Guernsey color and characteristics. DHIA records must average 13,000 pounds of milk on a 305-day $2\times$ mature equivalent basis, and the animal must classify Very Good or Excellent, or average 15,000 pounds of milk and classify at least 82.5 points. Females are eligible for a colored registration certificate, and male progeny of the fifth generation of the original nominee are eligible for registration.

During 1965, 12.8 percent of all registration applications sent to the Club were returned for corrections. Breeders need to complete and verify all applications before submitting them. Most errors were for omissions or incorrect registration numbers, or lacking the correct signatures.

JOINT OCCUPANCY

American Guernsey Cattle Club Executive Secretary Karl B. Musser suggested joint occupancy at some central headquarters by the dairy breed associations in 1947. He stated: "We would build one structure to house our five breed offices and do a great deal of our routine clerical work in cooperation. At the same time, we could have separate executive and promotional offices and remain strictly competitive."

The Jersey Board of Directors considered the joint occupancy proposal in 1960. A Guernsey-Jersey Joint Occupancy Committee met in Peterborough, New Hampshire, in July 1964. The Holstein-Friesian office was contacted; and Ayrshire, Brown Swiss, and Milking Shorthorn associations were invited to a joint discussion at Waterloo, Iowa, in October 1964. The Guernsey-Jersey committee considered space needs and practical office facilities for rental, lease, purchase, or new construction near transportation facilities. Considerable economy would be attained by the joint use of mechanical processing (computers and others) of breed records to benefit all breeders of registered dairy cattle. A central location would permit ready communication and a united front for the purebred dairy cattle industry, which is the source of seedstock for all dairy cattle. Members of the American Guernsey Cattle Club at the 1965 annual convention voted 701 to 35 in favor of joint occupancy of office facilities with one or more other breed associations.

EXTENSION SERVICE

Guernsey promotion began with activities of importers, owners, and the American Guernsey Cattle Club secretary. An Extension Service was organized in 1920. Eight fieldmen and breed representatives served Guernsey breeders in 1970.

GOLDEN GUERNSEY PRODUCTS

Guernsey milk and cream were marketed under a copyrighted Golden Guernsey trademark around 1920. Golden Guernsey, Incorporated, promoted sales as the activity increased. Licenses to use the trademark were granted to producers and distributors to advertise and sell Guernsey products meeting health, sanitary, composition, and other requirements. A small royalty promoted sales of the products.

A Gurn-Z-Gold trademark was approved early in 1953 for creamline milk with less than 4.4 percent butterfat, and homogenized milk below 4.0 percent fat, processed from Golden Guernsey milk. The Guernsey Royal trademark was adopted in 1960 for use when Golden Guernsey milk was processed to meet minimum state legal standards up to 3.7 percent of butterfat. Gurnskim was added in 1961. Nearly 265 million quarts of Guernsey products were merchandised under the trademarks during 1967. *The Story of Golden Guernsey Milk Products* booklet advertised Guernsey products as emblems of quality.

BREED PUBLICITY

The Guernsey Breeders' Association around Philadelphia established the *Guernsey Breeders' Journal* in 1885 with Willis P. Hazard as editor. The responsibility was great, and they requested the American Guernsey Cattle Club to assume the publication. The *Journal* was printed quarterly with sections of the *Herd Register*. It became a monthly magazine in 1910, and has been published twice monthly since July 1916.

Production records were published first in Advanced Registry volumes. These were replaced by the Performance Register, which also contained show winnings and type classifications. Volumes I to XX contained performance records from 1949 through 1966. A Production Leader booklet listed Gold Star Sires and Dams, summarized sires, and herd lactation averages for those with ten or more cows.

The Guernsey Club had the first breed booth at the National Dairy Show in 1907. Several books and breed brochures have been published. William H. Caldwell wrote three books: *Langwater Guernseys, The Glenwood Girls,* and *The Guernsey*—all published by the Guernsey Club. James E. Russell of Glenburnie Farm prepared *Heredity in Dairy Cattle* in 1944. This booklet described the art of breeding dairy cattle, based on successes and failures of four leading dairy herds. Written for Future Farmers and 4-H Club members, it was of equal interest and value to experienced breeders.

The American Guernsey Cattle Club owns the office building in Peterborough, New Hampshire 03458. Max Dawdy is the Secretary-Treasurer.

REFERENCES

Ames, F. Lothrop. 1918. Stock breeding as a science and an art. *Breeders Gaz.* 74:1108–9, 1221–22.

Becker, R. B., C. F. Simpson, and C. J. Wilcox. 1963. Hairless Guernsey cattle. *J. Hered.* 54:2–7.

Becker, R. B., C. J. Wilcox, and W. R. Pritchard. 1961. Progressive posterior paralysis (crampy) in mature cattle. *Florida Agr. Expt. Sta. Bull. 639.*

Becker, R. B., C. F. Simpson, L. O. Gilmore, and N. S. Fechheimer. 1964. Genetic aspects of actinomycosis and actinobacillosis in cattle. *Florida Agr. Expt. Sta. Tech. Bull.* 670 and *Ohio Agr. Expt. Sta. Res. Bull.* 938.

Berousek, E. R., J. A. Whatley, R. D. Morrison, S. D. Musgrave, and W. R. Harvey. 1961. Heritability and repeatability estimates of production and type of Guernsey cattle. *Guernsey Breeders' J.* 107:41.

Betts, Silas. 1900. The Guernsey cow. *Herd Register* 1:xvii-xxii.

Biddle, Craig. [Report of churn test of an Alderney cow.] *Memoirs of Philadelphia Soc. for Promotion of Agr.* 4:155.

Caldwell, William H. 1925. *Langwater Guernseys.* Peterborough, N.H.

————. 1941. *The Guernsey.* Peterborough, N.H.

————. 1941. *The story of the Glenwood Girls.* Peterborough, N.H.

Clark, John S. 1942. History of the score card. *Guernsey Breeders' J.* 61: 1069–71.

Codman, James M. 1900. The Guernsey breed of cattle. *Herd Register* 1:iii–v.

Colebank, Lawrence O. 1967. Can the colored breeds come back? *Hoard's Dairyman* 112:1205, 1224.

Copeland, Lynn. 1965. A study of Guernsey bulls. *Guernsey Breeders' J.* 116: 541.

Cummings, C. M. 1952. The Advanced Registry for 75 years. *Guernsey Breeders' J.* 84:611–17.

Dawdy, Max L. 1970. A breed priority—More sampling and proving of young Guernsey sires. *Guernsey Breeders' J.* 126:651–55.

Dickey, Howard C. 1970. A look at the future in dairy cattle breeding. *Guernsey Breeders' J.* 126:232–33, 292.

Eckles, C. H., and E. L. Anthony. 1956. *Dairy cattle and milk production.* 5th ed. Macmillan, New York.

Fitzpatrick, W. W. 1934. Golden Guernsey Milk as sold under the label of the American Guernsey Cattle Club. *Guernsey Breeders' J.* 45:169–70, 203–4.

Hazard, Willis P. 1872. *The Jersey, Alderney and Guernsey cow.* 10th ed. Porter & Coates, Philadelphia.

Hepburn, William K. 1943. Great brood cows are the intrinsic value behind great sires. *Guernsey Breeders' J.* 63:102–3, 109.

Hill, Charles L. 1917. *The Guernsey breed.* Kimball, Waterloo, Iowa.

Legates, J. E., and George Hyatt, Jr. 1956. Relationship between type and production. *Guernsey Breeders' J.* 96:910–11, 931.

Lounsbury, L. R. 1960. Memories of breed progress. *Guernsey Breeders' J.* 106:112–17, 152.

Munn, W. A. 1937. Restricted registration. *Guernsey Breeders' J.* 51:1133.

Prentice, E. Parmalee. 1942. *American dairy cattle.* Harper, New York.

Rakes, J. M., O. T. Stallcup, and D. F. Potts. 1962. Composition of Guernsey milk. *Guernsey Breeders' J.* 110:323–24.

Rice, V. A. 1947. A neglected field—cow families. *Guernsey Breeders' J.* 71: 289–93.

Robertson, James B. 1936. The Guernsey cow called May Rose II. *Guernsey Breeders' J.* 49:326–29, 343.

Russell, James H. 1944. *Heredity in dairy cattle.* Peterborough, N.H.

Shultz, Earl N. 1947. The first official classification. *Guernsey Breeders' J.* 71: 930–31.

————. 1948. A look at the breed. *Guernsey Breeders' J.* 73:1185–86, 1219.

Stewart, Robert D. 1958. "Gold Star" program to recognize sires, dams and breeders. *Guernsey Breeders' J.* 102:342–43.

————. 1958. AR-HIR testing. HIR-DHIA trial in Pennsylvania. *Guernsey Breeders' J.* 102:1408–9.

————. 1958. Report of study of skin secretion of Guernsey cattle in relation to milk color. *Guernsey Breeders' J.* 102:1410, 1414.

————. 1961. Guernsey classification program correlates type and production. *Guernsey Breeders' J.* 108:342–43.

Swain, James. 1868. Channel Island cattle. *Amer. Agriculturist* 27:135–36.

Woodward, Edwin G. 1942. Guernseys in America. In E. P. Prentice, *American Dairy Cattle*. Harper, New York. Pp. 375–83.

American Guernsey Cattle Club publications
  1878–1933. *Herd Register*. Vols. 1–45.
  1877–1927. The story of half a century.
  1939. Breeding better Guernseys.
  1941. *Performance Register*. Vols. 1–.
Guernsey Breeders' J.
  1935. The new Herd Test for Guernseys. 37:199–200, 207.
  1941. Great production sires. 60:554–57.
  1944. Hamilton County classifies. 66:9–12.
  1944. Herd classification. 65:12.
  1960. The Guernsey Breeders' Journal for half a century. 106:104–110.
  1961. Revision of the Constitution and By-Laws of the American Guernsey Cattle Club. 107:20–22, 42.
  1967. The American Guernsey Cattle Club type classification rules and regulations. 119:744.
  1970. The proposed Provisional Registration Program in dramatic form. 125:128–29, 508–9.

CHAPTER 12

FRIESIANS IN THE NETHERLANDS

THE BREED of cattle called Holstein-Friesians in the United States is known as Friesians in the native country and much of the British empire. The Netherlands comprise 12 provinces, including a province reclaimed from the Zuider Zee (South Sea). A 20-mile earth dike extending from North Holland, Wieringen Island, and the coast of Friesland was begun in 1927 and closed in May 1932. This made a freshwater lake of the brackish Zuider Zee, since renamed the Ijsselmeer (Ysselmeer). The northern provinces of Groningen, Friesland, and North Holland extend westward from the German border. Early development of the Friesian breed centered largely in Friesland and North Holland, and spread to other provinces.

The Netherlands lies between 52° and 53° 30″ North latitude, farther north than Newfoundland. The land extends from 17 feet below sea level to 300 feet elevation, with an extreme altitude of

Headpiece: Vignette of Friesian cow.

960 feet in the southeast. About 1,200,000 acres have been reclaimed from the sea, marshes, and lakes by building dikes and pumping off the water. The new province will include 553,500 acres reclaimed from the Ysselmeer. Winds off the Gulf Stream in the Atlantic Ocean moderate the winter rigors associated with a northern latitude. The moderately cool summer and moist climate favor grasses, root crops, and cereal grains—feeds adapted for dairy cows. Crops include rye, oats, fodder, sugar beets, wheat, peas, and lesser cereals in order of acreage. Farms and fields in the lowlands are separated by canals. Some electric fences are used. Most farms comprise 35 to 75 acres; the largest in 1926 was 270 acres. The soils vary between clay and sand in large part, with peat in low areas. Some soils are low in copper or magnesium.

Cows obtain about 75 percent of their nutrients from pastures and early-cut hays which grow on 55 percent of the farmland. Much of the grassland is on moist soils ill-adapted for cultivation, or on heavily fertilized light sandy soil. Rotational and strip grazing give more efficient use of the forages which are of high nutritive value. Hays are cured naturally on tripods or by artificial drying. Grasses are ensiled with some sugar-containing preservative or after wilting for 1 or 2 days in favorable weather. Concentrates are limited to 850 to 1,100 pounds mainly during the winter.

EARLY CATTLE

Early *Kjokkenmoddings* or shell mounds of the Old Stone Age along the coast contained bones of migratory wild fowls, stag, roe-deer, and wild boar. About 3 percent of other animals included *B. primigenius* Bojanus. Traces of a smaller ox were found. *B. longifrons* Owen was brought in under domestication late in the Old Stone Age. Professor Worsaae, Danish archaeologist, considered that Danubian immigrants brought domesticated cattle and horses north during the Neolithic era of the New Stone Age. Manmade mounds or terpens were built up in some lowlands of Friesland and Groningen for safety of people and animals during floods.

Little is known of cattle near the North Sea other than from fossil remains found in connection with burials, migrations, and conquests. Skulls of present-day Friesian cattle studied by D. L. Bakker in

1909 compared closely with *B. taurus brachyceros* Rutimeyer, which Owen called *longifrons*. Two outer horn shells of the aurochs in the Friesch Museum (Leeuwarden) were credited to the late Stone Age or early Bronze Age, from a mound near Oudaga that contained stone axes, early pottery, and a bronze "dagger." One horn shell exceeded 3.75 inches at the inside diameter of the base and 24 inches along the outer curvature. Bakker reported a "bastard" race, apparently crossbred between *B. primigenius* and *B. longifrons*. They most closely resembled the present Friesian cattle.

G. J. Hengerveld cited a legend that the Friesians brought cattle and cultures with them from the lower Rhineland about 300 B.C. and that the mother of their cattle—"a cow as white as snow"—came earlier from India. Batavians were said to have arrived two centuries later from Hesse near the Rhone headwaters, bringing black cattle. Circumstantial and genetic evidence points to other sources for color inheritance.

The Romans found the tribes of northern Europe to possess many cattle, as mentioned in *De Belli Gallico* (Book 6, Chapter 26) written about 65 B.C. A. Tacitus stated, "They [Friesians and Batavians] owned cattle, not excelling in beauty, but in numbers."

White cows were held in religious reverence, suggesting scarcity of this color. The Friesians paid tribute to the Romans in meat, cattle, hides, and horns. The Batavians furnished a contingent of soldiers instead. Later the Friesians and Saxons defied Roman rule. In the fourth century A.D., these northern tribes extended their influence along the coast to the Rhine River.

Two skulls of cattle of similar proportions were in the Friesch Museum, one dating to Roman times. This skull is long and narrow, with horns extending outward and slightly forward. Both skulls resemble present-day Friesian cows.

DISASTERS AFFECT THE REGION

The North Sea broke through the coastal dunes between 1219 and 1287, flooded the lowlands, and formed the Zuider Zee. This lake bottom is now being reclaimed.

Dutch cattle long have been recognized as dairy animals, with cheesemaking as an outlet for their product. A cattle market was operated in Utrecht in 660 A.D., another in Haarlem in 1266, one at Schiedam in 1270, and others later.

Diseases, floods, and wars decimated the cattle population in parts of the Netherlands at long intervals. Many cattle died in 376–395, 810, and even as late as 1782. Nearly 396,000 cattle died in the last great epidemic. Stronger dikes were built in the thirteenth century. The lowlands of Groningen, Friesland, North and South Holland, and Zeeland suffered most from inundations. Many people and cattle drowned in floods in 512, 516, 533, 570, 584, and 1170. Some 80,000 persons and animals drowned in Friesland alone in 1187; about 100,000 drowned in 1219. Other floods spread destruction in the thirteenth, fourteenth, and fifteenth centuries. In 1570, 9,000 people and 70,000 cattle were drowned in Groningen. Losses were suffered in 1686, 1717, and later, until stronger and higher dikes reduced floods after 1825. War breached the dikes of Walcheren Island in 1944 and the Germans breached dikes again before surrender. Gale winds pushed a high tide over the dikes in February 1953, drowning many people and cattle.

Contagious diseases swept Friesland repeatedly. In 1713 and 1714 with entry of rinderpest, over 300,000 cattle died in 13 months; two-thirds of the cattle died in 1744–56. Friesland lost 109,597 cattle in 11 months (1744–45) and 97,756 head in 6 months in 1768–69; other northern provinces had proportionate losses. Other outbreaks were less severe. Danish, Holstein, and small German cows from Oldenburg, Munster, and Hanover were purchased to replenish the herds following a rinderpest outbreak in 1769. By 1800 cattle had increased to more than 900,000 head in the Netherlands. About 18 million pounds of cheese were produced annually from 1801 to 1804. Armies of Napoleon I took large drafts of cattle in the beginning of the nineteenth century, yet cattle breeding prospered. More German cattle were brought into the northern parts of the Netherlands following drownings from great storms between 1820 and 1825 and losses from a severe lung disease between 1833 and 1849. Bakker wrote of these disasters:

> If one takes into consideration the great losses caused by drowning and contagious diseases combined with the recurrent wars, it cannot be wondered that cattle breeding was

found in a state of decadence at the end of the eighteenth century, and of the old superior class of cattle, few or none were to be noted long. It was necessary to buy smaller cattle from Denmark [which then included Schleswig-Holstein]. These were graded up however by the end of the eighteenth century to possess the qualities previously attributed to the native breed.

Professor G. J. Hengeveld mentioned that the soil contained more clay near the seacoast, and "consequently we find large cattle in the provinces of Groningen, Friesland, North Holland, and in some parts of North Brabant and Limburg."

Colors of Dutch Cattle

Written history yielded limited evidence concerning cattle in the historic period. Bakker examined 80 paintings by Dutch landscape artists in art galleries, which show colors of 220 cattle. A retabulation of Bakker's study is in Table 12.1 Bakker believed that the black-and-white colors came into the Netherlands with cattle brought from Jutland and elsewhere following outbreaks of rinderpest and other conditions.

John Speir observed Dutch landscape paintings in British and Dutch art galleries. Few black-and-white cattle were pictured before 1750. This color pattern presumably was possessed by animals brought as replacements from Denmark and Germany after 1750. Cattle on the Jutland peninsula were black-and-white, top-crossed later with Friesian bulls, and renamed the Black and White Danish breed.

Spahr van der Hook assembled records of cattle in public sales during the eighteenth century and earlier. He concluded from the color descriptions that some cattle possessed black color markings in the sixteenth century. Then the red coat color became more prevalent until 1750, when the proportion with black color pattern increased. His tabulations supported in part the observations drawn from artists' landscape paintings.

In 1905 K. Hoffman mentioned that red and red-spotted (with white) cattle had been recognized long in the Friesian breed. However, few farmers bred specifically for red-spotted cows. These animals were accorded a section in the early Friesian herdbook; 40

red-spotted bulls and 173 red-spotted cows were registered in 1903, compared with 2,889 bulls and 10,486 cows of the black-and-white pattern. Blue-spotted, dun-spotted, and mouse-gray cattle were in fewer numbers. Because red is recessive to black, this color sometimes resulted from mating of black-spotted parents carrying the hereditary gene for recessive red.

CATTLE SHOWS

The first cattle show in the region was held at Brake (actually in north Germany) in 1836, the practice being introduced from England, according to U.S. Consul John W. Wilson (1883). Solomon

TABLE 12.1
COLORS OF CATTLE IN THE NETHERLANDS, AS PICTURED BY FAMOUS ARTISTS

Period	Number of cattle	Numbers of cattle of various colors					
		Red	Fawn	Yellow	Dark	Others	Black-and-white
1550–1600	22	8	6	1	3	4	0
1600–1750	163	96	50	6	4	7	0
1750–	35						35

a. One white cow appeared with 6 fawn or yellow cows painted by R. Savery (1576-1637).
b. Bakker counted 35 black-and-white cattle in 20 paintings since 1750. He did not tabulate the cattle of other colors that appeared also.

Hoxie imported "the first prize cow" of Holland to the United States in 1880. By 1883 nearly every district held annual cattle shows, resulting in greater uniformity among the animals.

The Provincial Commission for Cattle Improvement (PCV), subsidized by government appropriations, conducted local bull shows in the spring. The *Friesch Rundvee Stamboek* (herdbook society) held a central spring show for registered bulls in Leeuwarden, and held another in early October for herdbook bulls. Judging was on conformation, with the bulls being rated as first-, second-, or third-class groups. A first-prize animal was designated from among those in the first class. A pamphlet from the export association commented that "the . . . excellence reached could be judged from the exhibits at the great National Exposition of the Royal Netherlands Agricultural Society held at The Hague in 1923 and at Utrecht in 1935 and the National Show in 1928."

Special shows in 1949 commemorated the seventy-fifth anniversary of the *Netherlands Rundvee Stamboek* and seventieth anniversary of the *Friesch Rundvee Stamboek,* cattle herdbooks of the respective provinces. Such shows were held again in 1954 and 1959.

Herdbook Associations in the Netherlands

The *Nederlandsch Rundvee Stamboek* was given legal status by King's decree on September 27, 1873, following establishment of the Holstein Herd Book in the United States in 1871. The NRS was organized subsequently in 1874 through a committee of the Netherlands Agricultural Society. Thomas E. Whiting, who imported Friesian cattle to Concord, Massachusetts, was a patron member. This herdbook was largely a list of inspected cattle approved by provincial committees, and their owners. Body measurements were added to a numerical score in 1897.

Such animals sold for export at higher prices than unlisted animals. Entries were grouped by provinces before 1902, and according to four local breeds thereafter. The NRS herdbook added pedigree details in 1907. Supervision by provincial committees gave way to control from a central office at The Hague in 1912. The NRS registered Friesians, the red-and-white Meuse-Rhine-Yssel, and the white-headed black Groningen breeds in separate sections of the herdbook.

An Advanced Registry was established in 1922. Admitted were cows that had good conformation and at least two lactations above a minimum requirement.

Breeders in the province of Friesland disagreed with methods of the early Netherlands herdbook. They met in "De drie Romers" Inn in Roordahuizum on July 12, 1879, and organized the *Friesch Rundvee Stamboek* for approved Friesian cattle in the province of Friesland. Black-and-white and red-and-white Friesians were entered by both associations. In the first year, 26 bulls and 203 cows were registered in the FRS herdbook. Entries had increased to 375 bulls and 2,523 cows by May 1884. After that date, the progeny of registered bulls and cows entered upon qualifying by a combination of inspection, measurements, and a conformation score.

Systematic shows for bulls were instituted in Friesland in 1896;

in 1901 a more detailed inspection of animals was required for admission to the herdbook. Animals from pedigree stock were entered as calves in a young stock register. Heifers then were inspected when they were one year old, and again after their first calving. If approved at the second inspection, they were entered in the register with their pedigrees, conformation scores, and production records. Since 1903, animals are required to score at least 70 points for conformation. They are measured for length of trunk, height at withers and rump, depth and circumference of chest, widths of chest, hips and thurls, and length of rump. The dam of a male calf now must score at least 16 out of 20 points for udder, teats, and milk veins before he can enter the young stock pedigree register. Bulls are inspected at 12 months and again at 18 months before full registration and service with herdbook animals. Herdbook numbers of fully registered animals were branded on the horns. All lactation records are published.

Scorecards used by the Friesian herdbook inspectors with bulls and cows were as follows:

	Bulls	Cows
Head (shape, eyes, muzzle, horns)	9	8
Horns	6	
Neck, chest, withers, shoulders	12	10
Ribs, back, flanks	10	8
Loin	8	8
Rump	10	12
Thighs	6	6
Tail	3	4
Legs, posture, gait	10	
Legs		6
Milk veins, skin, hair	6	
Udder, teats, milk veins		20
General appearance	20	
General aspect (including skin, hair), posture, gait		18
Total	100	100

The Netherlands Cattle Herdbook (NRS) used a scorecard from 1907 until it was revised by a committee on April 12, 1965. The new scorecard listed:

Head, eyes muzzle	Withers and shoulders
Horns	Ribs and flank
Neck and chest	Back and loin

Rump (pelvis)
Thighs and shanks
Tail and switch
Muscle covering
Posture, gait, balance
Hoofs
Color pattern
Hide, hair
Udder quality
Udder form and placement
Teat placement

Teat shape
Milk veins
Length of trunk
Height at withers
Height at tailhead
Depth of chest
Width of chest
Width of hips
Pelvis width
Length of rump
Feeding capacity
Dairy tendency

The scorecard for bulls differed in minor points.

A = excellent   90–100 points   B = quite acceptable   75–79 points
AB = very good   85–89 points   B — = acceptable   70–74 points
B + = good   80–84 points   BC = unacceptable   not recorded

Conformation	Development	Appearance	Milk indications	Skeletal structure	Muscle covering	General	Total score
AB	AB	A	AB	AB	A	A	92
AB	A	B +	AB	AB	AB	AB	87
B +	AB	AB	AB	B	B —	AB	85
B +	B +	B +	AB	B +	B +	B +	82
B +	B +	B —	AB	B +	B —	B	79
B —	B —	B —	AB	B	B —	B —	72

Animals scoring below 70 points were unacceptable in the herdbook.

An auxiliary book (*Hulpboek*) entered bulls and cows without respect to pedigree when such animals scored at least 75 points for conformation. Bulls were not eligible for entry in the auxiliary book after September 12, 1922. The number of animals entered in the Friesian herdbook is given in Table 12.2.

Herdbooks were maintained for Friesian cattle in western Groningen (began in 1880) and in Overjssel for only a few years.

The *North Holland Herd Book* was organized in November 1883 with a branch herdbook office in Dover, New Jersey. Three volumes were published in America between 1888 and 1892. Solomon Hoxie received a letter concerning formation of this herdbook to the effect that:

. . . The North Holland Herd Book was started in November, 1883, as a result of the growing American trade, and mainly influenced by B. B. Lord & Sons (American importers). They use a scale of points, . . . take no other colors than black and white, and then leave the final conclusion about the admittance or refusal of an animal with their inspectors, thereby facilitating matters. The main difference is found in the fact that the Netherlands Herd Book does not admit any female unless they have actually brought a calf and proved to be sound milkers, while the North Holland Herd Book only requires female animals to be two years old at the time of inspection, no matter if they are in calf or not. The North Holland Herd Book Association is originated and run by farmers, who are practical breeders, and I am very sure they aim to do what is right, desire to elevate the breed as well as the Netherlands Herd Book Association, but they do it solely in the hope to get a large share of the American trade. And this purpose, I also fully admit, is a legitimate and laudable one.

The North Holland Herd Book Society discontinued in 1907, the members joining the larger Netherlands Herd Book Society.

Hereditary Defects

Every bull submitted for registration is examined by a veterinarian to eliminate any bulls with defective reproductive organs or other abnormalities. Animals are rejected if the achilles tendon has been extended surgically; if the retractor muscle of the penis has been cut through; or if the bull calf has been affected with umbilical hernia.

TABLE 12.2
Registrations of Cattle in the Friesian Herdbook[a]

Year	Calves reported	Bulls registered	Purebred cows registered	Cows inspected and admitted to Hulpboek (auxiliary)
1928–29	14,408	778.0	3,635.0	1,402.0
1957–58	62,972	1,491.0	16,329.0	4,660.0
1969	59,054	1,675.0	78,002.0	591.0
Of these, red-and-white		30.0	511.0	35.0
Percentage with red		1.8	0.6	0.7

a. *Friesch Rundvee Stamboek.* Female calves from Hulpboek cows were eligible for registration as purebreds.

FARMING METHODS

A pamphlet by the Friesch Rundvee Stamboek described farming practices:

> During the time the cattle are in the open they feed exclusively on grass, the milk yield being highest during this period. During the winter months, which the cattle have to pass in the cow-house, the feeds on the pasture farms mainly consist of hay, ensiled grass and sometimes of artificially dried grass. On the mixed holdings besides hay and straw, tuberous plants and roots are also fed during the winter. These standard rations are in normal times supplemented by 650 to 880 lbs. of concentrates per milking cow. . . .
>
> Both the production of milk and sale of breeding cattle are aimed at. Milking is done twice daily at equal intervals.

Cows obtain about 75 percent of their nutrients from pastures and early-cut hays grown on 55 percent of the farm land. Moist soils ill-adapted for tillage, or heavily fertilized light sands, produce grasses. Rotational and strip grazing afford efficient use of the forages while of high nutritive value. Hays are cured on tripods or by artificial drying. Coarse grasses are ensiled with some sugar-containing preservatives or after wilting 1 or 2 days in favorable weather. Concentrates are limited to 850 to 1100 pounds mainly during the winter.

MILK RECORDS OF FRIESIAN COWS IN EUROPE

In 1833 the King of Wurtemberg brought foreign cows to his estate to test and compare them under similar conditions. John Klippart published their records in 1865:

Breed	Average annual milk yield	Average body weight
	(pounds)	
North Holland or Friesian	6,549	1,200
Swiss	5,764	1,225
Alderney, or Jersey	3,860	765

A herd of 42 to 61 Holland or Friesian cows in Saxony between 1852 and 1859 (386 cow-years) averaged 8,494 pounds of milk yearly. G. J. Hengerveld estimated that Friesian cows would average 3,500 kilograms (7,716 pounds) on milk. Further, "It is stated by many farmers that from time to time their productiveness amounts to 5,000 or 6,000 liters (11,360 to 13,631 pounds) of milk." The butterfat content was stated then as 3.0 to 3.5 percent.

Solomon Hoxie reported from an international meeting on May 30, 1882, concerning private milk records: "The publishing of milk records in connection with the descriptions in the herd books also was referred. The delegates from both European associations were unfavorable to such a measure on account of their want of confidence in records as they are ordinarily kept."

In 1895 J. Mesdag, Dairy Konsulent of Leeuwarden, began testing for butterfat yields with 49 Friesian cows in three herds; he was aided by the Friesian Agricultural Society. The Gerber test was applied to composite 1-day milk samples at 10-day and 15-day intervals. The average butterfat percentage of milk from individual cows was reasonably constant from lactation to lactation, and the sire exerted a great influence on butterfat yields. "Control" societies were similar to those in Denmark. The Friesch Rundvee Stamboek contributed to support the societies with the Leeuwarden branch of the Friesch Agricultural Society. Supervision was reorganized in 1907 under the Friesch Rundvee Stamboek.

Dairy Konsulents of North and South Holland supervised production contests in those provinces in 1900–1901, with prizes offered by the Holland Agricultural Society. The competition was repeated in 1901–1902. Prizes were based on a combination of production, body scores, and fattening ability. The Groningen division of the Netherlands herdbook supervised a similar production contest in 1901–1902, under the Dairy Konsulent. Payments for milk on butterfat tests by the cooperative creameries stimulated production testing. All cows in a herd were tested under these plans.

The governmental Central Milk Control Service was established on June 28, 1943. The respective Provincial Control Laboratories analyze milk samples taken by local association supervisors at 14-, 21-, or 28-day intervals from each cow in the herds. The Gerber method has been applied for butterfat percentage. Beginning in October 1957, protein in milk samples was determined by either the Kjeldahl or amido black pigment-binding method. An EL-X8 model electronic computer, installed late in 1967, calculates the production of individual cows.

The production records showed a remarkable increase in the average butterfat percentages in milk from registered Friesian cows. Manufacture of butter was the major outlet for milk in Friesland. Studies in Denmark and the Netherlands had shown a lower feed cost per kilogram of butterfat produced in high-test milk than in milk with a low butterfat test. Breeders then selected herd bulls from cows with a high average butterfat test. The effect has been a gradual increase in the average butterfat tests, as shown in Table 12.3.

TABLE 12.3
AVERAGE PRODUCTION, LARGELY BY HERDBOOK COWS, IN FRIESLAND

Year	Number of cows	Milk (lbs.)	Fat (%)	Protein (%)	Butterfat (lbs.)
1895	49	9,285	2.99		277.6
1905	829	8,826	3.17		279.8
1915	12,479	10,569	3.26		344.5
1925	15,012	9,938	3.54		351.8
1935	19,033	10,644	3.74		398.1
1945	Milk not recorded during World War II				
1955	46,601	10,340	4.04		417.7
1960	163,111 [a]	9,691	4.02		389.6
1967	174,557 [b]	10,016	4.13	3.36	413.7
1968	618,429 [c]	10,227	4.01	3.34	410.1

a. Included all Friesian cows milk recorded in Friesland.
b. About 31 percent were protein tested every 3 weeks.
c. Friesian cows in other provinces included.

Much cheese was made in other provinces. Less advantage existed for high-test milk in cheese than in butter manufacture, yet the tests in those provinces have increased substantially. The range in tests published in the herdbooks showed the trend for the Friesian breed. See Table 12.4.

TABLE 12.4
PERCENTAGES OF BUTTERFAT IN MILK OF FRIESIAN COWS IN FRIESLAND
AND IN OTHER PROVINCES

	Friesland	Other provinces
Records tabulated in 1937	8,368	13,477
Butterfat tests		
Range	2.83–5.62%	2.04–5.29%
Average	3.80%	3.54%
Records tabulated in 1967	174,557	618,429
Average	4.13%	4.01%

The solids-not-fat content was determined in mixed milk from selected herds of 18 cows or more in Friesland during 1928 and again in 1949. The average composition of milk from these farms at the 21-year interval was:

	1927–28	1948–49	Increase
		(percent)	
Butterfat	3.58	4.12	0.54
Solids-not-fat	8.52	8.90	0.38
Casein	2.38		

Since October 1957 a large proportion of the herdbook cows have been tested for protein in the milk, currently by the Kjeldahl method at three-week intervals. The provincial milk recording service supervises this work. The protein percentage of milk is highly hereditary, as is the butterfat percentage. R. D. Politiek of the Wageningen agricultural college found them to be inherited independently of each other.

Two desirable herdbook cows on pasture are shown in Figures 12.1 and 12.2.

FIG. 12.1. Four lactations by Wimpie 22 79412 F.R.S. averaged 12,008 pounds of milk, 4.19 percent, 503 pounds of butterfat. This typical Friesian cow was owned by J. M. Wasenaar, near Jelsum, Friesland.

GOVERNMENT INTEREST IN CATTLE IMPROVEMENT

The Netherlands and provincial governments have sponsored activities for improvement of breeding cattle. The general practice had been to use a young bull heavily during a single season (May to August) and then to slaughter or sell him for export before any knowledge of his transmitting ability was measured through his progeny.

The Minister of the Interior opposed a proposal to appropriate 10,000 florins for improvement of breeding cattle in 1896. However, 30,000 florins were appropriated for this purpose in 1897, and were renewed annually. State rules for its use were drawn as follows (translated freely):

Article I. Money for improvement of cattle on behalf of the state shall be used:
1. For maintenance premiums for bulls.
2. To contribute for purchase of bulls, and in case of necessity for females.
3. For maintenance premiums for female breeding stock served by prize bulls. Not over one-fourth of the total fund could be premiums for females.

Article II. Premiums and contributions under Article I, 1 and 2, were dispensed at special shows for a province or district. Only animals were considered that:
1. Appeared worthy of premiums.
2. Animals belonging in the district where the show occurred. Bulls must be used for breeding, and be registered in one of the Netherlands herdbooks.

Article III. In each province through the provincial committee [*Gedeputeerde Staten*] consisting of the agricultural society and herdbook society, a provincial commission is appointed whose function consists of:
1. To regulate use of state moneys (acceptable to the Minister of the Interior).
2. To name one or more prize judge commissions.
3. To administer and disburse the moneys.

Article IV. Owner of a prize bull must use him for service for a definite time at a fee approved by the provincial commission, and submit a list of the cows served.

Article VI. Premiums for females (Article I, 3) are awarded only to pregnant cows, or one calved within the date set, and retained for further breeding to a herdbook bull.

Article IX. Each provincial commission must report expenditures before March 1 to the Minister of the Interior, and report condition of breeding cattle in the province during the preceding year.

The appropriated moneys were prorated among the provinces in proportion to the number of breeding herds. Ten provinces contributed additional sums to these prize moneys.

PREFERENT BULLS

Examinations of progeny of bulls were begun soon after 1900 to determine transmitting ability. Since 1910 the Provincial Committee for Cattle Improvement (PCV) in Friesland made examinations at the request of the owner, interested breeders, or the herdbook society. Examinations began when a bull was four years old and were repeated until a decision was made. All available progeny were compared with their dams—sons as regarded conformation and daughters for production and conformation. The herdbook office supplied production records and conformation scores, and location of the animals.

Milkability of daughters was determined with a four-compartment milking machine to determine time and rate of milk letdown, proportion from fore and rear quarters, hand strippings, and response of the cow to machine milking.

FIG. 12.2. Juweel Vaan 76020 N.R.S. averaged 8,690 pounds of milk, 3.85 percent, 335 pounds of butterfat in 324 days for 6 lactations in a certified milk dairy. Two sons and a daughter were exported as breeding animals.

Results of examinations were published in the *Friesch Landbouw-blad* (Agricultural Gazette). If a bull improved his progeny sufficiently, he qualified as Preferent A or Preferent B. A classification for Preferent C was used for a few years. The NRS used a designation of Select Bulls that improved conformation and production of their progeny. These reports were assembled into book form by the Provincial Commission for Cattle Breeding. The book contained the pedigree, photograph when available, number of progeny, a daughter-dam comparison of milk yield, and protein and butterfat tests.

Over 70,000 bulls were entered in the FRS herdbook since 1910, 108 of which were declared Preferent. The herdbook society was responsible for examinations and declaring Preferent Bulls since 1955.

No *fixed* production standard was set for Preferent Bulls in Friesland. The Friesian Cattle Health Service had to certify that the bull did not transmit recognized hereditary defects.

The Board of the Netherlands Herd Book initiated the Preferent qualification in 1916 for Friesian bulls in other provinces. A committee that passed on the qualifications included the Head Inspector for the herdbook, a state consulting expert for cattle breeding in the respective province, a Board (NRS) member, and an inspector for the NRS. Animals were inspected on pasture.

Qualified bulls were declared Preferent 1st Class or Preferent 2nd Class. Reports by the committee were published in their journal *De Stamboeker* (The Herd Book Man). Reports in a booklet included a photograph of the bull and a typical daughter, his pedigree and conformation score, the committee's report, and an analysis of production records of daughters and their dams. Production standards for daughters to qualify their sires in 1937 were:

Age at calving	Minimum average dairy milk yield	Minimum milk	Minimum fat	Minimum butterfat
(years)	(pounds)	(pounds)	(percent)	(pounds)
2	22	6,618	3.40	278
5	32	9,596	3.40	403.7

The comparisons of daughter-dam production were presented graphically in an arrow chart by the *Friesch Rundvee Stamboek* and in a spot graph by the *Nederlandsch Rundvee Stamboek.* Examples of these are shown in Figures 12.3 and 12.4.

PREFERENT COWS

The rating "Preferent Cow" was initiated in 1947 in Friesland and in 1950 in the *Nederlandsch Rundvee Stamboek.* A cow must qualify in four categories. A score of 70 points for conformation is required to become registered. Within four years, she must drop at least three pedigree calves. Her milk production must average at least 3.80 percent of butterfat. Over half of her lactations in 330 days or less must equal at least the following amounts:

Age	Amount of milk
Up to 2 years 3 months	3,300 kilograms (7,275 pounds)
2–4 to 2–9 at calving	3,600 kilograms (7,937 pounds)
2–10 to 3–3 at calving	3,900 kilograms (8,598 pounds)
3–4 to 3–9 at calving	4,200 kilograms (9,259 pounds)
3–10 to 4–6 at calving	4,500 kilograms (9,921 pounds)
4–7 to 5–6 at calving	5,100 kilograms (11,243 pounds)
5 years 7 months or older	5,500 kilograms (12,125 pounds)

When a lactation period exceeded 330 days, it was converted by multiplying the average daily milk yield by 330. When the yield was insufficient, 50 kilograms of milk, up to 300 kilograms, were deducted from the requirement for each 0.1 percent of fat above 3.80 percent. Allowances of 10 and 15 percent were made for periods of extreme shortage of feed during World War II. When the cow had grazed on certain recognized "light soil," the milk requirement was reduced by 5 percent.

The fourth category required that at least four direct descendants of the cow must have been approved, including one female with a production record. Conformation of the progeny must merit at least ten points based on the scores:

Female		Male	
78–79	2 points	72	1 point
80–82	3 points	73–74	2 points
83–84	4 points	75–77	3 points
85 and higher	5 points	78–79	4 points
		80 and higher	5 points
		Preferent son	10 points

At least one female progeny was required to meet the production requirements of her dam, as above. If there were more than one female progeny, one of them need not necessarily meet the production requirements.

Preferent Cows must pass the milkability test with a milk flow of at least 4.74 pounds per minute.

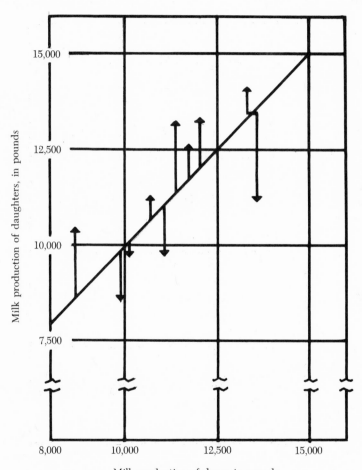

Milk production of dams, in pounds

FIG. 12.3. In an arrow chart, the dam's production is indicated on the abscissa above the base line. The daughter's production is indicated by the length of the shaft of the arrow with relation to the scale on the ordinate at the left.

To be recognized as a *High Lactation Cow,* she must have at least seven milk records, meeting 110 percent of requirements in all but two of them. Since 1959, 48 Friesian cows each had a lifetime yield of at least 100,000 kilograms (220,600 pounds) of milk.

The Central Milk Recording Service, established in 1943, is a national government organization cooperating with the agricultural and herdbook societies. The recording service also analyzes trans-

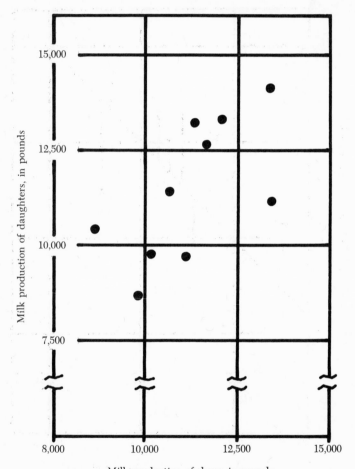

Milk production of dams, in pounds

FIG. 12.4. On a spot graph, each dot is located at the point where a line parallel to the base would intersect a vertical line from the base, indicating the respective levels of production of the daughter and the corresponding dam.

mitting ability of bulls having 15 or more milking daughters. Dairy products are marketed as butter, cheese, and dried and fresh milk.

Milking qualities are determined on at least 20 daughters of bulls in natural use, or 25 to 30 daughters by artificial service. Some 159 sire-groups were tested for milking qualities during 1967–68. Rate of letdown and percentage of milk from the two fore quarters is particularly important, as is the behavior of the cow during machine milking. The four-cup milking machine takes milk from the separate quarters. The first such machine was made by the Delaval company for research at the Missouri Agricultural Experiment Station before 1934. Perhaps the first such determinations, by hand milking, were made at the New York (Geneva) station in 1888.

STATE PREMIUM BULLS

An animal must have scored at least 75 points and have a good production pedigree to qualify as a State Premium Bull. Anna's Bertus 15025 FRS illustrated the conformation selected (Figure 12.5). The designation was replaced by Recommended Bull, or Select Bull, for transmitting improved conformation and production.

FIG. 12.5. Anna's Bertus 15025 FRS was the type desired in a prize winning young Friesian bull.

The Health Service for Cattle was instituted in Friesland in 1918 by the *Friesch Rundvee Stamboek*, cooperating later with the Co-operative Dairy Factories. The Health Service was in charge of eradicating tuberculosis from Friesland in 1951 and for the Netherlands by May 1956. Bang's disease (brucellosis) and vibriosis were eradicated similarly. Heifers and cows are still inoculated each spring against aftosa (foot-and-mouth disease). Since the close of World War II, over 75,000 breeding animals have been exported to 41 countries on four continents. None went to Canada or the United States although many were exported there before 1905.

George Reed of Valley Farm, New Hampshire, described conditions and practices in the Netherlands in 1949 as follows:

Nearly all the cows we saw had excellent top lines and rumps, some of them were a little heavier over the withers than we like to see in good Holstein dairy cows but their average test is up around 4% and their production seems to be quite high for the amount of concentrates they feed, because their feeds are mostly . . . grass silage, hay, beets, and potatoes, and not as much concentrated feed as we feed in the U.S.

This was in addition to pasture grasses under a system of rotational grazing. He continued:

Their classification is very strict. The highest classification for a bull is 88 points. . . . We saw one cow that was classified 93 and I think the best classification we saw on a bull was 86, although I would easily classify "Excellent" in every bracket in our type of classification. In some of the herds . . . our classifiers would find almost half of the animals classifying in the top brackets because they were very uniform in top lines, rumps, legs, and udders. I believe that their animals may be a little shorter in the neck than ours on the average; they weigh over 200 pounds more than our average Holstein will. They have very alert heads, well-dished, with a broad muzzle and strong jaw. All the female animals have their horns left on and most of the bulls.

Their herds were not depleted much from occupation by the Germans. Only one farm that we visited had had to turn in one animal, the poorest in the herd, for food purposes. That was more or less in the form of a fine because the farmer was ordered to turn over a certain amount of his milk to the occupation army and this man disregarded the order and sold all the milk in town. . . . But they were very particular not to destroy or harm any of the registered animals because they wanted to have the best breeding to carry on with after they conquered these countries and they also wanted the people to be friendly toward them.

One place we visited in Leeuwarden was the weekly Friday sale where they sell between 2000 and 3000 Black and White cattle every Friday at private treaty with no auctioneer. They come along and slap each other's hands and the seller keeps shaking his head until the buyer reaches a price that he is satisfied with and then he nods his head in the affirmative and the animal is bought. At these sales the inspectors are present (to judge the young milk cows, whose calves they want to keep in the Herdbook, while registration of the cow may be cancelled after the sale).

When animals are selected for inspection and are not already registered from a registered dam, the inspectors are very rigid, and they measure the rump, the body, the shoulders, the head, neck, feet and legs, and the udder. They certainly are very strict on registration of animals they enter in the Herd Book in this manner. (They must score at least 75, rather than 70 points).

The sales grounds are maintained by Leeuwarden, and a fee is charged for animals that come to the stalls. The grounds have brick pavement; water and eating facilities are available.

Artificial Breeding

Artificial breeding began on an experimental basis in 1936 in the Netherlands, and in limited practice 2 years later. The bulls were owned mainly by farmer cooperatives. Provincial supervising committees established minimum standards for milk yields and butterfat tests of dams and granddams of the bulls, but the cooperatives prefer higher standards. Reports of the Central Committee for Artificial Insemination showed that 62 percent of the cows and about 40 percent of the heifers were bred artificially in 1968. Some 695 Friesian, 389 Meuse-Rhine-Yssel, 11 Groningen, and 1 Jersey bull were in artificial service in the Netherlands in 1970.

Bulls in artificial service are blood-typed for antigen pattern to facilitate identification of their progeny when necessary. Formerly a commission of experts inspected the first 100 calves sired by a bull in the so-called "100-calf-test," which considered body conformation. This method has been discontinued.

Older bulls with recognized transmission are used on about four-fifths of the cows, while selected young bulls are being sampled for future replacements. Their progeny are analyzed for conformation, production, milkability, and any recognized hereditary defects. Breeding efficiencies are estimated on 60- to 90-day nonreturns, which were found to be 4.1 percent higher than actual results. Semen is prepared in ampules, tablets, or pellets formed and stored in liquid nitrogen (−320° F.).

When yields of 15 daughter-dam pairs are recorded, these are published by the herdbook societies. A program of measurements and weights deals with ultimate carcass quality and values.

The headquarters of the two herdbook societies are: H. de Boer, Friesch Rundvee Stamboek, Zuiderplein 2–6, Leeuwarden, The Netherlands; and Drs. A. Koenraad, Stradhoudersplantsoen, 24, 's-Gravenhage, The Hague, The Netherlands.

REFERENCES

Bakhoven, H. G. A., et al. 1953. Het gehalte van de melk van Friese koein san vetfrije drogstof en het verband tussen vetgehalte en vetvrije drogstof. Officieel orgaan. *Koninkl. Ned. Zuivebond* 45(4):53–57.

Bakker, D. L. 1909. Studies uber die Geschichte den heutigen Zustand und die Zukunft des Rindes und seiner Zucht in den Niederlanden mit besonderer kritischer Beruchsichtigung der Arbeitweiss des Niederlandischen Rindviehstammbuches. Inaug. Diss., Univ. Bern. Martricht.

Bakker, D. L., et al. 1948. [The history of the Dutch cow.] In *Rindvieh*. Amsterdam. Ch. 1.

Blink, H. H. 1922. The Friesian Herdbook Association. *Holstein-Friesian World* 19:2033–35.

Clemons, G. M. 1937. Holsteins in Holland. *Holstein-Friesian World* 34:1115–16, 1143.

DeLeeuw, P. 1925. Improving the Holstein-Friesian cow in Holland. *Holstein-Friesian World* 22:1264–65, 1278, 1280.

Dijkstra, J. M., and E. T. Roelofs. 1959. *Preferent Stieren in Friesland, 1910–1959*. Leeuwarden.

Hauser, E. O. 1953. The Dutch strike back against the sea. *Sat. Eve. Post* 225(45):19–21, 118.

Hengerveld, G. J. 1880. Origin and purity of the breed. *Dutch-Friesian Herd Book* 1:10–19.

Hoekstra, P. 1966. Het niewe keuringsrapport van het Nederlands Rundvee Stamboek. *Tidschr. Diergeneesk* 91:1265–70.

Hoffman, K. 1905. *Monographien landwirtschaftlicher Nutztiere.* Vol. 4. *Das Hollander Rind.* Leipzig.

Houghton, F. L. 1897. *Holstein-Friesian cattle.* Brattleboro, Vt.

Janse, L. C. 1950. Composition of Friesian milk. *Neth. Milk Dairy J.* 4(1):1–9.

Klippart, John W. 1865. Report of an agricultural tour in Europe. *Ohio Agr. Rept.* Ser. 2:17–280.

Lubbock, Sir John. 1872. *Prehistoric times.* Appleton, New York.

Moscrip, W. S., and F. L. Houghton. 1926. The Holland system of selective registration. *Holstein-Friesian World* 23:43–44, 66.

Politiek, R. D. 1957. Het eiwisgehalte in de melk. Thesis, Landnouwhogeschool, Wageningen.

Reed, George. 1949. A New Englander visits Holland—Part Two. *Holstein-Friesian World* 46:2442–46.

Rumler, Robert A. 1960. Abroad with Secretary Rumler—Part Two. *Holstein-Friesian World* 37:480–81, 562–63.

Stanford, J. K. 1956. British Friesians—A history of the breed. Max Parrish, London.

Tacitus, A. About 65 b.c. *De Belli Gallico.* Book 6, Ch. 28.

U.S. Consular Repts. Cattle and dairy farming. 2 vols. GPO, Washington, D.C.

Van den Bosch, I. G. J. 1930. The scoring system in the Netherlands. *Holstein-Friesian World* 27:337, 352, 356.

———. 1930. The measuring system in Holland. *Holstein-Friesian World* 27: 381–82.

———. 1930. Preferent bulls in Friesland. *Holstein-Friesian World* 27:426.

Van den Hoek, Spahr. 1952. *Geschiedenij van de Friesa.* [History of Friesian agriculture.] *Landbrouw I.*

Van Welderen, E., and H. G. A. L. Bakhoven. 1932. *Preferent stieren in Friesland 1910–1932.* Leeuwarden.

Herdbooks

L'essential de functionnement du Herd-Book. Leeuwarden. 1937.

Friesch Rundvee Stamboek. Leeuwarden. 1879.

Nederlandsch Rundvee Stamboek. The Hague. 1874.

North Holland Herd Book. Dover, N.J. 1888–1892.

Centrale Melkcontrole Dienst. Stichting. 1968.

The Herd Book in the Netherlands. The Hague. 1969.

The Black and White Friesian Cattle. Leeuwarden. 1969.

CHAPTER 13

HOLSTEIN-FRIESIANS IN THE UNITED STATES

CATTLE WERE introduced into the United States from the Netherlands in 1621 by the early Dutch colonists. Definite records of their
importations are not available. The diary of Samuel S. Forman of
Cazenovia, New York, mentioned eight Dutch cattle imported in
1795 by Mr. Lincklean, agent of the Holland Land Company: "The
cows were the size of oxen; their colors were clear black and white,
not spotted, but large patches of the two colors; very handsome
bodies and straight limbs; horns middling size, but gracefully set;
their necks were seemingly too slender to carry their heads; their
disposition mild and docile."

Small importations into Vermont about 1810, into New York
about 1825, and into Delaware later were not kept pure. A cow that
Winthrop W. Chenery of Belmont, Massachusetts, bought from a
sailing vessel was so good that he imported a bull and two cows in
1857 and four more cows in 1859. An outbreak of disease caused

these cattle and all descendants except a bull to be destroyed. Mr. Chenery imported a bull and a cow in 1861 which were among the early cattle registered in this country. The cow Texelaar yielded 1,704½ pounds of milk in June 1863. She was fresh as a 4-year old. Gerrit S. Miller of Peterboro, New York, made the second importation which became registered. Many prominent cattle traced to Mr. Miller's herd. Superiority of these cattle as milk producers made them popular.

During 25 years (1852–76), 182 animals entered the country that were in the herdbooks later. Some 6,927 cattle were imported in the 10-year period between 1877 and 1886. A total of 7,757 animals came before 1905. Further importations were barred because of foot-and-mouth disease quarantine. Only Canadian cattle have entered the United States since that time.

HERDBOOKS IN THE UNITED STATES

HOLSTEIN HERD BOOK

Six breeders met in Boston on March 15, 1871, and organized the Association of Breeders of Thoroughbred Holstein Cattle, which was to publish the *Holstein Herd Book*. They recognized as "thoroughbred Holsteins . . . those large improved black-and-white cattle imported from the provinces of North Holland, Holstein, or intermediate territory, or . . . traced in direct line, on the side of both sire and dam, to animals of undoubted purity of blood of said importations."

Nine volumes of the *Holstein Herd Book* were published.

DUTCH-FRIESIAN HERD BOOK

The Unadilla Valley Stock Breeders Association in New York was organized in 1874 and purchased cattle imported by Thomas E. Whiting of Concord, Massachusetts. More of his cattle were bought in 1876. After Mr. Whiting died in 1877, the Association purchased his herd, records, and a manuscript intended as a "Register of Thoroughbred Dutch Cattle." The papers included an article by G. J. Hengeveld, head inspector of the Nederlandsch Rundvee Stamboek.

Solomon Hoxie and his associates, who owned the Whiting cattle, met at Brookfield, New York, on December 8, 1877, and organized the Association of Breeders of Pure Bred Friesian or Dutch-Friesian Cattle. The Whiting manuscript became the nucleus of the *Dutch-Friesian Herd Book*.

FIG. 13.1. Solomon Hoxie incorporated production records and a system of type classification based on body measurements and scores into the Main or Advanced Registry of the Dutch-Friesian Herd Book. The purposes were "to increase and maintain public interest in the breed; to inaugurate improvement in the breed; and to collect observations upon which a science of cattle culture might be built."

The *Dutch-Friesian Herd Book* was divided by Secretary Hoxie (see Fig. 13.1) into an Appendix Registry and a Main Registry. Spotted black-and-white Dutch-Friesian cattle from Friesland and North Holland were entered in the Appendix Registry on purity of breeding. Animals were advanced to the Main Registry by the executive committee under either of two conditions: (a) imported animals and progeny of Main Registry cows, sired by Main Registry bulls, must have had well-developed escutcheons under the plan of the Pennsylvania Commission (Guenon's scheme of escutcheon classification); and (b) females not fulfilling the requirements (a) must have had a milk record according to age, as follows: at least 6,000 pounds if under 2½ years old at date of calving. At least 7,000 pounds if over 2½ and under 3½ years old. At least 8,000 pounds if 3½ and under 4½ years old. At least 10,000 pounds if over 4½ years old at date of calving.

Bulls were measured, and their conformation was judged by the inspector. Bulls had to score 80 points or more, based on a Scale of Points, after 1881. Measurements were taken of height at shoulders, height at hips, length of body from shoulder to pins, length of hips, girth of fore chest, and width of hips and of thurls. The measurements of 55 cows from a group of 185 cows appeared in Volume 1 of the *Dutch-Friesian Herd Book*. Maid of Twisk No. 1 was credited with producing 15,960 pounds of milk in 336 days.

HOLSTEIN-FRIESIAN ASSOCIATION OF AMERICA

The two herdbook associations differed over the correct name of the breed and the source of pure breeding stock; they also were apprehensive that production records would "set up an aristocracy within the breed." Two attempts failed to unite them in 1880. A proposal prevailed to adopt the name Holstein-Friesian Association of America and maintain an Advanced Registry. The archives, herdbooks, and assets were combined. Ninety-one members of the Dutch-Friesian and 206 of the Holstein associations joined the new association in Buffalo on May 26, 1885. Thomas B. Wales of the *Holstein Herd Book* became president, and Solomon Hoxie became superintendent of the Advanced Registry. After the animal was registered, it might qualify for the Advanced Registry by type inspection, body measurements, and milk or butter production. Bulls entered the Advanced Registry on production of progeny and on inspection. The new organization published the *Holstein-Friesian Herd Book.*

AMERICAN BRANCH OF THE NORTH HOLLAND HERD BOOK

The Holstein and Dutch-Friesian associations barred registration of animals imported after March 18, 1885, unless owned by members in the United States or Canada. A. C. and N. F. Sluiter, Hollanders, had an importation in quarantine for which registration was refused. After losing a suit to become members, they established the American Branch of the North Holland Herd Book. The parent body had been founded in that province in November 1883. Three volumes of their herdbook between 1888 and 1892 registered 293 bulls and 860 females. The Holstein-Friesian Association accepted them for registration in 1892.

WESTERN HOLSTEIN-FRIESIAN ASSOCIATION

A $100 membership fee in the Holstein-Friesian Association of America, and the need to travel as far as New York to attend annual meetings, created a feeling of disproportionate representation. An agricultural depression around 1890 accentuated dissension. The Western Holstein-Friesian Association was formed in 1892 with a Nebraska charter. Volume 1 of their herdbook in 1895 contained

2,100 entries. In 1898 the Western Association reunited with the parent organization, which accepted the Western herdbook and assets.

The term *thoroughbred* in the bylaws of the Holstein-Friesian Association of America was changed to *purebred* before 1890.

THE HOLSTEIN-FRIESIAN REGISTRY ASSOCIATION, INCORPORATED

During a depression, and amid other circumstances, Howard C. Reynold appealed for membership in a new herdbook society, with lower membership and registration fees. The Holstein-Friesian Registry Association organized at Harrisburg, Pennsylvania, on July 31, 1925. Proxies were voted. Low registration fees precluded promotional activities except for a modest breed magazine. A committee was appointed " . . . to formulate a Scale of Points to be recognized by this Association and gather data and determine the minimum requirements for registration in this Association after January 1, 1928, as to size, conformation, physical development and dairy qualities. . . . This Association shall recognize only such milk and butter records as represent normal hereditary, economical production."

No herdbook was published, and the breed magazine was discontinued. The association discontinued by proxy vote of active members, and records were transferred to Brattleboro in September 1966. Delegates to the parent association voted at the 1966 annual convention to readmit cattle from the Holstein-Friesian Registry Association up to March 1967, entering dead animals at a nominal fee to complete pedigrees. Some 8,106 dead animals were registered as ancestors of 4,786 living animals, and with 154 memberships.

MEMBERSHIP PARTICIPATION IN ASSOCIATION AFFAIRS

Bylaws of the Holstein-Friesian Association of America provided for a president, four vice-presidents, six directors, a secretary, a treasurer, and a Superintendent of Advanced Registry. These 14 officers constituted the Board of Directors who managed Association affairs. They were elected at annual meetings, where many members voted proxies of absent members. A smaller Executive Committee from the Board functioned after 1895.

The Association reincorporated in 1913, enabling annual meetings to be held outside of New York. A president, vice-president, and nine directors took the place of the previous officers. State and local breeders' clubs were represented by delegates at annual meetings. Revision of the bylaws changed procedures. The number of directors was increased to 16 in 1919 based on increased membership.

Solicitation and voting of proxies was abused on occasions. This problem had been discussed but not solved. Honorable Frank O. Lowden was nominated for president in 1921. He accepted nomination under condition that the entire Board support him, and a delegate system of representation be established for annual meetings. That autumn a called convention met these provisions, which still stand. Delegates are nominated by local members, and are elected in proportion to active membership. Each state has one delegate, plus one additional delegate for each 150 active members. (An active member has registered or transferred an animal within 2 years). These delegates discuss proposals one day and transact business the next day at the annual meeting. The secretary is elected by the directors, instead of by members. There are no proxy votes; instead an alternate acts for an absent delegate. Effective in 1967, a delegate who has served 3 consecutive years, is ineligible for re-election until a year has passed.

In 1948 the Secretary became an Executive Secretary, with more responsibilities. The Association office was reorganized for efficiency into the Extension Service, Registry, Advanced Registry, Office Management, and Accounting Departments. Also a Computer Operations Department has been added.

Individual life membership was at a $25 fee. Starting in 1961, new memberships were for a 10-year term at $20, and were renewable for 10 years at $10. Active members were required to select a herd prefix name before 1965, which was reserved for 20 years, to use in registering homebred animals. In 1969 there were 26,056 active members, 44,403 total members, and 13,600 junior members under 21 years old.

COLOR MARKINGS

Friesian cattle in the Netherlands were predominately black-and-white in large patches. Red-and-white Friesians were registered there in small numbers. Dun color also occurred but was not registered in the herdbooks.

Color standards were adopted in the United States to exclude color markings observed often among grade cattle. Color markings that barred registration, as revised in 1950, were: solid black, solid white, black in switch, solid black belly, one or more legs encircled with black that touched the hoof at any point, black on a leg beginning at the hoof and extending to or above the knee or hock, black and white intermixed to give a gray appearance, and colors other than distinctly black-and-white. Some Holstein-Friesians and other black breeds carry the recessive gene for red hair color. On the average the red color crops out in one of four progeny when both parents carry the gene. Solid color is dominant over broken or spotted color. A fine black-speckled pattern tends to be recessive to large black spots and is seen less frequently. "Brockle face" is dominant.

Two changes concerning coat color of Holstein-Friesians were approved in 1969. Females born since July 1969 with black in the switch or beginning at the hoof became eligible for the herdbook with the suffix OC to the name and B preceding the registration number. A separate red-and-white Holstein-Friesian herdbook accepted females of proved registered ancestors. A 6-month moratorium applied to females over 2 years old, born before January 1970.

THE SHOW RING AND TYPE

Cattle from the Netherlands were exhibited soon after importation and before a breed association was organized. Shows brought the breed before the public and stimulated type improvement. Winthrop W. Chenery exhibited "Dutch" cattle in 1864. Gerrit S. Miller's cattle competed with W. A. Russell's at the New York State Fair in 1873. Importers Smith & Powell of Syracuse, New York, showed in 1876 and later. The "Code of Show Ring Ethics," recommended by the PDCA was adopted in 1966.

NAME OF THE BREED

Dutch cattle were exhibited under different names. Chenery wrote of his "Dutch" cattle for the *Report of the Commissioner of Agriculture* in 1864, which was edited to "Holstein" without his knowledge and was published. He adopted the name and used it when forming the Association of Breeders of Thoroughbred Holstein Cattle, of which he became president.

Scales of points were published in 1880 by the Dutch-Friesian Breeders' Association and used for type inspection of animals for the Main or Advanced Registry. These scales contained 22 articles for bulls and 25 for cows and totalled 100 points for perfection. Revised scales gave more attention to mammary development. Their members exhibited "Dutch-Friesian" cattle at shows. Rivalry was keen between the societies over the name; they tried to have the others' cattle barred from some shows.

Holstein-Friesians competed at the World's Columbian Exposition in Chicago in 1893 but did not enter the milking contest where the Babcock test was used for butterfat determinations. Seven herds competed at the Pan-American Exposition at Buffalo in 1901. More than 100 head were at the St. Louis World's Fair in 1904, where Sarcastic Lad 23971 was made champion bull.

The National Dairy Association held its first National Dairy Show at Chicago in 1906 and soon became the dairy "court of last resort." It followed the state fairs and three regional shows: Eastern States Exposition, Springfield, Massachusetts (1917–); Dairy Cattle Congress, Waterloo, Iowa (1910–65); and the Pacific International Exposition, Portland, Oregon (1911–). The National was not held in 1915 due to foot-and-mouth disease in 1914, a financial depression in 1932–34, and World War II in 1942–45. The National Dairy Association disbanded in 1946. Charles L. Hill, organizer and long-time president, was honored for leadership by the Purebred Dairy Cattle Association, which represents the breeders of purebred dairy cattle. Until 1965 the Dairy Cattle Congress succeeded as the major show in many dairy activities—educational exhibits, junior and collegiate dairy cattle judging contests, displays of agricultural equipment, and some breed meetings. The Dairy Shrine

Club honored dairy leaders, and had headquarters nearby until 1967. The national show for Holsteins was replaced by three regional shows in 1966, anticipating more breeder participation and increased ringside audiences.

SHOW CLASSIFICATION

The show classification dividing cattle into groups according to sex and age includes calves; junior and senior yearlings; 2-, 3-, and 4-year old males and females; get of sire (junior and senior gets at some shows); produce of dam, best uddered cow; junior, senior, and grand champion male and female; premium exhibitor; and premium breeder awards. Some shows do not have a class for 4-year old bulls. Some shows have group classes—young herd (male and three females), dairy herd (four cows in milk), county and state herds comprising eight to ten animals of several ages. Although the Grand Championships were the climax of the show, the most coveted prizes were the Get of Sire and Premier Breeder awards because these indicated quality of the herd from whence the winners came.

The 1-day "Black-and-White Show" renders great service to the cattle industry. It originated with Holstein breeders around Richmond, Utah, in 1915 with initiative of fieldman H. A. Mathieson. Jersey "parish" shows were patterned after it in 1928, and Brown Swiss "canton" shows started in 1938. Many junior dairy shows follow the 1-day plan.

An annual "All-American" contest has been sponsored by the Holstein-Friesian World since 1922, based on photographs of leading prize winners in the United States and Canada. Animals were placed by a panel of leading judges. This contest recognized top animals even though the animals may have been shown at only a single show. A "Junior All-American" contest was established in 1952.

TRUE TYPE

Diverse ideas of ideal type existed among breeders and by some leading judges, as evidenced by placings of some animals at successive shows. Axel Hansen suggested in the *Holstein-Friesian*

World that the officially recognized judges meet to attain more similarity of judgments.

The Executive Committee passed a resolution January 10, 1922, for such a conference. Secretary F. L. Houghton invited about 40 leading breeders, showmen, and judges who might plan some definite action. They met in Philadelphia on March 20, 1922, with Fred Pabst presiding. After much discussion, Pabst suggested that clay models of an ideal type be prepared to scale. The group suggested paintings also, agreed on general procedures, and selected ten persons as a subcommittee to carry out the details. The Executive Committee named this group as an official committee and allotted them $50,000 to study and establish "true type."

The committee included W. S. Moscrip (chairman), T. E. Elder, R. E. Haeger, Axel Hansen, H. H. Kildee, A. C. Oosterhuis, Fred Pabst, W. H. Standish, Ward W. Stevens, and F. L. Houghton (secretary). They engaged Edwin Megargee, animal painter, and Gozo Kawamura, sculptor, to convert their ideas into specific illustrations. After they studied photographs of leading animals and scored and criticized many desirable animals on farms in Waukesha County, Wisconsin, the artists prepared the clay models and paintings to scale.

On June 8, 1922, the committee and artists met in Kansas City with many of the original invited group at the Heart of America Holstein sale and first Association convention under the delegate system. Minor changes conformed with the height of withers and fullness of fore udder of two excellent show cows consigned by George B. Appleman of Kansas. The models and paintings received unanimous approval at the directors' meeting in St. Paul in October 1922.

Duplicate metal models, painted to conform with the approved paintings, were loaned to land-grant agricultural colleges. Smaller scale models were sold. The Committee on Type (W. S. Moscrip, chairman) revised the Scale of Points to conform with the models. These models and President Frank O. Lowden are shown in Figure 13.2.

Spring Brook Bess Burke 2d 131387 was a prominent cow for production and type. Her daughters Bess Johanna Ormsby and

Bess Mercedes Ormsby, and progeny of her sons Sir Pietertje Ormsby Mercedes 37th, Creator, King of the Ormsbys and Winterthur Bess Burke Best exerted a wide influence on production and Holstein type (Fig. 13.3).

The last scale of points was superceded by the unified dairy scorecards, approved by the American Dairy Science Association and copyrighted by the Purebred Dairy Cattle Association. The dairy cow and bull scorecards, first drafted in 1943, were revised in 1957 with descriptions reworded and points redistributed. These scorecards were divided into four major headings: general appearance, dairy character, body capacity, mammary system (for cows), and feet and legs (for bulls). Separate characteristics were listed for each breed concerning color, size, and horns. The ideal cow and bull of each breed were shown in colors, and charts illustrated the parts of the dairy bull and cow.

FIG. 13.2. President Frank O. Lowden and the true type model Holstein-Friesian bull and cow. These models were developed by a special True Type Committee of ten members. Other dairy breeds followed the example with less extensive projects.

Type Classification and Sire Recognition

After the true type models were completed in 1922, discussion arose that the show ring benefited only a few breeders who exhibited their animals. How could these benefits be applied to other cattle at home, and herds not competing? Could a judge visit a herd and constructively criticize the individual animals? Could type be recorded permanently as production records had been, for use in breed programs?

Few people recalled that Solomon Hoxie devised an inspection system for the Main or Advanced Registry of the *Dutch-Friesian Herd Book* before 1880 based on body measurements and scores. His plan was used by the Holstein-Friesian Association of America from 1885 to 1893, and was discontinued during a depression.

Type Classification

A plan was developed to classify animals on farms into ratings according to conformation scores, recognizing the sires that transmitted desirable type. Thus Herd Classification and Sire Recogni-

FIG. 13.3. Spring Brook Bess Burke 2d 131387 was noted for type and production. She transmitted these characteristics through 2 daughters and 4 sons.

tion began in 1929. Nine part-time inspectors were appointed to classify animals. They met at Silver Glen Farm, St. Charles, Illinois, to unify methods. The scorecard soon was revised, as follows:

Classification	Original score	Revised score
	(points)	
Excellent	85 or more	90 or more
Very Good	77.5–84	85–89
Good Plus	72.5–77.4	80–84
Good	67.5–72.4	75–79
Fair	60–67.4	65–74
Poor	Below 60	Below 65

The "breakdown" method of scoring according to anatomical parts was devised by F. W. Atkeson for Brown Swiss cattle, based on divisions of the unified dairy scorecards. The Holstein-Friesian Association adopted a similar plan in 1944. Twelve inspectors classified between 1929 and 1970, some being repeated. Some 1,239,864 animals were classified. Those of 99,361 Holsteins classified in 4,048 herds during 1970 are given below.

	Percentage of animals rated		
Rating	1929–68	1969	1970
Excellent	1.2	1.6	1.2
Very Good	12.8	13.1	12.8
Good Plus	43.1	41.4	42.8
Good	35.9	36.1	36.0
Fair	6.8	7.4	6.9
Poor	0.2	0.4	0.3

Breeders applied for type classification on a voluntary basis. All cows that had calved and bulls over 3 years old were submitted for inspection. Registration certificates of animals classified Poor were cancelled. Only female progeny from cows or bulls classified Fair were registered. This regulation was dropped in 1959. Effective in 1966, cows may be rated Excellent up to four times, starting at any age after the second calving. An additional E designation may be earned within 6 to 8 years, 9 to 11 years, and 12 years or older.

SIRE RECOGNITION

A percentage of all available daughters (minimum of ten) and all available dams were required to have been inspected. Three sire ratings were recognized. A Bronze Medal Preferred Sire had improved daughters over their dams. A Silver Medal Preferred Sire was strikingly prepotent for herd improvement. A bull qualified as a Gold Medal Proved Sire when as a Silver Medal Approved Sire, 50 percent (at least six) of his daughters exceeded Advanced Registry requirements by 50 percent. The average butterfat test must have been between 3 and 5 percent. The plan became effective in 1929. Later changes adapted the program to bulls in artificial service.

A bull earned the Silver Medal Type rating when at least ten (75 percent of his registered) daughters averaged 81 points; 50 to 75 percent averaged 82 points; 25 to 50 percent averaged 83 points; or less than 25 percent classified 84 points or above. For bulls in artificial service, all daughters in at least ten herds selected by the Executive Secretary must be classified. At least 50 must have been inspected, less than half of them being in a single herd. If the average classification score of the daughters equaled 81 points, the sire became a Silver Medal Type Sire.

The Silver Medal Production Sire rating was earned when 50 percent of the daughters (at least ten) qualified in the Advanced Registry or Herd Test with production 50 percent over the Advanced Registry requirements according to age, with not below a 3.4 percent butterfat test. The records must average at least the breed average of 430 pounds of butterfat (January 1954) on a 305-day $2\times$ milking mature equivalent basis. The daughters' average production must exceed "expectancy" by at least 40 pounds of butterfat. Expectancy is the amount halfway between the breed average and the average production of the dams. As the breed average increased, requirements for the recognition programs also increased. Since January 1964 a bull's daughters must average at least 13,800 pounds of milk and 510 pounds of fat on a 305-day $2\times$ mature equivalent basis. The lactation average for registered Holsteins in 1965 was 15,114 pounds of milk, 3.67 percent and 555 pounds of fat.

A bull became a Gold Medal Sire when his daughters qualified him for Silver Medals in both type and production. One of the great Gold Medal Sires is shown in Figure 13.4.

The average type score at each age was determined from 24,075 classification scores in 1956. The Breed Average Age (BAA) score at each age was regarded as 100 percent, and the scores of individual animals computed on a relative percentage basis. Records of bulls with at least ten classified daughter-dam pairs were reviewed periodically since 1959. To become a Silver Medal Type Sire under the program, the average score of the daughters must exceed expectancy (midway between breed average and dams' average) by at least 1.00 points if a sire has 40 pairs or more; 1.25 points for 30 to 39 pairs; 1.50 for 20 to 29 pairs; 1.75 points for 15 to 19 pairs; or 2.00 points for 10 to 14 pairs. A sire then was recognized as a SMT sire when all available daughters were classified and had a BAA percentage of at least 103 points. A point was 1 percent above 100 percent.

In January 1968 standards were increased and Silver Medals changed to temporary Type Qualified and Production Qualified designations, based on the current status. The sire's ten pairs then must have a BAA expectancy of at least 101.00 points. The standard for daughters to exceed their dams then became 1.00 points if he has 100 or more daughter-dam pairs, 1.25 points for 50 to 99 pairs, 1.75 points for 25 to 49 pairs, 2.25 points for 15 to 24 pairs, or 2.75

FIG. 13.4. Wisconsin Admiral Burke Lad 697789 V.G. More than 100 daughters produced an average of over 14,000 pounds of milk and 500 pounds of butterfat. His 163 classified sons and daughters had an average score of 85.6 percent. He was among the leaders in numbers of Gold Medal sons and daughters.

points if he has 10 to 14 daughter-dam pairs. The system was automatic and without a fee. Special applications, however, required a $50 application fee plus the regular fee for individual animals in the applicant's herd, and $5 per head for animals outside the herd.

The temporary designation Production Qualified was based on the most recent daughter-herdmate comparison in the USDA Sire Summary. The sire must have at least 10 daughters summarized. The predicted difference must at least equal +200 pounds of milk with a 30 percent or more repeatability, or it could be +300 pounds with a repeatability of 20 through 29 percent. At least 51 percent of his daughters must be registered, or identified by blood typing (unregistered daughters) if less than 30 daughters are included. The designation must be rated as each new sire summary is computed.

A sire designated simultaneously as Production Qualified and Type Qualified is automatically recognized with the permanent Gold Medal award.

Records useful for herd improvement are obtainable from electronic tapes filed in the Association office. Breeders may request information from this "Holstein Fact Finder" source, including: 1. descriptive data of a bull's classified daughters; 2. production records of all tested daughters; 3. production and type classification of individual Holsteins; 4. functional strengths and weaknesses transmitted by a sire.

Since 1971, mechanized office procedures necessitated assignment of an official ID number to each breeder registering a Holstein-Friesian animal. The ID number is a combination representing the owner's surname and is in sequence of assignment. Official ID numbers appear on the more recent registration certificates.

Revised Classification Program

The type classification program was revised for 1967, as coordinated by George W. Trimberger. Thirteen columns were added for "description of animal," using keyed numbers for descriptive details. These numbers referred to:

Stature. 1. Upstanding. 2. Intermediate. 3. Low-set.

Head. 1. Clean-cut, well proportioned with style and strength. 2. Strong—lacking style. 3. Short. 4. Plain and/or coarse. 5. Weak.

Front end. 1. Shoulders smoothly blended. 2. Medium strength and width. 3. Coarse shoulder and neck. 4. Narrow and weak.

Back. 1. Straight, full crops, strong, wide loin. 2. Straight, weak crops. 3. Low front end. 4. Weak loin and/or back.

Rump. 1. Long and wide, nearly level. 2. Medium width, length, or levelness. 3. Pins higher than hips. 4. Narrow, especially at pins. 5. Sloping.

Hind legs. 1. Strong, clean, flat bone, squarely placed, clean flat thigh. 2. Acceptable. 3. Sickled and/or close at hock. 4. Bone too light or refined. 5. Post-legged (too straight).

Feet. 1. Strong, well-formed. 2. Acceptable, with no serious faults. 3. Spread toes. 4. Shallow heel.

MAMMARY SYSTEM

Fore udder. 1. Moderate length and firmly attached. 2. Moderate length, slightly bulgy. 3. Short. 4. Bulgy or loose. 5. Broken and/or faulty.

Rear udder. 1. Firmly attached, high and wide. 2. Intermediate in height and width. 3. Low. 4. Narrow and pinched. 5. Loosely attached and/or broken.

Udder support and floor. 1. Strong suspensory ligament and cleanly defined halving. 2. Lack of defined halving. 3. Floor too low. 4. Tilted. 5. Broken suspensory ligament and/or weak floor.

Udder quality. 1. Soft and pliable. 2. Intermediate. 3. Could not determine. 4. Meaty.

Teat size and placement. 1. Plumb, desirable length and size. 2. Acceptable with no serious faults. 3. Rear teats back too far. 4. Wide front teats. 5. Undesirable shape.

Miscellaneous. 1. Winged shoulders. 2. Front legs toe out. 3. Weak pasterns. 4. Crampy. 5. Small for age.

SELECTIVE REGISTRATION OF MALES

Resulting from a resolution in 1926, pedigree ratings of two classes were established for Holstein male calves based on transmitting ability of sires and type and production of dams. Lack of popular use led to the lower rating (Selected Pedigree) being discontinued in 1958, and the Preferred Pedigree registration certificate before 1962.

PRODUCTION RECORDS IN AMERICA

Three cows imported by Gerrit S. Miller from West Friesland yielded an average of 9,597 pounds of milk yearly in 1870–72. Private records of Holsteins cited in Volume 3 (1873) of the *Holstein Herd Book* showed that "76 cows have given 60 lbs. . . . or more per day. . . . We particularly hope attention will be given to butter tests during the next season." The Association adopted a resolution in 1880 to publish private records of 6,000 pounds for 2-year olds and up to 10,000 pounds or over for cows past 4½ years old at a $2 fee.

Solomon Hoxie of the Dutch-Friesian Association devised rules for a system of recording production in the Main or Advanced Registry. Records were attested by a committee representing some local agricultural society or stock-breeders' association. Requirements for entry in the Advanced Registry of the *Dutch-Friesian Herd Book*, and later in Volume 1 of the *Holstein-Friesian Association of America's Herd Book* were:

Age	Milk in 10 months	Milk in 7 days before, and after 8 months after calving		Butter in 7 days
		Before	After	
(years)	(pounds)	(pounds)		(pounds)
2	6,500	354	118	9
3	7,900	432	144	11
4	9,300	511	170	13
5	10,700	589	197	15

Lady DeVries 689 was credited with producing 18,848¼ pounds of milk in 365 days (Volume 3, *Dutch-Friesian Herd Book*).

Holstein-Friesian cows competed in public milking contests in 15 states and Ontario between 1883 and the 1890s. They did not enter the World's Columbian Exposition trials at Chicago in 1893, where the Babcock test was first used officially. Popularity of public tests waned with development of systems of production testing on farms.

Solomon Hoxie proposed in 1892 that the Advanced Registry be based on butterfat yields, determined by the Babcock test, which the Association approved in 1894. Inspection for type and body measurements under the Advanced Registry was discontinued as an economy measure at his suggestion. Prizes were awarded for Advanced Registry records in 1894. High private records had less credence.

Secretary F. L. Houghton prevailed upon Colleges of Agriculture to supervise the records and apply the Babcock test (Fig. 13.5) in "authenticated fat tests" of 7-days' duration beginning on May 30, 1894. Such tests were initiated in 1895; each milking was supervised and the Babcock test was applied to milk samples taken by representatives of these state institutions. The peak of 7-day records was reached in 1920–21 when 14,099 records averaged 437 pounds of milk, 3.57 percent and 15.4 pounds of butterfat. Colonel G. Watson French of Iowana Farms, Davenport, Iowa, lacked confidence in short-time tests, and advertised persistently that "It is the other 358 days that count."

The public slowly gave greater attention to yearly records when buying breeding animals, and short-time testing declined. The 7-day tests were voted out as of December 31, 1932, after 158,000 official 7-day butterfat tests had been conducted.

Yearly semiofficial testing began in 1908. Daily milk records were kept by the owners; 2-day butterfat tests were supervised officially during each month of the record, and qualifying records were published by the Association. A division for 305-day production records was instituted in 1916. In 1919 a cow was required to drop a living calf within 14 months of previous calving. A "dairyman's" 305-day division was added in 1920, based on two milkings daily after the first 45 days.

GOALS OF PRODUCTION TESTING

Yearly records of selected cows were intended to indicate maximum capacity for production. More was learned about feeding and management. Cows were given every opportunity to attain maximum production; they were fitted in the dry period and bred to calve in autumn so the highest milk yields occurred when cool weather favored high fat tests. Quality pasture in spring and summer tended to sustain yields late in lactation. Milking three and four times daily stimulated production. Feeders catered to each cow, and box stalls with water cups added comfort. High records had advertising value.

The average production of Holstein-Friesians under selective testing from 1914 to April 1928 was computed by M. S. Prescott (Table 13.1). He commented that although some breeders had given attention to higher butterfat tests, the influence had not spread and the average hovered around 3.4 percent butterfat.

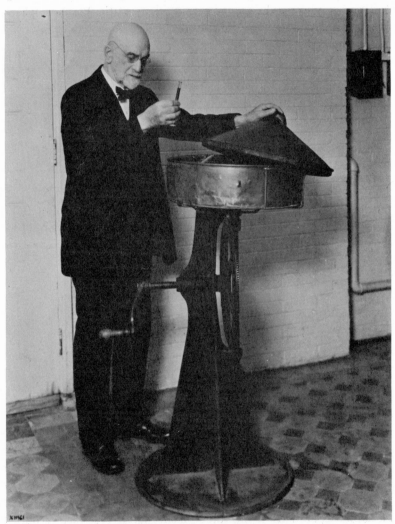

FIG. 13.5. Dr. S. M. Babcock and the original centrifuge used in the butterfat test which bears his name. The invention enabled breeders to determine quality of milk, in addition to quantity, when selecting breeding animals.

Selective testing located many transmitting animals but the records failed to measure the productive range of progeny or the persistent production of a cow through successive lactations. Average fat percentage in the milk of well-fitted cows may have been above that obtainable under good management of entire herds. Selective testing served the industry of breeding dairy cattle. Its place was to be supplemented by the Herd Improvement Registry.

TABLE 13.1
AVERAGE PRODUCTION OF HOLSTEIN–FRIESIAN COWS UNDER SELECTIVE TESTING, 1914–28

Years	Lactations	Milk (lbs.)	Test (%)	Butterfat (lbs.)
1910–14	1,367	14,422	3.43	495
1915–19	3,548	14,921	3.42	510
1920–24	14,384	16,026	3.38	541
1925–28	9,188	16,753	3.39	567
Average	28,481	16,046	3.39	544

The early trend had been to milk many cows three or four times daily on test, to obtain maximum capacity of cows in milk and butterfat. The trend has been toward two milkings daily, as shown below.

		Milkings daily while on test	
Year	Two	Three	Four
		(Percent)	
1937	18.3	61.2	20.5
1946	45.8	50.2	4.0
1955	63.5	36.5	(discontinued)

HERD IMPROVEMENT REGISTRY

The plan to test entire herds evolved from the Cow Testing Association idea in Denmark, via the Scottish Milk Records Association with Ayrshires and the Herd Improvement Registry of the Ayrshire Breeders' Association in the United States. When proposed for Holstein-Friesians, Malcolm H. Gardner believed it might lower safeguards and public confidence in the records. F. N. Strickland described operations of the Rhode Island Herd Test Plan requiring every cow in the herd to be tested.

The Herd Improvement Registry with Holstein-Friesians began

in January 1928 to improve management of entire herds. Also "testing all the daughters of a sire is a real check of his ability to transmit the factor for high production and this will be one of the outstanding features of the Herd Improvement Test." Testing increased in the HIR division. See Table 13.2.

Selective testing located high-producing individuals for advertising and publicity purposes. Testing every cow in the herd, on the other hand, measured transmitting ability of sires and dam and facilitated herd management.

DAIRY HERD IMPROVEMENT REGISTRY

In 1956 the Board of Directors established rules under which DHIA records might be recognized. The Purebred Dairy Cattle Association studied modifying supervision of DHIA records so that they would be acceptable for breed use. The rules included approval of supervisors by State Superintendents of Official Testing, identification of purebred cows from registration certificates, surprise check tests with a preliminary dry milking, method of paying supervisors, and IBM computation of records. The modification was tried in Pennsylvania in 1958–59, then applied nationwide. Only DHIR testing was conducted since 1967.

TABLE 13.2
A COMPARISON OF THE TRENDS IN AVERAGE PRODUCTION IN SELECTIVE
ADVANCED REGISTRY AND WITH ENTIRE MILKING HERDS

Year	Number of records	Average production		
		Milk (lbs.)	Test (%)	Butterfat (lbs.)
Advanced Registry				
1910–28	28,481	16,048	3.39	544
1928–29	1,927	17,050	3.54	579
1949	2,346	15,658	3.64	571
1959	1,355	15,733	3.83	603
1962 (2×)	192	13,753	3.90	536
1962 (3×)	250	18,719	3.79	711
Herd Improvement Register and Dairy Herd Improvement Registry				
1929	4,834	10,864	3.34	366
1939	10,315	11,354	3.46	393
1949	32,145	11,220	3.58	401
1959	78,635	13,621	3.62	502
1969[a]	130,585	15,435	3.68	568

a. Lactation averages on a 305-day 2× mature equivalent basis.

COMPOSITION OF HOLSTEIN MILK

Milk of Holstein cows in 28,481 yearly records between 1910 and 1928 averaged 3.39 percent fat. Although some breeders had begun to use herd bulls from higher testing families, little effect had been noted. Their example spread and bulls from such lines were used more widely. By 1960 the 90,056 10-month lactations under HIR test averaged 3.70 percent butterfat. Herd sires had not been selected for protein content or solids-not-fat in their dams' milk. One leading herd produced milk averaging over 4.0 percent fat, yet their milk contained over 8.5 percent of solids-not-fat only during the cooler months of the year.

R. D. Politiek in the Netherlands and the Milk Marketing Board showed that with Friesian cattle the percentage of butterfat and of solids-not-fat (SNF) were inherited independently even though the average trend was for them to vary together.

TOTAL NUTRIENT TESTING

The Association inaugurated a Total Nutrient Program in July 1962. Some 23,454 lactations with SNF data completed through 1964 averaged 12,603 pounds of milk, 3.67 and 8.54 percent, 462 pounds of butterfat and 1,077 of SNF on a 305-day $2\times$ mature equivalent basis. Bulls in the 1965 volume of the Green Book (names that begin with A, B, and C) were tallied for daughters with lactations up to 305 days. Some 4,791 records averaged 8.55 percent SNF, with a standard deviation (about 68 percent of them) between 8.20 and 8.90 percent, and an extreme range of 7.20 to 10.20 percent SNF. Some 6,432 lactation records of milk from Holstein cows summarized by the Inter-Regional Research Committee on Milk Composition averaged 3.67 ±0.38 percent fat, 8.45 ±.30 percent SNF, and 3.12 ±.25 percent of protein. The standard deviations included all records in the computations.

The Association sponsored research by the Michigan, North Carolina, and Wisconsin stations. Preliminary study of 2,500 lactations showed a range of 2.9 to 5.0 percent of butterfat. Lactations of 267 Holstein cows with an average of 3.7 percent of butterfat (current breed average) ranged between 7.8 and 9.3 percent of SNF, with

75.0 percent of them between 8.3 and 9.0 percent SNF. Seventy percent of them were at 8.5 percent or above. Butterfat tests were not a reliable index of low, medium, or high contents of SNF.

At the Guernsey annual meeting in May 1962 A. W. Hobler reported results of tests by Rutgers University with nine herds of four breeds. He stated, "Each cow is tested monthly for protein, SNF and fat. I find that we have many cows that may test 3 or 4 points more in fat that test less in protein than the cows with less fat. This is not a general rule, but there is a big variation between the herds. . . . It is so much more desirable to have the tests on protein than for SNF."

Solids-not-fat contents of milk fluctuate daily and seasonally, with stage of lactation, advancing gestation, age of cow, and particularly with individuals and between breeds. Although SNF tests vary less widely than fat tests do, they can form the basis for selective breeding as is done with fat tests. H. T. Thoele stated, "To breed selectively for changes in milk composition, we need individual cow records on milk yield, per cent fat, per cent solids-not-fat and possibly protein."

Many experiment stations have been investigating this characteristic cooperatively since 1959 (as noted above), improving methods with experience. In 1960 the Agricultural Extension Service in the state of Washington began to measure SNF in DHIA records. Two artificial breeding studs in California began in 1961 to evaluate daughters of their bulls for SNF.

President Leon A. Piquet addressed the annual convention in 1961, stating "The Special Committee has recommended that testing of individual cows in a herd monthly for solids-not-fat be made a part of our Association's testing program as fast as methods are developed and can be placed in effective use."

RECOGNITION OF TRANSMITTING DAMS

A cow became known as an Advanced Registry dam when three progeny, male or female, entered the Advanced Registry with 7-day or semiofficial records of females. Three or more A. R. daughters qualified a bull. *All* production records now are published in the Advanced Registry, Herd Improvement Register, and consolidated to the Dairy Herd Improvement Registry in 1967.

A Committee on Brood Cow Recognition studied records of 1,000 cows with 100,000 pounds of milk in lifetime production. Over 600 of them had one or more classified daughters. Less than 1 percent would classify as Gold Medal Dams under a plan then considered. Breeders were solicited for their "best brood cows." Sixteen among 50 would meet the requirements under consideration.

REQUIREMENTS FOR GOLD MEDAL DAMS

A dam may or may not have had a production record or type classification. If tested or classified, her requirements were the same as those of her progeny. Production records were computed to a 305-day $2\times$ mature equivalent basis. A cow must have produced 100,-000 pounds of milk, or all records average 13,000 pounds of milk and 525 pounds of butterfat. This was increased in 1965 to 14,400 pounds of milk and 555 pounds of fat. A daughter must have scored 80 points at 2 or 3 years of age, 81 points at 4 or 5 years, or 82 points at 6 years or older; and a qualifying son at least 82 points or be a Silver Medal Type Sire. A son may have rated Silver Medal Production Sire, or all tested daughters average at least 14,900 pounds of milk and 580 pounds of butterfat. At least three progeny must have qualified for type and three for production. The dam may be qualified by three progeny out of four, four out of five or six, five out of seven, or six out of eight or nine progeny. Production requirements for daughters increased steadily to 14,900 pounds of milk and 580 pounds of fat in 1968. Between 1957 and 1969 some 1,078 Gold Medal Dams had qualified.

UNDESIRABLE RECESSIVE FACTORS

An amendment to the Association's by-laws in 1957 stated "The Executive-Secretary shall receive and keep on file information concerning the inheritance of any Holstein-Friesian animal concluded by him on evidence to be a carrier of an undesirable recessive factor which may affect its use or value for breeding purposes."

The recognized "undesirable recessive factors" include red hair factor, bulldog (short legs, bulging forehead, and undershot jaw), prolonged gestation, hairlessness, imperfect skin, muscle contraction, mule foot, and dwarfism. The Board may add others. This recognizes and tends to reduce distribution of undesirable recessive characters. Other factors reported include ataxia, porphyria (pink tooth with loss of red pigment in urine), and external hydrocephalus. Of 437 animals reported through 1967, 324 carried red factor. Recessive red factor is being tested voluntarily under the Secretary's office. Bulls are mated with eight or more red cows, and their successive calves reported. Seven consecutive black-and-white calves gave a probability of 1 in 128 that the bull did not carry red factor. A certificate was issued to this effect. Five bulls were known carriers of the mule foot recessive gene.

An ineligibility color survey during 1962 and 1963 found that 0.12 percent of calves from 1,848 herds were red-and-white, and 4.2 percent of herds reported at least one red calf.

STATEMENT OF POLICY

A Statement of Policy adopted in 1961 recommended that a buyer be informed when purchasing an animal or semen of a carrier with an undesirable recessive character. The Executive-Secretary supervises the official color-carrier test given above, to deal with this character.

ARTIFICIAL BREEDING

Artificial breeding of dairy cattle in the United States apparently began in 1917 on a between-herd basis when King of the Ormsbys 178078 was owned jointly by Allamuchy, Tranquility, and Winterthur Farms in Delaware and New Jersey. Superintendent Arthur Danks carried fresh semen from a natural service between farms for selective matings. Carnation Farm in Washington practiced artificial insemination on a within-herd basis in 1920, multiplying services of Matador Segis Walker 148839 and his full brother Segis Walker Matador 166136 late in their useful lifetimes.

C. L. Cole collected semen by massaging the ampullae of an albino Holstein bull at the University of Minnesota in May 1936 and inseminated a heifer successfully. He inseminated cows with fresh semen in several herds near Grand Rapids, Minnesota. Some 105 among 121 cows were pregnant in the spring of 1938.

Enos J. Perry observed the first farmers' cooperative artificial breeding association for dairy cattle in 1937 on the island of Samsoe, Denmark. He organized a similar association in New Jersey in 1938 through the cooperation and interest of Guernsey and Holstein breeders. A similar organization was set up in Missouri the same year. The movement spread gradually.

H. W. Norton, Jr., commented on artificial breeding in the 1938 annual report of the Association:

> Artificial insemination offers tremendous possibilities for breed improvement if properly managed and safeguarded. Under this plan the services of our best bulls can be made available for use with choice females located in widely separated areas. . . . Cooperative associations make great savings possible to breeders . . . and allow them to breed their herds collectively to better sires than they might afford individually.
>
> At the same time this method offers very great danger if careless or unscrupulous methods are followed. Properly safeguarded it may prevent the spread of disease but carelessly handled it may spread disease in every herd in the group. While we should favor artificial insemination because of its possibilities for improvement of the breed, we should protect our breeding records and have assurance of their accuracy and correctness in every detail.

In 1941, 1,976 applications for registration in this breed were for calves resulting from artificial inseminations. The reduced demand for bulls because of artificial breeding and increasing numbers of cows per herd, is suggested from the change in proportion of males to total registrations. Reduction in numbers of small herds also requires fewer bulls (see Table 13.3). One calf was produced from semen frozen nine years previously. Some 1,429 bulls were in A.I. studs in 1969.

The Association has supported research on blood antigens at the University of Wisconsin since 1942. This method of verifying parentage went into use in 1943. Sires in artificial service now must be blood typed. This has been done under contract between the Purebred Dairy Cattle Association and the University of California since June 1955. A Canadian court accepted blood antigen tests in 1954 to verify parentage of Holstein-Friesian cattle.

Uniform rules for artificial breeding were drafted by a committee of the American Dairy Science Association and adopted through the Purebred Dairy Cattle Association by the respective breed associations. A revision in 1955 and 1960 covered storage and use of frozen semen. Some 69.8 percent of Holstein-Friesians registered during 1970 resulted from artificial inseminations.

The Research Committee sponsored 7 active projects in 1961: inheritance of solids-not-fat in milk; chemical study to determine

TABLE 13.3
PROPORTION OF MALE REGISTRATIONS OF HOLSTEIN–FRIESIANS

| Year | Registrations | | Ratio males to females |
	Males	Females	
1900–1909	42,066	88,973	1:2.1
1910–19	233,216	436,470	1:1.9
1920–29	299,551	812,276	1:2.7
1930–39	182,538	683,174	1:3.7
1940–49	301,569	1,069,174	1:3.5
1950–59	277,973	1,817,486	1:6.5
1960–69	199,520	2,473,336	1:12.4
1970	20,324	261,250	1:12.8

carriers of recessive red hair color; heritability of milk and butterfat production; genetic relationship between milk, fat yields, and increases in production with age; evaluation of genetic changes in some bull-producing herds; evaluation of sires with progeny based on herdmates in a single herd; and prolonged gestation.

SIRE EVALUATION

Transmitting ability of a sire was estimated by computing the average production of his registered daughters on a 305-day $2\times$ milking mature equivalent basis. This was compared with production of the tested dams or as a daughter-dam comparison (since the 1930s). Some bulls selected on such comparisons failed to maintain such production levels in subsequent artificial use. An adjusted daughter-herdmate comparison was developed, as discussed elsewhere. This method was projected further for a *Predicted Difference* from the average production of the breed. Concerning the latter method, E. L. Corley of the USDA Dairy Cattle Research Branch at Beltsville, Maryland, stated in 1967: "The USDA production summary represents a new and improved version of the herdmate comparison that has been used since 1960. It is the most reliable index of dairy sires' breeding value yet utilized."

The production records of both registered and grade daughters on DHIR and standard DHIA were combined. "By expressing breeding value on all bulls in terms of Predicted Differences, it is now feasible to evaluate and directly compare all bulls included in the summary as to their transmitting ability for production."

REGISTERED HOLSTEIN SIRE PERFORMANCE SUMMARIES

The Association began to publish a sire guide with the ratings effective in 1968. The latest USDA sire summary for an active bull with ten or more daughters cited the average production per daughter, and number of herds represented. Production was stated as average milk, test, and butterfat yield of daughters, adjusted herdmate average, repeatability, and predicted differences. The average type classification was given next in three parallel lines, representing all daughters, daughter-dam pairs, and the dams, respectively.

The descriptive type classification adopted in 1967 was given as stature, head, front end, back, rump, hind legs, feet, fore udder, rear udder, udder support and floor, udder quality, and teats. The breed age average (BAA percent) score also was presented. Such detailed comparisons were almost impractical prior to use of electronic computers.

PROGRESSIVE BREEDERS' REGISTRY

Balanced achievements in programs for breed improvement were recognized with the Progressive Breeders' Award plaque awarded for the first time in 1938. Bronze bars were added for qualifying in succeeding years.

In 1970 a breeder qualified for the Progressive Breeders' Registry award by meeting eight requirements. All eligible animals over 8 months old must be registered. At least 20 registered females must have freshened, and 75 percent must have been bred by the applicant. The herd must be on DHIR test, and average at least 700 pounds of milk and 25 pounds of butterfat in 305 days above the

lactation average of the breed for the past 3 years (16,000 po
of milk, 575 pounds of fat in 1970). The herd shall have been cl
fied in the last group or special classification, and the females in
the herd have a Breed Age Average percentage not less than 102.
Two-thirds of eligible animals shall have an average classification
score of at least 81 points. They shall be federally accredited as free
from tuberculosis or in a modified accredited area; under state or
federal supervision and free from brucellosis (Bang's disease). The
breeder must be a member of the Holstein-Friesian Association of
America and active in the state or local club. Satisfactory evidence
must be in the application for the Progressive Breeders' Registry.
There were 118 Progressive Breeders' Registry awards granted in
1969.

BREEDING HOLSTEIN-FRIESIAN CATTLE

The conformation of Holstein-Friesian cattle has improved gradually since the true type model bull and cow were developed and
since type classification was applied.

Although no Holstein cows entered the milking trials at the
World's Columbian Exposition in 1893 where the Babcock test for
butterfat was applied officially, the Association favored its use the
next year. A definite step toward a higher butterfat test was initiated when the owner of Dutchland Farm declared at the dispersal
sale that he wished to retain a small nucleus because of friendly
relation with the breeders. He withdrew the Changeling and Den-
ver families from the sale. Their milk exceeded the herd average in
butterfat percentage. The average butterfat content of Holstein
milk has been increasing gradually since then. A similar trend occurred in the Netherlands after it was found that butterfat was
produced more economically in higher testing milk. R. D. Politiek
showed that fat and protein contents of milk are inherited independently, even though they usually increase or decrease similarly.
A long breeding trial at the New Jersey station increased the average butterfat percentage to above 4.0 percent, even though the
solids-not-fat frequently averaged below 8.5 percent during the
warmer months.

Recessive red-and-white Friesians, always registered in the Netherlands herdbooks, were accepted for a separate herdbook by the delegates at the 1970 convention. The polled character occurs in some herds, and a slight tendency toward twinning has been reported. Flexed pasterns in newborn calves nearly always are outgrown before the yearling stage. Abnormal skin and four lethal types of "bulldog" seldom occur.

Syndactylism (mule foot) resulted from mating certain bulls with known female carriers. Crampy or progressive posterior paralysis is a recessive in grown cattle descended from known carriers. Lack of resistance against the bacterial and fungus types of lump jaw (actinobacillosis and actinomycosis) has allowed cattle to contract the infections as young as 9 months, or later. Wry-tail is a common hereditary skeletal defect of the sacral vertebrae. Albinism is caused by complete lack of the gene(s) for color pigmentation in the body. Frequency of these hereditary defects can be reduced from knowledge of the genetic makeup of both parents and their more distant relatives.

Less is understood of such complex hereditary characters as the percentages of butterfat, proteins in milk, and the persistency of lactation. The total nutrient testing program since 1962 is a beginning toward further improved nutritional values of milk.

JOINT OCCUPANCY

In 1947 Executive Secretary Karl B. Musser of the American Guernsey Cattle Club proposed that the five dairy breed organizations relocate at some midwest area, with one structure to house their offices, and do a great deal of routine clerical work in cooperation. The Board of Directors of the American Jersey Cattle Club joined in discussion at Peterborough, New Hampshire, in July 1960. The latter group also visited the Holstein-Friesian office. The latter advised that they would consider joint occupancy only at the home office in Brattleboro, Vermont. This proposal did not fulfill some joint occupancy goals, so the proposal still is in flux.

The Holstein-Friesian Association of America has headquarters at Brattleboro, Vermont 05302, with Robert H. Rumler as Executive Secretary.

BREED PROMOTION

Early promotion of Friesian cattle was by exhibition at shows, reports of private production records, and public milking contests between breeds. Promotion became more effective after the Holstein and Dutch-Friesian herdbook associations united in 1885. A meager budget was appropriated to a Literary Committee, with Solomon Hoxie as chairman. Advertising space has been bought in farm magazines since 1905.

BREED MAGAZINE

The *Holstein-Friesian Register* was founded by E. P. Beauchamp at Terre Haute, Indiana, on March 15, 1886. F. L. Houghton purchased it 2 years later and continued publication until he died in 1928. The *Holstein-Friesian World* appeared at Lacona, New York, under ownership of C. C. Brown and E. M. Hastings in 1904. The *Black and White Record* began in 1916 and consolidated with the *World* in 1918. The *Holstein-Friesian Register* consolidated with the *World* in 1928. Maurice S. Prescott is editor. The Association supplemented space in these magazines with news releases, articles, and advertising in the farm press.

EXTENSION SERVICE

The Association established the Extension Service in 1917. Activities included advertising, news releases, assistance with junior club activities, and booths at fairs. Cooperation is given to 48 state Holstein Clubs and over 450 county and district clubs. Brochures and pamphlets have been published concerning Holstein-Friesian cattle.

Activities of breed association fieldmen are coordinated with state and regional activities at annual regional extension conferences. An activity index compares regional achievements with national averages on registrations, transfers of ownership, active membership, new members, new buyers, herds on production test, herds classified, state association members, and new junior members. There are 15 national and state Holstein fieldmen in addition to the type classifiers.

REFERENCES

Armstrong, T. V. 1959. Variations in the gross composition of milk as related to the breed of the cow. A review and critical evaluation of literature in the United States and Canada. *J. Dairy Sci.* 42:1–19.

Barrett, George R. 1958. A new Association program. Automatic recognition of Silver Medal Sires. *Holstein-Friesian World* 55:2980, 2984.

———. 1963. An Association research report—Is classification of bulls worthwhile? *Holstein-Friesian World* 60:819–20.

Briquet, J. Jr., and J. L. Lush. 1947. Heritability of amount of spotting in Holstein-Friesian cattle. *J. Hered.* 38:99–105.

Chenery, Winthrop. 1864. Holstein cattle. *Rept. Comm. Agr., USDA,* pp. 161–67.

Cole, C. L. 1938. Artificial insemination of dairy cattle. *J. Dairy Sci.* 21:131.

Dunn, L. C. 1923. The inheritance of spotting in Holstein cattle. *Holstein-Friesian World* 20:1681–82, 1702.

Eldridge, F. E., W. H. Smith, and W. M. McLeod. 1951. Syndactylism in Holstein-Friesian cattle. Its inheritance, description and occurrence. *J. Hered.* 42:241–50.

Gardner, Malcolm H. 1925. The herd test for Holstein-Friesians. *Holstein-Friesian World* 22:1721, 1734.

Hengeveld, G. J. 1880. Origin of Dutch cattle. *Dutch-Friesian Herd Book* 1:10–18.

———. 1882. A review of the origin and history of the breed. *Dutch-Friesian Herd Book* 2:7–17.

Houghton, F. L. 1897. *Holstein-Friesian cattle.* Brattleboro, Vt.

Hoxie, Jane L. 1923. *Solomon Hoxie.* Little & Ives, New York.

Moscrip, W. S. 1923. Report of W. S. Moscrip, chairman of the committee having to do with revision of Scale of Points, preparation of pictures and models of perfect animals, school for judges, etc. *Holstein-Friesian Herd Book* 51:302–3.

Norton, H. W., Jr. 1954. Milestones along the way. *Holstein-Friesian World* 51:83–87.

Prescott, M. S., et al. 1930. *Holstein-Friesian History.* Lacona, N.Y.

———. 1960. *Holstein-Friesian History.* Diamond Jubilee Ed. Lacona, N.Y.

Reaman, G. E. 1946. *History of the Holstein-Friesian breed in Canada.* Collins, Toronto.

Stoehr, H. R. 1962. C. L. Cole began in 1936. *Hoard's Dairyman* 107:419.

Strickland, F. M. 1926. The Rhode Island herd test plan. *Holstein-Friesian World* 23:151–52.

Thoele, H. W. 1961. Further development in solids-not-fat research. *Holstein-Friesian World* 58:1721–23.

Warner, Fred B. 1966. Service to breeders improved with description type classification. *Holstein-Friesian World* 63:2754–57.

Herdbooks

Dutch-Friesian Herd Book. 1880–85. American Association of Breeders of Pure Bred Friesian or Dutch-Friesian Cattle. Vols. 1–4.

Holstein Herd Book. 1872–85. Holstein Breeders' Association of America. Vols. 1–9.

Holstein-Friesian Herd Book. 1886–. Holstein-Friesian Association of America. Vols. 1–.

Western Holstein-Friesian Association Herd Book. 1895. Vol. 1.

Holstein-Friesian World
   1954. Significant dates in Holstein history. 51:212, 214, 216, 218.
   1958. Rules for red factor test. 59:956.
   1967. USDA-Holstein agreement provided broader base of sire information for production. *Holstein-Friesian World* 64:2042, 2052.
   1970. Holstein reclassification. A major advance in Holstein's classification program. 67:2052–55.
Holstein-Friesian Association of America publications
   Annual reports.
   Color markings of Holstein-Friesian cattle.
   The Holstein Story.
   1941. Inspection and classification of herds and the recognition of sires. Rules and regulations.
   1945. *Holstein-Friesian Type and Production Year Book.* Vols. 1–.
   1953. The judging manual.
   1954. The Holstein handbook.
   1954. Type classification and recognition of sires.
   1956. Progressive Breeders' Registry. Rules effective Jan. 1, 1956. Reinstated for 1970.
   1958. Type classification—Rules and regulations.
   1966. Descriptive type classification.
Jersey J.
   1967. PDCA showing code of ethics. 14:109.

CHAPTER 14

CATTLE ON THE ISLAND OF JERSEY

T HE NATIVE home of the Jersey breed is the Island of Jersey, the
most southern of the Channel Islands, in St. Michael's Bay on the
coast of Normandy and Brittany, 30 miles north of the old feudal
town of St. Malo. The rainfall averages 34 inches yearly on the Island's 45 square miles; frosts are few, and snow seldom falls on the
Island. It is in the same latitude as Labrador, but prevailing ocean
currents and winds contribute to its mild climate. The mean annual
temperature is 51° F.

The native islanders were fishermen in part during Roman times
when the Emperor Antonius (A.D. 138–161) knew the island as
Caesarea. St. Magliore, a Breton, introduced Christianity about
A.D. 565, establishing a Christian school (later an abbey) near the
present Elizabeth Castle.

The Channel Islands were ruled by Rollo, the Norseman, when

Headpiece: Vignette of Jersey cow.

he became the first Duke of Normandy. The islands still are subservient to the Duke or Duchess of Normandy, now rulers of the British empire. Jersey has enjoyed special privileges as an independent state under the British Crown since 1279, enacting its own laws, levying taxes and duties, and coining at least a part of its money. Because the 12 parishes touch the English Channel, parishioners have had the right to harvest seaweed or "vraic" for fuel and fertilizer since ancient times.

EARLY CATTLE

Part of a skull of *B. longifrons* was found at St. Ouen's Bay on the Island of Jersey. Other ox bones occurred in Neolithic kitchen middens on nearby Guernsey. The early cattle doubtless were brought by the Neolithic settlers, and in later movements from the French provinces of Brittany and Normandy.

Under the Norman abbots, the feudal lands of Jersey were divided into 52 cueilettes and vingtaines, with several good manors. The abbey of St. Heliers was a priory amalgamated with Cherbourg on Normandy. There was a considerable ducal domain on the island about 1100, accounting for early trade with that province.

Early mention of cattle on Jersey was incidental to other events. During a political factional disturbance in 1643, soldiers representing the belligerent civil authorities (parliamentarians) sallied forth from the fortified Castle Elizabeth and made a foray upon cattle grazing in neighboring enclosures. The cattle and horses of a royalist were confiscated when he fled to St. Malo during this period (1643–49). After the royalists got the upper hand, troops had difficulty in buying straw and forage for their horses. The natives claimed they had not more than enough for their own cattle.

EARLY AGRICULTURE

From an unpublished manuscript *Les Chroniques de Jersey* by Phillipot Payn (1585), Reverend Phillip Falle drew parts of his early *Caesarea, or, An Account of Jersey*, published in 1694 and revised in 1734. In the 1734 edition he stated: "About 150 years ago the Island lay pretty much open under the feudal allotments and use of lands, but when the humour of planting seized our people, they fell to enclosing, for shelter and security to their fruit." He mentioned the great increase of enclosures, fences, hedgerows, and highways as being out of proportion to size of the island. He also described the gathering of seaweed or vraic twice yearly for fuel and fertilizer.

John Shebbeare (1771) mentioned the Jersey cows as one of the chief products of the island. The Reverend Mr. Valpy (1785) listed knitted woolen goods and cows as the main exports from Jersey. Culley (1786) stated that Alderney cows were being used about British estates. Thomas Leyte, a military surveyor (1808), wrote: "The cattle of this island are generally small; very few are bred for the market, except calves. Bullocks are universally used in the cart, and killed when past their labour. A sufficient number of cows are kept for use of the inhabitants in milk and butter; the remainder are exported to England, where they sell at a high price."

Thomas Quayle's *General View of the Agriculture and Present State of the Islands on the Coast of Normandy, subject to the Crown of Great Britain* (1815) described the Jersey cow as the possession highest in the Jerseyman's estimation. He told of tethering cows on pasture, and moving them five or six times daily. Colors of the cattle ranged between cream (often mixed with white) and black, or black-and-white, but mainly from fawn or red mixed with white. Cows yielded about 10 quarts of milk daily, with an extreme of 22 English quarts, some churning 12 to 14 pounds of butter per week. Thirty pounds of parsnips with meadow hay were fed daily in winter. Lucerne or alfalfa was introduced, and its regular culture occurred in 15 years.

Jersey cows were called Alderneys in England, according to W. Pless (1824). George S. Syvret (1832) wrote that Jersey cows generally were small but gave very fine butter, and were esteemed in England. Henry D. Inglis visited Jersey in 1833 and 1834, and wrote "Milk is scarcely at all used in a Jersey menage; it is all wanted for butter, either for the market, or for indoor consumption." Potatoes were increasing in acreage, and lucerne yielded four cuttings during the year. Inglis recognized that ". . . from the system pursued by the agricultural society of Guernsey, the breed of that island now differs in many essential particulars from the breed on Alderney and Jersey. . . . All over England, the Alderney cow—as it is generally called—is celebrated not only for its beauty, but for the richness of its milk, and the excellence of the butter made from it."

Colonel J. Le Couteur (1845) stated that in winter a cow on Jersey was fed some straw, " . . . from ten to twenty pounds of hay, with about ten to twenty pounds of parsnips, white carrots, turnips, or mangel-wurzel."

Cows calved mainly in early spring in time to go on pasture (usually late April). They were tethered on pasture with a 6- or 8-foot chain attached to a halter, and connected with a swivel and ring to a stout iron stake driven into the ground. The cows were moved three to six times daily after grazing each small semicircle clean (Fig. 14.1). Further:

> In form the Jersey is deer-like, and small in size; the colors mostly prized are the light red and white; the brown and the fawn; brindled specimens are rarely seen. . . . A cow is at her prime at six years of age, and continues good until ten years old. Many are kept that are much older but then they begin to fall off. . . . A good cow on the average gives fourteen quarts of milk per day, or eight to nine pounds of butter per week. . . .
>
> A great improvement has taken place in the breed of cattle within the last twelve or fifteen years, which is attributable to the formation of agricultural societies in the island. . . . Bulls are seldom kept after their second year, for they become extremely wild and troublesome."

A letter from Le Couteur to the secretary of the American Jersey Cattle Club in 1869 mentioned that "Guernsey cattle are not deemed foreign [on Jersey] but there are scarcely over a dozen of that breed in our island. They are of larger bone, and carcass, considered to be coarse, though famous milkers, requiring much more feed than the Jerseys. Our judges at the cattle-shows have discarded both them and their progeny."

H. G. Shepard wrote recently:

> The practice of tethering cattle at pasture on Jersey has continued up to the present day, though for economic reasons some use is now being made of electric fencing. The fact that the

pastures are part of the farm, or never very far from it, makes easier the task of moving the animals several times a day. Owing to the mildness of the climate, it is possible for cattle to go out to pasture nearly every day of the year (unless it is very cold or wet), even for a few hours in the winter. The "tethering" system, of course, allows the fullest conservation of pasture, and thus the Island pastures can carry a higher proportion of livestock than would otherwise be the case."

During the present century cereal acreage has been reduced in Jersey, and more early potatoes are grown for export to England. Tomato growing in the open field became second in importance about 1900. As the land slopes gradually to the southward, tomatoes can follow potatoes on suitable soils. Tomatoes were grown extensively until overproduction restricted the crop to best-suited land. There were many orchards a century ago, when apples and cider were exported in quantities. In 1961 only 141 acres were in orchards.

The Colorado potato beetle gained a foothold near the end of the

FIG. 14.1. Cattle on the Island of Jersey have been tethered ever since the enclosure of land.

German occupation in World War II. The potato export trade with England was restricted immediately after the liberation on this account. Potato beetles are on the neighboring Cotentin peninsula and a few beetles find their way to island beaches nearly every year if the wind blows from the continent. Strict control and crop spraying have kept the potato crop from becoming infested. Broccoli, cauliflower, and flower production for export tended to reduce acreage of early potatoes during recent years.

The cattle population usually numbers between 6,000 and 7,000 cows and bred heifers, nearly 3,000 younger females, around 150 bulls over 12 months old and fewer under that age, part of which are intended for export. Cattle numbered 9,178 animals in 1962, all but 166 of them being *Jersey Herd Book* stock. Five females entered the *Herd Book* as Foundation Stock in 1962. Limited numbers of calves were raised during the German occupation, while acreage of wheat was increased temporarily. There were 267 electric fences and 414 milking machines on the Island in 1962.

Some 21.2 percent of the Island lands were grazed in 1962 and 8.1 percent were harvested as hay for livestock—mainly Jersey cattle. The cultivated area is reported in vergees; an acre equals 2¼ vergees.

ACTS OF THE STATES

On July 16, 1763, the States (Island legislature) at the request of the Deputy Attorney General passed an Act forbidding import of cattle, sheep, hogs, fowl, eggs, meat of any kind, butter, and fat from France, under penalty of confiscation of the vessel and cargo.

A stringent law was passed on August 8, 1789, forbidding importation from France into Jersey of any cow, heifer, calf, or bull. Bullocks could enter certain ports but had to be reported to the parish constable. Masters of vessels had to give an affidavit of the origin of any cattle in shipment. Prohibited animals were to be slaughtered. Masters of vessels had to account for the source of animals shipped out, before clearance of the vessel, with a signed certificate from the seller of each animal as being the breed of the Island, together with the port of intended delivery.

When meat for the military garrison could not be obtained wholly from England, an Act was passed on March 18, 1826, which permitted cattle to be brought from the other Channel Islands and bullocks to be brought from France only into the harbors of St. Helier, St. Aubin, or Mount Orgueil. Notice had to be given before landing them, and the harbor master had to keep a register of cattle that embarked.

On September 8, 1864, an Act permitted importation of any cattle from France for slaughter or re-exportation, but all animals disembarked had to be reported within 24 hours to the Agent, taken to a place provided by the Harbor Committee, branded with the letter F, 3 inches square, and held there until slaughter or re-embarkation under permit.

A law of September 19, 1878, guarded against importation of animals from areas known to have an infection of certain contagious diseases. Every precaution was taken to keep the Island breed pure and free from diseases.

COLORS OF JERSEYS

Thirteen Jerseys in the first show sponsored by the Jersey Agricultural and Horticultural Society in 1834 were described. Seven were red and white and one was brown and white; colors of five animals were not stated. Dr. L. H. Twaddell mentioned in 1865:

> The Jerseys are of all shades of color, from a pale yellow fawn, running through all the intermediate hues, even occasionally to a red, an intermixture of black or grey, known as French grey, that merged into black with an amber colored band along the back, the muzzle invariably shaded with a lighter color; and individuals are often seen black and white or pure black, unrelieved by any other color. A yellow brindle is sometimes seen, but this is by no means a favorite.
>
> The darker colors are the most popular in England, from the belief that they are hardier in constitution and bear the climate better, but this opinion does not accord with our experience in America.

George E. Waring, Jr., visited the Island in 1873 and noted that most Jerseys had some white markings and switches, white legs and feet.

A tabulation of the color markings of Jerseys registered in Volume 1 (1873) of the *Jersey Herd Book* showed:

Foundation Stock	Solid color	Broken color
Bulls	86	64
Females	243	481
Pedigree Stock		
Bulls	17	30
Females	4	18

A fad for solid-colored Jerseys which began in England harmed the breed by overlooking meritorious animals with white markings. The fad became a craze with some, the accusation being made that some American importers gave more attention to color than to dairy qualities. Even the secretary of the American Jersey Cattle Club wrote in protest. Most Island breeders were not led astray by the fad. There was a craze for "fancy colored" Jerseys—silver gray, mouse gray, lemon fawn, etc.—during the 1880s. Had this fad persisted, the breed would have been harmed. Again, Island breeders were not led astray. The 1904 report of the Island society stated, "The colours most prevalent are fawn, light to dark brown, rusty red, some few grays, some blacks. The secondary colour is white. No Scale ever carried points for colour." Jersey bulls registered in 1937 on the Island ranged in color from fawn to dark brown. Forty-five of them were solid-colored and 40 broken-colored animals. This showed the slow gradual spread of the dominant "self" or solid color in the breed.

AGRICULTURAL SOCIETIES

Garrard, an English writer about 1800 to 1803, gave a brief description of "Alderney" cattle which he obtained from the Secretary of the Agricultural Society of Jersey existing in 1799. No evidence was found as to when it was organized. Quayle mentioned in 1815 that "several years ago, the gentry of this Island set on foot an institution of this nature [an Agricultural Society], which has corresponded with the Board of Agriculture. It is said to be still in existence; but certainly not with many signs of animation." The secretary, Philip Durell, gave the seal of the earlier Society to the new Society organized in 1833.

The Governor of the Island published notice on August 24, 1833, of a meeting to form a "Society of Agriculture and of Horticulture in Jersey . . . for the improvement of the farms and gardens, also that of the race of cattle." A committee represented each parish, with Colonel J. Le Couteur as secretary, to draft rules and bylaws. The Society formed on September 28, 1833. The States granted an annual subsidy of £100 until 1841. The name adopted was "The Jersey Agricultural and Horticultural Society." King William IV graciously consented to become Patron in January 1834, whereupon the prefix "Royal" was added. This Society and its committees function today, supervising shows and other activities under patronage of the successive English Sovereigns. There were 672 senior and 46 junior members in 1967.

CATTLE SHOWS

Michael Fowler, an English cattle buyer, was one who encouraged forming the Society as a means of improving the Island cattle. Five cattle dealers, in the presence of the president and Society committee, drew up two scales of points for judging cattle at the shows, the method being: "Two of the best cows on the Island were selected as models; one was allowed to be perfect in the forequarters and barrel, the other in her hind-quarters. With the help of the best breeders and judges, a scale, for governing the judges at the cattle shows, was drawn up." Colonel Le Couteur described the method in 1845. The scales were divided into seven articles, allowing 25 points for perfection in bulls and 27 points for cows and heifers. The scales were used in judging the first show at Easter, 1834.

The judges wrote that the cattle were very much out of condition, too slightly formed behind, gait unsightly, udders ill-formed, tail coarse and thick, hoofs large, and heads coarse and ill-shaped. Many lacked the golden tinge within the ears which denoted a property to produce yellow and rich butter. Some females had short bull-necks, too much flesh or dewlap under the throat, or were heavy in the shoulders. It was indicated that these faults could be corrected in the breed.

Three pence were added to the prize money of a bull for each subscriber's cow with calf by such a bull. Prize moneys earned were withheld until a bull had been in service for a season. Prize-winning cows or heifers had to drop one calf on the Island. Entries increased to 71 animals at the show in 1836.

In 1838, when Queen Victoria ascended the throne, she continued as patroness, and 153 animals were exhibited. Progeny of previous prize winners were recognized in the 1838 show. Premiums were forfeited if an owner withheld service of a prize-winning bull from the public.

Colonel Le Couteur reported in 1841: "Among the members of your Society, seven years attention to breeding have almost caused the ancient characteristic defect of dropping hind quarter of the Jersey cattle to disappear, besides several minor defects, and it only remains to give squareness to the hind quarter and roundness to the barrel to render it a most beautiful animal."

The Scale of Points of 1844 had 26 articles for bulls and 30 for females; it was drafted again by the cattle dealers.

A group of breeders and officers revised and enlarged the Scale of Points in 1849. Colonel Le Couteur wrote in 1846 that " . . . the fixing of points and pedigree to cattle have established the fact that a cow may be equally good as beautiful; and on many farms, including that of the writer, two cows may be found with prize points, each producing 14 pounds of rich butter (per week) in May and June." Le Couteur prepared two drawings in 1849 to illustrate the points of perfection of a Jersey bull and a Jersey cow according to the revised Scale of Points. These drawings were approved by the Society on January 5, 1850. Copies of these original drawings are shown in Figure 14.2.

The first direct notice of milk production occurred in 1838, when a second sweepstakes prize was offered "for the Cows, the property of one person, that shall produce the finest coloured and richest cream, at the Spring Show, the trial to be made in lactometer."

In 1860, President Edward Mourant donated a prize to be awarded the owner of the best cow in milk 3 to 5 years old. The judges were to disregard the Scale of Points but to pass on the superiority of each cow generally, and also "the milk of the best cows, when selected from the general lot, shall be tested by means of lactometers, and its quality, together with the form of the udder, shall be considered points of great weight and importance."

FIG. 14.2. Colonel Le Couteur's drawings of the ideal Jersey bull and cow, approved by the general meeting of the Royal Jersey Agricultural and Horticultural Society in 1850.

Some 179 animals out of 205 exhibited in 1862 were decorated. The annual report commented: "Beauty of symmetry alone cannot ever be the acme of perfection; the latter can only be attained when goodness and beauty are equally combined."

Colonel Le Couteur donated a prize for the Island cow in May, 1869—"Cow giving the richest milk." A class was arranged at the show on May 22, 1879 for "cows giving the richest milk on trial, the same animal having calved since January 1st, 1879 and producing 2 pots of milk at one milking." One pot was equal to 2 imperial quarts (2.5 pounds per quart).

In 1885 Dr. John S. Linsley criticized Island showyard practices for failure to take producing ability of Jerseys into account more fully. When a production class was added to the show of May 27, 1885, the entry rules specified: "To the cow giving the richest milk on trial, the said animal having calved since January 1, 1885, and producing two pots of milk at one milking."

An entry rule set up in 1894 for the *Herd Book* specified that "any cow or heifer in milk, not registered in the *Herd Book*, obtaining either first, second, third, or fourth prize, or 'Certificate of Merit' at any of the Departmental Shows or Butter Tests, shall be eligible for qualification in the *Herd Book* as Foundation Stock [subject to payment of fee]." This was an outcome of the 24-hour Butter Tests started in 1893 when the English Jersey Cattle Club offered three medals in the milking contests at the Society's three shows. Shows for bulls were held separately from 1837 to 1919, when they were combined with the Spring Show for cows and heifers.

Jersey cows on the Island were relatively small animals; mature cows weighed 640 to 800 pounds. Imported Lady Viola 238437 typified the popular "Island type" cow with extreme refinement of the period. See Figure 15.1 for her picture. John Perree, a leading breeder, exporter, and long an officer of the Society, stated in 1938 that foreign demand led to breeding larger animals than formerly (Fig. 14.3).

An Autumn Show was instituted in 1928. Classes for Junior and Senior Yearling bulls and heifers divided after July 1 of the previous year. Provision was made in 1950 for 1,000-gallon cows (10,000 pounds of milk in 361 days, Class AA or AAA) with at least 5 percent butterfat. Get of Sire classes consisted of five female progeny, the sire of which must be *alive on the Island* but not necessarily shown. Junior Progeny consisted of five uncalved females.

There were usually two judges for cow classes, two for bulls, two for uncalved females, and separate judges for the Championships and Challenge Cups. A referee judge with long experience also was elected.

The Spring Show classification for 1966 was as follows:

Bulls: Junior Yearling bulls, born on or after July 1, 1965; Senior Yearling bulls, born January 1 to July 1, 1965; bulls born May 30, 1963 and before January 1965; aged bulls, born before May 20, 1963; Senior Progeny bulls, based on merits of five registered female progeny; and Junior Progeny bulls, based on five uncalved female progeny.

Females: Junior Yearling heifers, born on or after July 1, 1965; Senior Yearling heifers, born January 1 and before July 1, 1965; heifers in calf, under 3 years old; heifers in milk, under 3 years old; young cows, 3 to 6 years old; aged cows; 1,000-gallon cows (produced at least 10,000 pounds of milk and 500 pounds of fat and qualifying for Class AA or Class AAA); and Cow and two progeny, one to be a female.

The Scale of Points was revised in 1956, omitting weights for bulls and females. Mammary development was described:

Article	Points
10. Udder well developed, not fleshy or divided, well balanced; teats of uniform length and size, squarely placed	15
11. Fore-udder running well forward, and well attached	10
12. Rear-udder wide and well attached, not rounding abruptly at the top	10
13. Milk veins large, prominent, and tortuous in their course	3

Secretary H. G. Shepard observed in 1939: "These Shows provide the means of estimating progress in breeding."

Junior members held their first Junior Cattle Show in 1962. Junior entries also showed in the open classes. Junior members were 8 to 20 years of age, and owned up to six herdbook animals. Two members were added to the Central Committee in charge of shows. Type classification of entire herds began in October 1967. Twenty-two animals owned by G. A. Richardson had an average score of 86.2 points.

PARISH SHOWS

The first parish Agricultural Society on the Island was organized in St. Peter's parish in 1846. All parishes organized locally, the latest being St. Helier's Agricultural Society in 1902. Jerseys winning at parish shows usually competed in the Island shows. Both Island and Parish shows were held regularly prior to German occupation on July 2, 1940, in World War II. Following the German surrender to the Royal Navy, the Three Parish Show was held on September 26, 1945, after a five-year interruption.

JERSEY HERD BOOK

Though "pedigree points" were listed in the revised Scale of Points in 1849, the Society's Agricultural Department established the herdbook 17 years later (1866). Colonel C. P. LeCornu mentioned in 1867 that some people opposed a herdbook, since all cattle on the Island were "without any cross with foreign stock." He mentioned

FIG. 14.3. The bull Observer was being transported to the show. He was a larger bull than many used earlier.

also that wide variations occurred in dairy conformation and milking ability, and that by study of pedigrees some poor strains could be avoided. He addressed several parish societies and convinced many farmers of the need of a herdbook system.

Animals were first examined for entry in the *Herd Book* as Foundation Stock on April 4, 1866. Owners of approved bulls kept a complete record of qualified cows served. Calves of approved parents could be entered as Pedigree Stock, subject to inspection and qualification at the proper age. Bull calves were not allowed to serve until examined and approved at 1 year of age. Heifers were examined after the first calving. Volume 1 of the *Jersey Herd Book* was published in 1873. Show winnings of animals appeared first in Volume 15. Bulls could be approved in 1960 when 10 months old.

Secretary H. G Shepard described (in a letter) the current routine for registration of young stock on the Island:

The system of preliminary registration and subsequent qualification which was evolved when the *Jersey Herd Book* was founded in 1866 is still, in essentials, carried on today. Such a system can operate only in a compact community such as is Jersey, where the distances from farm to farm or from farm to St. Helier are at the most nine miles. It works in this way. Calves by a qualified bull and out of a qualified cow or registered heifer-in-milk must be registered within 8 days of birth if they are to be retained. Very few bull calves are registered. When a heifer has dropped her first calf, she is inspected by a panel of judges, and "qualified" as Commended or Highly Commended. Her calf is then eligible for registration. The produce of animals rejected (nowadays less than 1 per cent) is ineligible for registration, but a rejected animal may be resubmitted after next calving and, if then qualified, calves born subsequently are eligible. The small number rejected is due to present day breeders' reluctance to bring out an animal which they know would not be accepted. They are usually slaughtered before calving again. Bulls must be a year old before being "qualified," and are not allowed to serve until then. The dam of the bull must also be brought up for inspection.

All calves intended for registration are earmarked within two or three weeks of birth by an official of the *Jersey Herd Book*, and not left to the breeder to do at his convenience.

The Earmarking Scheme for calves was adopted January 12, 1946.

Five *Herd Book* judges were elected for a 5-year term by the Jersey Herd Book Committee. One man retired each year, and retirees were ineligible for re-election for 2 years. Two or more judges passed upon heifers after first calving, and bulls at 1 year old, for entry in the *Herd Book*. These animals had undergone preliminary registration at birth, the record involving parentage and purity of pedigree. Unworthy animals were rejected. The initials H.C. or C. added after the name and registration number indicate the respective rating earned at inspection.

A rule adopted in 1872 by the American Jersey Cattle Club refused entry to an imported animal in their Herd Register unless that individual or its sire and dam had been entered in the *Jersey Herd Book*. This rule stimulated much interest in registration of Island cattle.

Since 1891 the dam has had to be inspected along with her bull calf in order for the latter to be approved. Exceptions were made for dams that had been exported or died but which had satisfactory show winnings. Butter tests of dams also have been considered for the same purpose since 1893. The judges now take into account the yearly records of the dam. Nevertheless, a bull is approved on his own appearance and that of his dam. There have been occasions, according to H. G. Shepard, when a bull was only "Commended" because his dam was not quite good enough in type even though she had high production records.

Ph. L. S. Mourant stated that although the Jersey breed was pure before 1866, " . . . it is universally admitted that the *Herd Book* has had for effect the weeding out of undesirable stock. Slowly but progressively the type has improved."

Volume 40 of the *Herd Book* was published during the first year of the German occupation, after which printing was limited until after the liberation. Examinations for entry in the *Herd Book* were conducted every 6 weeks during all of World War II.

PRODUCTION RECORDS ON JERSEY

In early days, few private records of milk or butter production were published for cows on the Island of Jersey, but the statements were made frequently that a good cow should produce a pound of butter per day, and perhaps up to 14 pounds in a week when fresh and on good pasture. Butter tests were started by the Society in 1886. A bronze and a silver medal were donated by P. C. Hanbury for winners of a 2-day milking show supervised by the Society on May 15–16, 1889. Mabel 6th No. 112 and Mabel 13th No. 114 won the awards with yields of 36 pounds 6 ounces and 38 pounds of milk, respectively, after allowances had been made in the scores for the number of days since calving, for butter analysis, and for solids-not-fat. The trials were repeated in 1890 and again in 1892, when 17 cows competed. The English Jersey Cattle Society offered Gold, Silver, and Bronze medals to winners of 1-day butter tests on the Island, beginning in 1893.

Ernest Mathews, a judge of the butter tests from England, concluded from a study of the butter contest records over 18 years that "richness of milk appears to run in certain families, and to be transmissible through both sire and dam equally as quantity of milk." A number of butter test winners were taken to the United States by T. S. Cooper & Sons prior to 1906. Some 904 butter test records of 24 hours' duration were conducted between 1889 and 1914, when the show yard was put to military use during World War I. These tests called attention to several top-producing families on the Island —Fancys, Gamboges, Oxfords, and others.

Yearly butterfat tests were initiated in December 1912: "The aim of the Committee in introducing these records is to enable breeders to breed from the best milking cows, and to use bulls from the best milking strains only, and thereby to increase the milk and butter producing qualities of the Jersey cow." The Society's committee sent inspectors every fortnight to weigh and sample milk of each cow on the farm, samples being tested by the Gerber method, and total production reported at the end of the year. One of the first cows to produce above the specified requirements was LaFosse Lady 14009 whose published record was: 1913—103 days . . . 1,656.5 lbs. milk, 5.52%, 103.83 lbs. fat; 1914—214 days . . . 3,145.0 lbs. milk, 5.62%, 199.75 lbs. fat. This was a creditable record for a cow mainly tethered on pasture, with a little hay and a few roots, and which was milked twice daily during lactation. Few cows receive much concentrates or are milked thrice daily. In 1924, 140 herds were testing nearly 1,000 cows. Later rules allowed as few as 4 cows in a herd to be placed on test at a time. Milk recording ceased during the German occupation in World War II, and was resumed on November 1, 1945, after five years' interruption.

Interest developed after World War II in knowing the production of cows in successive lactations, as well as of other cows in the herd. Production was recorded for 361 days, or the owner might choose a 305-day record by notifying the recorder on his last visit prior to that time.

The provision in 1954 that an owner must have three-fourths of his milking cows recorded for production was revised in November 1955 to include the entire herd. Cows that were ten years old or older, or which had been recorded for six consecutive lactations, were permitted to be recorded at the option of the owner.

A registered bull was awarded a "Star" for each 100 pounds of butterfat that his dam produced above the minimum requirements for her age. Medal of Merit, Gold, and Silver Medal certificates were awarded to cows that met the same production and calving requirements adopted previously by the American Jersey Cattle Club for similar awards. A bull received the medal award based on three daughters winning the medal, out of different dams.

Six "Goddington" prizes were awarded to the leading Certificate of Merit producers each year to encourage lactation records. The highest production before World War II was that of Juillep's Nice Girl 42002, with 17,895.5 pounds of milk, 5.49 percent and 983.47 pounds of butterfat in 1935.

Jerseys have been famous for rich milk. The average butterfat tests of Certificate of Merit lactations on the Island ranged between 3.72 and 7.53 percent fat in 1961. The average production of 1,806 cows during 1963 was 9,089 pounds of milk, 5.34 percent and 435 pounds of butterfat in 311 days. Records were discontinued for 5 years during World War II but resumed in November 1945. Production of Island cows should not be compared with that of cows receiving different feed and care in other countries.

TON OF GOLD COWS

Patterned after the American Jersey Cattle Club's plan, Ton of Gold certificates were issued to cows producing 2,000 pounds of butterfat or more in four consecutive years recorded in the Milk and Butterfat Record Scheme.

EXPORTS OF JERSEY CATTLE

R. M. Gow cited a sworn statement dated September 1, 1741, that the sloop *Jane* was chartered to take eight cows from Jersey to Southampton at 9 shillings freight per cow. This was the earliest exportation of cows from the Island known to him. Reverend Mr. Valpy mentioned export of cows from Jersey in 1785: "The other article of export which deserves to be mentioned (in addition to knit woolen stockings), is cows. They are well known in southern England under the denomination of Alderney, or Norman cows. In the month of June last, no less than 120 were exported to England, and all are the produce of the Island."

A little cow purchased by Michael Fowler on return from Barnet Fair produced an average of 14¼ pounds of butter weekly over 17 weeks. He discovered that she had come from the Channel Islands. Subsequently, he and three sons shipped 7,330 cows and bulls to England, America, Australia, France, and New Zealand in 1873–79.

IMPACT OF WORLD WAR II

Annual exports of cattle from Jersey ranged from 269 head in 1804 to 2,483 in 1867. More than 819 bulls and 11,139 females went to the United States before World War II. Exports of 662 Jerseys left in 1946 to seven countries.

Islander described conditions of the Island cattle during World War II:

> From June 1940 Jersey cattle had to be slaughtered to provide meat for the population, but until 1944 the number about equalled the normal annual export. Eventually French cattle were imported for slaughter. Thus after many years, foreign cattle did land alive in the Island, for necessity knows no law, but they went straightway to the slaughterhouse. At the liberation, every animal, French or Jersey, which had been in German ownership, was brought in for slaughter. After June 6,

1944, cattle could no longer come from France and after the reserves drawn from there had been exhausted the Jersey cow (or bull) had to supply the meat for the Islanders. Then in December last, the Germans having used up their resources, including horses, demanded up to 50 head of cattle to be killed weekly for their rations. Therefore, to supply this demand and the civilian ration, over 1,000 head were slaughtered from Dec. 1 to April 30. Besides this, 200 head were requisitioned by the military for shipment to Germany, going forward in two consignments. It now appears that 130 of these exported Jerseys are in Alderney.

An Island committee allotted cows for slaughter to the various parishes and farmers according to the number owned. The farmer was permitted to select or buy a cow for this purpose, as required. The young stock situation was less favorable.

Farmers could only keep a small quantity (of milk) for calf rearing and the fewer calves they reared the less milk they needed to retain. It was therefore decided, as from December 1941 to severely restrict the number of heifer calves allowed to be reared and, therefore, registered. Bull calves and surplus heifer calves were slaughtered at three weeks of age (for veal). The number of heifer calves exempted varying from 60 to 100 per month. The number of heifer calves registered in pre-war days averaged 2,200 a year. . . .
The selection of the calves to be retained was the thankless job of another committee which evolved an ingenious system of points. Ideally, the dam of the heifer it was desired to keep should have been inspected, but lack of transport made this impossible. A limited number of bull calves was allowed to be retained, about 36 per annum. Before deciding on these the selection committee invariably inspected the dams of the calves which it was desired to keep and studied also the official pedigrees of them.

Cattle shows were not held, despite endeavors by the German military. Periodical examinations for qualification of animals in the *Herd Book* continued, but production recording was discontinued during the 5 years of German occupation. The Society officers constituted an emergency committee to continue the *Herd Book* duties. Five cases of foot-and-mouth disease occurred just before and during the occupation. Infected herds were destroyed promptly, movement of cattle prohibited, and no further outbreaks occurred. There was a marked decrease in bulls and replacement females. However, most of the effects on cattle numbers were overcome by 1948.

Tractors replaced 61 percent of the horses on farms from 1953 to 1963. Electric fences nearly doubled in number, while milking machines increased from 213 to 711. Less labor is employed on farms. Acreages of early potatoes, tomatoes, cauliflower, and silage crops displaced some grasslands for grazing and hay.

JERSEY BREED CONFERENCE

The Royal Jersey Agricultural and Horticultural Society was the first British breed organization that called an international breed conference on the Island, in May 1949 which was attended by the Society delegates and delegates from most other Jersey breed societies. The delegates considered import and export requirements, recording production, value of Jersey milk, and related subjects. The World Jersey Cattle Bureau was formed in 1951, and held conferences in Canada in 1954, South Africa in 1958, and Columbus, Ohio, in 1968. The aim was for more uniformity in breed improvement programs—record keeping, scales of points, and similar programs.

Before retiring as Secretary after 44 years of service with the Royal Jersey Agricultural and Horticultural Society, H. G. Shepard observed the change in attitude toward Jersey products: "The solids-not-fat content of milk is now assuming equal if not greater importance than the butterfat percentage." The Jersey Milk Marketing Board is reporting the solids-not-fat contents in the daily intake of milk processed on the Island.

ARTIFICIAL BREEDING

All bulls on the Island are at public service, under regulations. Artificial insemination had been used if a cow did not conceive to natural service. The Jersey Artificial Insemination Center, Limited, was organized as a private company with ten directors and the Earl of Jersey as chairman. C. Gruchy is veterinary officer at the stud on Val Poucin Farm. The organization holds membership in the Society. No case of brucellosis, foot-and-mouth disease, tuberculosis, or vibriosis has occurred on the Island during the past ten years. Semen, frozen in straws or ampules, is being transferred internationally.

BREED PUBLICATIONS

Three breed journals have been published on the Island. Two were sponsored privately: *The Island Cow* (1928–34) and *The Jersey Breeders', Growers' and Merchants' Gazette* (1936–39). Newspapers on the Island of Jersey carried some news of agricultural activities. *The Jersey at Home* began publication as a quarterly journal in 1951; it was the official organ of the Royal Jersey Agricultural & Horticultural Society. The cover page bore the photograph of a typical Jersey cow and Jersey bull on an outline map of the Island of Jersey.

A. Dodd is Secretary of the Society, at Springfield, St. Helier, Jersey, Channel Islands.

REFERENCES

Anonymous. 1937. The Channel Islands and Parts of Britanny and Normandy. London.

Boston, Eric J. 1954. *Jersey cattle*. London.

Bree, Ronald. 1946. What I saw on Jersey Island. *Jersey Bull.* 65:16–17, 22, 24.

———. 1946. Difficulties of Jersey Island. *Jersey Bull.* 65:82–83.

———. 1946. Procedure of Jersey. *Jersey Bull.* 65:116, 200–3.

Culley, George. 1786. *Observations on live stock*. G. G. J. & J. Robinson, London.

Duncan, Jonathan. 1836–1837. *Guernsey and Jersey Mag.* Vols. 1–4.

———. 1841. *The history of Guernsey, with occasional notices of Jersey, Alderney and Sark*. London.

Falle, Rev. Phillip. 1696. *Caesarea: Or, An Account of Jersey*. 1st ed. (Rev. ed., 1734.)

Garrard, George. 1800–1803. Different varieties of oxen common to the British Isles. (Portfolio.)

Gow, R. M. 1936. *The Jersey. An outline of her history during two centuries— 1784 to 1935*. American Jersey Cattle Club, New York.

Haskins, S. E. 1854. Charles the Second in the Channel Islands. Vols. 1–2. London.

Hazard, Willis P. 1872. *The Jersey, Alderney and Guernsey cow*. 10th ed. Porter & Coates, Philadelphia.

Inglis, Henry D. 1844. *The Channel Islands of Jersey, Guernsey, Alderney, Sark, Herm and Jethou*. 5th ed. Whittaker & Co., London.

"Islander." 1945. How conditions are on Jersey Island. *Jersey Bull.* 64:956–57, 1000–2.

Jacob, John. 1830. *Annals of the British Norman Isles.* J. Smith, London.

LeCornu, C. P. 1859. The agriculture of the Islands of Jersey, Guernsey, Alderney and Sark. *J. Roy. Agr. Soc. Engl.* 20:32–67.

Le Couteur, Col. J. 1845. On the Jersey, misnamed Alderney, cow. *J. Roy. Agr. Soc. Engl.* 5:43–50.

LePatourel, J. H. 1937. *The medieval administration of the Channel Islands.* Oxford Univ. Press, Oxford.

Leyte, Thomas. 1808. *A sketch of the history and present state of the Island of Jersey.* London.

Linsley, John S. 1885. *Jersey cattle in America.* Burr Printing House, New York.

Low, David. 1842. *On the domesticated animals of the British Isles.* Longmans, Green & Co., London.

Marett, J. R. de la. 1932. The origin of the Jersey cow. *Island Cow* 17:369–71.

Mourant, Ph. L. S. 1907. *The dairy queen. The Jersey cow.* Jersey.

Payn, Phillipot. 1585. Unpubl. manuscript cited by Rev. Phillip Falle.

Pless, W. 1824. *An account of the Island of Jersey.* T. Baker, Southampton.

Prentice, E. Parmalee. 1940. *The history of the Channel Island cattle. Guernseys and Jerseys.* Harper, New York.

Prentice, E. Parmalee, and contributors. 1942. *American dairy cattle, their past and future.* Harper, New York.

Quayle, Thomas. 1815. *General view of the agriculture and present state of the Islands on the coast of Normandy, subject to the Crown of Great Britain.* London.

Renouf, Thomas. 1887. Jersey cattle. In Cattle and dairy farming. *U.S. Consular Repts.* Part 1, pp. 205–7.

Shebbeare, John. 1771. *An authentic narrative of the oppression of the islanders of Jersey.* London.

Shepard, H. G. 1929. *Island Cow* 4:17–19.

———. 1934. *One hundred years of the Royal Agricultural and Horticultural Society, 1833–1933.* Jersey.

Stapleton, H. E. Autumn, 1949. Butter-fat percentages in milk of Island cows.

Syvret, George S. 1832. *Chroniques des isles de Jersey, Guernsey, Auregny et Serk.* Guernsey.

Thornton, John. 1879. History of the breed. Jersey. *English Herd Book of Jersey Cattle* 1:1–88. (Contains Acts of the States 1765, 1787, 1826, 1864, and 1878.)

Tubbs, L. Gordon. 1939. *The book of the Jersey.* Beech House, London.

Twaddell, L. H. 1865. In *Philadelphia Society for Promoting Agriculture.*

Valpy, Rev. R. 1785. A tour of Jersey. *Ann. Agr.* 4:268–76, 436–46, 511–52.

Wallace, Robert. 1893. *Farm live stock of Great Britain.* 4th ed. Oliver & Boyd, London.

Waring, George E., Jr. 1876. *A farmer's vacation.* J. R. Osgood & Co., Boston.

Wilson, James. 1909. *The evolution of British cattle.* Vinton & Co. London.

Royal Jersey Agricultural and Horticultural Society publications
   1872–. Jersey Herd Book, Vol. 1–.
   1833–. Annual Report (Agricultural Department); Catalogues of the Spring, Summer, and Autumn Shows; Jersey Herd Book Milk and Butter Fat Records.
   1949. Jersey breed conference.
   1951–. The Jersey Cow at Home. Vols. 1–.

CHAPTER 15

JERSEYS IN THE UNITED STATES

S HIPS TRAVELING to the Channel Islands often sailed from Jersey to Guernsey and finally to Alderney before returning to the English ports. When asked whence they came, it was natural for the sailors to reply that they came from Alderney. The cattle brought by these vessels commonly became called Alderneys in England. Likewise cattle from the Channel Islands first entered the United States under the name of Alderneys. One of the first ones came to the attention of the Philadelphia Society for Promoting Agriculture. A cow was imported by Maurice and William Wurts and was owned by Richard Morris in 1817 and by Reuben Haines in 1818. Richard Morris reported 8 pounds of churned butter from 1 week's milk. The next year, 16 pounds 14 ounces of butter were churned from 14 days' milk production. The rich yellow color of her butter was mentioned by Reuben Haines when on October 20, 1818, he sent a sample to Richard Peters, secretary of the Society. The island of her origin is not known.

IMPORTATIONS TO AMERICA

Captains of sailing vessels sometimes brought a few cows aboard ship to supply milk for the officers' table and for passengers during the voyage. Even as late as 1875, Secretary T. J. Hand entered such a cow in the American Jersey Cattle Club Herd Register, as these cows usually were sold here after the ship arrived.

A group of wealthy men about Hartford, Connecticut, sent John A. Tainter to the Island of Jersey in 1850, at which time he brought over about a dozen animals. These included the bull Splendens 16, and the cows Dot 7 and Violet 23. At least 13 clipper ships brought Jersey cattle to America in the early days. Tainter imported over 100 Jerseys on other trips from 1851 to 1861, the famous cow Flora 113 being among them in 1853. She was reported to have produced 511 pounds 2 ounces of churned butter in 50 weeks. The Fowlers, English cattle dealers, brought 601 animals between 1869 and 1881. T. S. Cooper & Sons, Coopersburg, Pennsylvania, imported 3,824 registered Jerseys between 1876 and 1928, including many famous show winners, producers, and transmitting animals. Seven Grand Champions at the National Dairy Show were among them. Other extensive importers were Meridale Farms, W. R. Spann & Sons, Frank S. Peer, Edmund Butler, and B. H. Bull & Sons of Brampton, Ontario.

Early popularity of the Jersey was indicated by the comment of Francis M. Rotch in 1861: "The cross of the Jersey upon the native has been tried in a few dairies in Otsego county, New York, and the results thus far are as promising. The quality of the milk in every instance is much improved, closely resembling that of the pure breed, whilst in quantity it is but slightly diminished from the native yield. The rich color of the milk, cream and butter is transmitted to the cross-bred very strongly."

Two English breeders—Philip Dauncey, who was an enthusiast for solid-colored Jerseys, and his follower William C. Duncan— bred some of the prominent ancestors behind the St. Lambert family, Pogis 99th of Hood Farm 94502, Rosaire's Olga Lad 87498,

St. Mawes 72053, and seven of the first ten cows awarded the President's Cup for high production.

The extent of importations was as follows:

	Bulls	Cows
Jerseys imported, 1850 to 1867	61	244
Imported Jerseys registered from 1868 to 1942, including animals in dam	4,607	22,148
	4,668	22,392

The peak of importations was reached around 1910. Direct importations in limited numbers were resumed in 1946, after the close of World War II. Fifty animals were brought from the Island to the United States in 1969.

COLOR MARKINGS OF JERSEY CATTLE

Early Jerseys brought to America were quite heterogeneous in hereditary makeup. Their "purity" had been maintained for years by law, precedent, and prejudice, except for a few Guernseys not deemed "foreign" a century ago. The latter had been brought as dowries by brides from the bailiwick of Guernsey. Color descriptions served as an index of this heterogeneity, though subject to the frailty of human description. Volume 1 of the *Herd Register* gave the color markings of registered Jerseys as light cream, silver, yellow, silver gray or drab, steel, steel gray, gray, French gray, dun, mouse, dove, squirrel gray, deer color, light fawn, fawn, bright salmon, orange fawn, red fawn, dark fawn, roan, red, brick dust red, light brown, brown, mulberry brown, dark brown, mink, chocolate, and black. Most of these colors were mixed with white markings. It was stated in Volume 1 that "the colors most prized are fawn and white."

Volumes 1, 2, and 3 cited in Table 15.1 were assembled under sponsorship of the Association of Breeders of Thoroughbred Neat Stock. Volume 100 was published by the American Jersey Cattle Club and contained the color markings of Jerseys registered around 1920. Color markings of a few animals were not stated.

Spread of the dominant character for solid color ("self") over a half-century is noted and, further, the shades of brown, cream, fawn, and gray vary less widely. Solid black or mahogany color is less common in the breed. Broken color is not discriminated against in the show ring. In fact it is often desired by many breeders. Three of the first four aged Jersey cows at the National Dairy Show in 1939 had white markings, while Lonely Craig 1075153—the Grand Champion female—was about 40 percent white.

The small Island type Jersey in favor earlier, was typified by Imported Lady Viola 238437 that was imported in 1911. She is shown in Figure 15.1.

TABLE 15.1
COLOR MARKINGS OF JERSEYS IN THE UNITED STATES BEFORE 1870 AND THOSE ENTERED ABOUT 1920 IN THE *Herd Register*

Herd Register Volume	Males		Females	
	Solid color	Broken color	Solid color	Broken color
1	0	60	0	202
2	18	116	32	316
3	9	76	19	169
Total	27	252	51	687
100	394	106	385	115

HERDBOOKS AND BREED ASSOCIATIONS

The Association of Breeders of Thoroughbred Neat Stock appointed a committee to compile pedigrees of Jersey cattle about 1857. Ten years later their first herdbook was published listing Jerseys under 68 different owners. Known later as the *American Jersey Herd Book*, the last volume (Volume 6) appeared in 1878. A single volume of the *Bristol Jersey Herd Book* of cattle in the vicinity of Bristol, Connecticut, was compiled by S. R. Gridley and W. Barnes and published in 1869. The Maine State Jersey Association published eight volumes of their herdbook between 1876 and 1898.

Correspondence among C. M. Beach, T. J. Hand, S. M. Sharples, and Colonel George E. Waring, Jr., suggested the need "to compile a trustworthy herd book." This resulted in formation of the American Jersey Cattle Club at a meeting in Philadelphia in July 1868. Volume 1 of their *Herd Register* appeared in 1871.

The Club incorporated April 19, 1880, in the state of New York and is the present breed organization. They published 117 volumes of the *Herd Register*. Publication was discontinued during a financial recession in 1931. Thenceforth a short pedigree was included on the registration certificate of each animal.

CLUB POLICIES

The policy of the American Jersey Cattle Club once was to have a small select membership, and to handle registrations and transfers of Jersey cattle for nonmembers at only a slightly increased fee than for members. The opinion prevailed that breed affairs could be conducted best by a limited membership among the leading breeders. The incorporated Club met outside of New York state for the first time, after a drive to secure the necessary 1,000 members in 1923 made such a meeting legal. The membership fee was reduced from $100 to $50 on August 30, 1933.

Five members must sign with the applicant before his membership will be considered by the Board of Directors.

FIG. 15.1. Imported Lady Viola 238437. First prize cow over the Island of Jersey in 1904 and 1905 and where shown in England in 1906, 1907, and 1908. She was purchased by Elmendorf Farm, Lexington, Kentucky in 1911. She typified the popular Island type Jersey of that period.

In 1941 George W. Sisson, Jr., long a leader in the Club, spoke of the nonmembers thus:

> Taxation without their enthusiastic cooperation with us in what this Club and what the cow can do. How can we change this situation? I am not going to formulate it, but I do hope that this Board of Directors will take that matter under advisement and see if they cannot develop some plan. We are a sort of benevolent autocracy in this Club. I have sat with you for 45 years and more and we have done the best we could for the Club, but we haven't reached the heart and soul of these 60 or 70,000 . . . whose money we take.

The reduced membership fee and an increase in the number of members contributed to a more democratic policy, as Sisson envisioned. Later changes aided also.

Plans to divide the United States into districts for representation by directors and a reform in the method of voting proxies were approved by a majority in a mail ballot of members in 1944. With increased membership, the directors considered regional representation as early as 1943. The setup of the Board of Directors was changed to regional representation in 1952.

The states were grouped into nine districts with relation to the number of registered Jerseys, and accessibility for travel to directors' meetings. These districts in 1959 were:

1. Maine
   New Hampshire
   Vermont
   Massachusetts
   Rhode Island
   Connecticut
   New York
2. Pennsylvania
   Maryland
   New Jersey
   Delaware
   West Virginia
   Virginia
3. Michigan
   Indiana
   Ohio

4. Kentucky
   Tennessee
   North Carolina
   South Carolina
5. Georgia
   Florida
   Alabama
   Mississippi
   Louisiana
6. Missouri
   Arkansas
   Kansas
   Oklahoma
7. Texas
   New Mexico
   Arizona

8. Minnesota
   Wisconsin
   Illinois
   Iowa
   North Dakota
   South Dakota
   Nebraska
9. Montana
   Wyoming
   Colorado
   Utah
   Idaho
   Nevada
   Washington
   Oregon
   California

One director was elected by the members residing in each district, with a director-at-large from each group: Districts 1, 2, and 3; Districts 4, 5, and 6; and Districts 7, 8, and 9. Directors served a 3-year term and may be elected for a second 3-year term. The President was elected annually and could succeed himself. The Board divided their duties into six committees: on performance and type, registration and breeds relations, information and extension, *Jersey Journal*, milk, and finance. The president, vice-president, and three directors constituted the executive committee. An executive secretary was employed by the Board, and his staff carried out the policies outlined by the Board of Directors.

The Board of Directors proposed a plan to divide the United States into 12 districts, each to be represented by a director. Members tabled the proposal at the annual meeting in 1958, and remanded the recommendation to the Board for further study and consideration later.

THE SHOW RING AND SCALE OF POINTS

The fourth national exhibition of the United States Agricultural Society held in Philadelphia in October 1856 was the first major exhibition at which Jersey cattle were shown in the United States.

The scale of points, adopted on the Island in 1849 and amended in 1866, was published in Volume 1 of the *Herd Register* in 1871. Some 31 "articles" for bulls and 34 for cows and heifers were weighted equally at one point each, regardless of their anatomical or physiological importance. In 1875 the scale was revised with proportionate points for each article, totaling 100 points for perfection. Eight points were allowed for escutcheon. When Guenon's theory that a person could forecast productive capacity of the cow by the pattern of the escutcheon was disproved, the scale was revised in 1885. The 32 points for total mammary development (udder, teats, and milk veins) were increased to 39 points. These scales of points guided the judges in evaluating parts of an animal when placing cattle in the show ring.

A committee revised the scale of points in 1903, dividing it into items for head, neck, body, tail, udder, teats, milk veins, size, and general appearance with appropriate descriptions under each item.

Concerning the selection of a Jersey cow, Secretary R. M. Gow stated in 1907: "The scale of points drawn up and adopted by the American Jersey Cattle Club will prove an excellent and instructive guide in selecting a Jersey cow. The best way, however, of determining the merits of any dairy cow is to use a pair of scales to ascertain the quantity of the milk, and a Babcock tester to ascertain its quality, or percentage of fat."

The committee on Scale of Points was continued, and produced a pictorial scorecard. The "Ideal Dairy Type" cow approved by the American Jersey Cattle Club was accompanied by a new scale of points in 1913.

At the fiftieth anniversary of the Club, in 1918, Secretary Gow stated: "I have before me the first volume of the *Herd Register*, published in 1871. . . . The first and second volumes of the *Herd Register* contained a number of photographs of Schreiber's of the presumably leading Jerseys of that day. If anyone doubts whether or not the Jerseys have been improved in conformation as well as in production, he need only glance at these pictures. Hardly one of the animals would obtain any place in the modern show ring."

James E. (Jimmy) Dodge, late manager of Hood Farm, presented a resolution before the Board of Directors from a nonmember of the Club in 1922 that the ideal weights of Jersey cows in the Scale of Points be increased by 100 pounds, and of bulls 200 pounds. The Directors arranged for weights or measurements of about 10,000 cows with Register of Merit production records. John W. Gowen analyzed the records and found that average production correlated directly with size of the cows. A committee appointed to make a study, also reported that cows winning at the shows must be of good size to appeal to dairy farmers. These cows must have capacity and other qualities indicating good producing ability.

The directors revised the Scale of Points in 1927 based on these findings to be: mature cows, 900 to 1,100 pounds—4 points; and mature bulls, 1,300 to 1,600 pounds—5 points.

A list of interpretations guiding action of judges in the shows supplemented the Scale of Points in 1929. These interpretations dealt with defects, injuries, udder defects, over-condition, and uniformity within group classes. Points for general appearance were increased in 1938 with slight reduction in points for head, rump, size, and fore udder attachment.

DAIRY COW UNIFIED SCORE CARD

A committee of the Purebred Dairy Cattle Association with representatives of each dairy breed association, prepared a Unified Dairy Cow Score Card in 1942. It was divided into four main headings: general appearance, dairy character, body capacity, and mammary system with their respective descriptions. Color, size, and horns were listed separately as breed characteristics. A similar scorecard was prepared for dairy bulls. The scorecard for cows was revised in 1957.

The National Dairy Association sponsored a National Dairy Show as the court of last resort in show ring competition from 1906 to 1941. An outbreak of foot-and-mouth disease in 1914 caused the show to be omitted in 1915. A recession cancelled the shows in 1932–34, and the Association disbanded after World War II. Its place was taken by an All-American Jersey Show at Columbus, Ohio, in 1946–48 and at other major shows designated annually by the Board of Directors. Junior owners competed in an All-American Junior Show, and in the open classes. The North American Dairy Cattle Show organized at Columbus, Ohio, in 1967 with Ayrshires, Brown Swiss, Guernseys, and Jerseys, and as a regional show for Holsteins and Milking Shorthorns. National Collegiate and 4-H Dairy Cattle Judging Contests and the annual meeting of the Dairy Shrine Club were held during the show.

PARISH SHOWS

A system of 1-day "parish shows" has been developed since 1928, dividing states into districts for them. Breeders assembled their Jerseys at a central place for judging and returned home the same day. Better animals often went to state fairs.

State fairs and large regional shows—Eastern States Exposition, Dairy Cattle Congress, and Pacific International Exposition—gave regional competitions as feeders to the National Dairy Show. Although organized independently, these shows were the show window of the breeding industry, rendering great educational and advertising service.

The first annual Jersey Pictorial Parade was supervised by the Club's Type Advisory Committee. Exhibitors of Jerseys submitted photographs of 132 entries that were successful during the 1963 show season. A panel of 12 approved judges placed awards. Seventeen of 30 animals had exhibited at the All-American Jersey Show. Some animals were recognized that had been at only one show.

TYPE CLASSIFICATION

The Directors approved a plan on May 31, 1932, for type classification of registered Jersey herds. The plan was developed by a committee with C. H. Staples (chairman), H. H. Kildee, Jack Shelton, and Lynn Copeland. J. B. Fitch filled the vacancy when Kildee withdrew. The Club adopted the plan at the annual meeting in June 1932. All females of milking age and bulls over 15 months were submitted for classification by official judges appointed by the Club.

The classification ratings were:

	Originally adopted	October, 1936		January, 1968
	(points)			(points)
Excellent	90 or more	90 or more		90 or more
Very Good	85–90	85–90		85–90
Good Plus	80–85	80–85	Desirable	80–84
Good	70–80	75–80	Acceptable	75–79
Fair	60–70	70–75	Poor	74 or below
Poor	Under 60	Under 70		

Classification ratings were published with the Register of Merit, Herd Improvement Test, or Tested Sire and Dam volumes in connection with the production records. Approved judges classified 14 herds during the first year. Registration certificates of Poor animals were cancelled, and only females from Fair cows were registered. This rule eliminated some registrations from the lower ratings. Frequent conferences of type classifiers tended to unify practices and decide on problem situations. Type ratings entered into requirements for qualifying "Star" bulls and Superior Sires, to be discussed later.

Status of the first 335,839 Jersey classifications are summarized in Table 15.2. Jerseys classified in 1948–49 averaged 83.36 points. Numerical scores were reported since 1963. Jerseys classified in 1969 averaged 82.71 points, with the changed emphasis on stature and udder form.

A revision was adopted in 1968. The terms Good Plus and Good were replaced with Desirable and Acceptable. The Fair rating was discontinued, and Poor was set at 74 points or below. Registration certificates of animals scoring 74 points or below were stamped in

TABLE 15.2
AVERAGES OF JERSEY TYPE CLASSIFICATIONS UNDER THE ORIGINAL AND REVISED METHODS

Original classification	Animals Classified (No.)	(%)	Revised classification	Animals Classified[a] (No.)	(%)
Excellent	15,192	4.5		366	2.02
Very Good	132,134	39.4		5,586	30.78
Good Plus	142,147	42.3	Desirable	8,358	46.07
Good	40,776	12.1	Acceptable	3,386	18.66
Fair	5,581	1.7	Poor (below 75)	449	2.47
Total	335,830			18,145	

a. Classified during the first 12 months of the revised program.

the Club office that male progeny were ineligible for registration. Stature was considered under general appearance, with emphasis on reasonably more scale. Medium suspensory ligament and udder quality were added to the description of the mammary system. Six hereditary defects—chronic edema, eye defects (crossed and pop eye), front feet toe out, parrot mouth or weak jaw, wry-face, and wry-tail—were recorded by the classifier when observed. All cows under six years old must be scored at each herd inspection, and older animals at the owner's option. Scores may be raised or lowered at later inspections. These changes would reduce the average score of the breed and would give opportunity to strive for improvement through selective breeding. All scores available from 1968 are used in Sire Summary Averages.

Defects observed under the revised Type Classification program included 35.46 percent toe out, 14.85 percent wry-face, 15.95 wry-tail, 22.42 percent eye defects, 9.48 percent weak jaw, and 1.84 percent with chronic edema among 18,145 Jerseys inspected in 498 herds in the first 12-months period. Arlis Anderson retired as classifier in 1967. John W. McKittrick succeeded as chief classifier, assisted by W. E. Weaver and later by A. D. Meyer. A. D. Meyer became chief classifier in 1969.

JERSEY PRODUCTION RECORDS

Memoirs of the Philadelphia Society for Promoting Agriculture (Volume 4, p. 155) published Richard Morris's letter about his 3-year old "Alderney" cow that yielded 8 pounds of churned butter in a week. Whether she was from Guernsey or Jersey is unknown. In the next lactation she produced 14 quarts of milk daily from which 16 pounds 14 ounces of butter "of so rich a yellow" were churned in 14 days.

Secretary George E. Waring, Jr., of the American Jersey Cattle Club, cited Flora 113, imported in 1851 for Thomas Motley. Her milk was churned separately. She produced 511 pounds 2 ounces of butter between May 18, 1853, and April 26, 1854. Her persistent production was notable as compared with common cows at that time.

J. Milton Mackie believed that milk and butter records should be used along with the scale of points in judging the merits of a Jersey cow. A. B. Darling of New York City requested in 1882 that the Club send a man to supervise a disinterested record of his cow Bomba 10330. Edward Burnett volunteered his services. She produced 205 pounds 6 ounces of milk in 7 days from which he churned 21 pounds 11½ ounces of butter. The Board of Directors placed his report in the minutes—the first official *production* recognition by the Club.

Major Campbell Brown of Spring Hill, Tennessee, collected 189 private butter test reports, mainly for 7 days, and published them in the *Country Gentleman* in 1882. Joining with Thomas H. Malone, William J. Webster, and M. M. Gardner, he published *Butter Tests of Jersey Cows* in 1884, assuming that the records would be of value to breeders. They hoped further to " . . . arouse such an interest as will force the American Jersey Cattle Club to elaborate a system which for the future may give more satisfactory results."

The Club appointed Major Henry E. Alvord as official tester in 1885. He supervised two butter tests that year. Several private tests reported to him were accepted by the Executive Committee on

January 4, 1887. The first production records published by the Club were *Butter Tests of Registered Jersey Cows* in 1889. The volume contained 1,693 private and 36 official records. Five volumes appeared, the last being in 1902.

REGISTER OF MERIT

Some 6,200 private and 597 authenticated and confirmed production records had been published by the Club. On March 12, 1902, the Executive Committee wrote:

Resolved, that it is the sense of this Committee that the Club shall receive, preserve and publish milk and butter-fat records of Jersey cows, provided that all seven-day records shall be authenticated by a representative of any State, Provincial or National experiment station or agricultural college, and that all yearly records shall be verified by a monthly check test of at least two-days duration conducted by such representative.

The committee of George S. Peer, H. S. Redfield, and George W. Sisson, Jr., drew up the rules that were approved by the Club on May 6, 1903.

The Register of Merit was established: " . . . with the object of raising to a higher standard the average excellence of the Jersey breed and to secure an authenticated and permanent production record to which reference can be made in selecting animals for breeding purposes."

Countess Matilda 74928 produced 270 pounds of milk containing 16.96 pounds of butterfat in 7 days, yielding 19 pounds 11 ounces of churned butter. Some yearly tests supervised at Hood Farm in Massachusetts were accepted, including Sophie 10th of Hood Farm 143234 with 8,683 pounds of milk, 5.51 percent and 478 pounds of butterfat in a year. Dollie's Valentine 105049, with a confirmed butter test at the Kentucky Agricultural Experiment Station in 1899, was recognized as the highest producer of the breed then. She produced 10,218 pounds of milk and 578 pounds of butterfat in 1 year, and was scored 96.5 points by John A. Middleton.

Register of Merit records were divided into 365-day (Class A and AA) and 305-day (Class A and AAA), double and triple letters indicating that a living calf had been dropped within the prescribed time.

Production records of 7, 14, and 30 days were accepted in Class B until 1917 when the short-time division had fallen into disuse. Production requirements for yearly records to qualify a cow were:

Age of cow	Milk	Butterfat
Under 30 months	8.000 pounds or 260 pounds	
5 years or older	10,000 pounds or 400 pounds	

A cow that also scored 80 points for conformation qualified for entry in Class A. Bulls entered Class A when three daughters from different dams qualified in production, and the bull scored 80 points. Without scoring, he entered Class B. Owners could select the cows they placed on test. Of the first 55 records, 54 cows scored by the 1903 Scale of Points averaged 91.48 points. Judges probably varied in their estimates of an ideal Jersey, compared with current classification.

Butterfat requirements soon were changed to 250.5 pounds of butterfat at 2 years old, up to 360 pounds at 5 years of age. A 305-day division was added in 1921 with the above requirements, and the 365-day division raised to 290.5 and 400 pounds of butterfat at the respective ages. Production records were quoted in terms of milk and "butter," harking back to the churn test before the Babcock test was perfected. The fiftieth annual meeting of the Club (1918) resolved " . . . that advertisements relative to the official records be written exclusively in terms of butterfat."

Sophie 19th of Hood Farm 189748 (Fig. 15.2) was among the great producers and transmitters of production. Four sons were used in large herds. One of them—Pogis 99th of Hood Farm 94502—was regarded as the greatest living dairy bull in his day in transmitting butterfat production and persistent milking ability.

TESTED SIRES

The Board of Directors amended the rules in September 1931 to eliminate minimum requirements for the Register of Merit and to publish all records of cows on test 270 days or longer. The number of registered daughters past 4 years old was published as an indication of selectivity in testing. A bull having ten or more daughters with records became known as a Tested Sire.

Production records were computed to a uniform 365-day mature equivalent basis prior to 1943 when computing the average production of progeny of Tested Sires. About 80 percent of the cows had been milked twice daily, with the 305-day division gaining in popularity.

The Board of Directors voted to publish the production records for Tested Sires on a 305-day 2× mature equivalent basis since 1943. During the 1955–56 fiscal year, 1,212 Jersey cows completed 305-day records in the Register of Merit averaging 9,045 pounds of milk, 5.52 percent and 492 pounds of butterfat. The trend was toward less selective testing and more use of the Herd Improvement Registry.

HERD IMPROVEMENT REGISTRY

An advance toward breed improvement came with a program for testing all cows in the herd for production. Following the Ayrshire Breeders' Association, the Board of Directors approved the Herd Improvement Registry which began operation in July 1928. The

FIG. 15.2. Sophie 19th of Hood Farm 189748 represented the rugged American type Jersey popular for size and butterfat production. She held the highest butterfat record over all breeds in 1914, and transmitted high production through 4 sons.

herd test was based on 1-day butterfat tests each month of all cows in the herd, over a 12-month period. Certificates of production of individual cows were issued on request and payment of a fee since March 1935. Only cows over 12 years old could be omitted at the owner's option if they had a previous Register of Merit or Herd Improvement Registry record. The average production of cows in the Herd Improvement Registry is listed in Table 15.3. The HIR division terminated in December 1965 for a single production program —the Dairy Herd Improvement Registry.

DAIRY HERD IMPROVEMENT REGISTRY

Supervision of DHIA records was modified in accordance with a study made by the Purebred Dairy Cattle Association and American Dairy Science Association so that the records were acceptable to the dairy breed associations. The DHIR was adopted on July 1, 1959. Records made under the new rules for supervision are known as Dairy Herd Improvement Registry (DHIR) records. They are used by the breed associations for the same purposes as the R. of M. records and the HIR. Some 28,613 production records in HIR and DHIR in 1960–61 averaged 8,048 pounds of milk, 5.3 percent and 427 pounds of butterfat. All production testing was consolidated as DHIR in 1966.

RECOGNITION OF PRODUCTION

A system of medal awards was revised in 1919 to recognize outstanding production with reproduction. Butterfat requirements for these awards were:

	305 days (Class AAA)	365 day (Class AA)
	(pounds)	
Silver Medal (under 5 years old, 2 years 95 days at start)	410	500
Increase in fat per day older at start	0.2	0.2
Gold Medal, any age	610	700
Medal of Merit, any age	740	850
President's Cup, to highest producer of the year, at least	870	1,000

A bull qualified for the corresponding medal since 1921, when three daughters from different dams earned that medal. The Silver

Medal requirement has been revised to 420 pounds of butterfat at 2 years old, with 0.175 pounds of fat added for each day older at start of the record. Silver Medals are awarded only to females under 5 years old, but to bulls of any age.

Since 1941 a cow rated a Ton of Gold certificate when she produced 2,000 pounds of butterfat in four consecutive years in HIR test.

TABLE 15.3

Average Production of Jerseys in Herd Improvement Registry, 1929–58, and in DHIR in 1970

Fiscal years	Number of records	Milk (lbs.)	Test (%)	Butterfat (lbs.)
1929–39	36,145	6,854	5.28	362
1939–49	131,899	7,091	5.35	379
1949–58	203,127	7,309	5.33	389
1970	28,026	9,914	5.05	501

A Gold Star Herd Award was earned in the HIR with an average production according to the number of cows, as follows:

Size of herd	Average butterfat production
	pounds
5 to 10 cows	475
10 to 50 cows	450
50 cows or more	425

Recognition of Transmitting Ability

Transmitting ability of Jersey bulls was recognized with the compilation of private *Butter Tests of Jerseys* in 1884, when the production records were assembled according to sires. Stoke Pogis III 2238 had nine daughters, including the famous Mary Anne of St. Lambert 9770 credited with 27 pounds 9¼ ounces of churned butter in seven days.

The Register of Merit recognized Jersey bulls having three daughters qualified for entry from different dams. Bulls with three Gold Medal daughters (producing 610 pounds of butterfat in Class AAA or 700 pounds in Class AA) from different dams, were rated as Gold Medal bulls beginning in May, 1911. The double- and triple-letter classes indicated that the cow carried a living calf for a definite time during the record. This recognition included Silver Medal and Medal of Merit bulls similarly in 1921.

The tendency toward linebreeding handicapped selection of a particular bull for exerting a wide influence across the breed. No single bull held this distinction since Golden Lad on the Island. Sybil's Gamboge 174663 ranked among the leaders, with 216 registered daughters in natural service, 88 of which yielded an average of 12,234 pounds of milk, 5.12 percent and 626 pounds of butterfat on a 365-day 3× mature equivalent basis in the Register of Merit. Born in 1914 on the Island, he died in Vermont when past 10 years old. Though generations have passed, size and udder quality characterize many of his descendants in natural and artificial service. His picture as an old bull is shown in Figure 15.3.

At least ten daughters must be tested and ten classified, these being at least 50 percent of all registered daughters 4 years or older.

A Jersey cow became a Tested Dam when three progeny qualified for production records. A Tested Sire son counted the same as a daughter, based on the average production of his progeny. Xenia's Sparkling Ivy 837775 was a Tested Dam that combined persistent

FIG. 15.3. Sybil's Gamboge 174663 sired 88 daughters whose Register of Merit records averaged 12,234 pounds of milk, 5.12 percent and 626 pounds of butterfat on a 365-day 3X mature equivalent basis.

production with type and transmitting ability. She is shown in Figure 15.4.

The Register of Merit rules were amended in 1931 to publish *all records* for 270 days or longer regardless of production. This was believed to present the transmitting ability more nearly when computing the average mature equivalent production of daughters of a Tested Sire and Tested Dam.

SUPERIOR AND SENIOR SUPERIOR SIRES

A Tested Sire qualified as a Superior Sire since 1932 when half of all daughters over 4 years old were tested and their average production exceeded 600 pounds of butterfat on a 365-day mature equivalent basis. Also at least ten classified daughters (40 per-

FIG. 15.4. Xenia's Sparkling Ivy 837775 was a Gold, Medal of Merit, and Ton of Gold cow, and classified Excellent when 15 years old. Her 10 progeny averaged 488 pounds of butterfat in 305 days, 2X mature equivalent basis, including a Superior and 4 Senior Superior Sires. Her daughter Dandy Sparkling Zinnia 1039863 has a lifetime record of 142,876 pounds of milk, 7,915 pounds of butterfat, and 11 progeny averaging 10,460 pounds of milk, 569 pounds of butterfat.

cent of those over 4 years old) were required, with 90 percent of them rating Good or higher. Requirements were changed in 1935 to require an average type classification score of 82 percent and, in 1943, 450 pounds of butterfat on a 305-day 2× mature equivalent basis.

A Superior Sire advanced automatically to a Senior Superior Sire in 1945 when at least 20 daughters averaged 450 pounds of butterfat and the classification score averaged 83 percent or higher. Some 37 Jersey bulls then had qualified. These standards overlooked some outstanding bulls in small herds. A new standard established in 1952 recognized milk and butterfat production, as follows:

	Superior Sire	Senior Superior Sire
	(pounds)	
Average production of all daughters on a 305-day 2X mature equivalent basis		
Milk	8,400	9,500
Butterfat	470	510
All classified daughters		
Average score, percent	83	84

At least ten daughters must be tested and ten classified, these being at least 50 percent of all registered daughters 4 years or older.

The Sire Award program was revised by the Board, effective in January 1965. The Tested Sire Award was discontinued and replaced by an Approved Sire Award. Production records continued on a basis of two milkings daily for 305 days on a mature equivalent basis. Milk production of all daughters of an Approved Sire must exceed the yearly lactation herd average of the herd(s) in which these records were completed. Requirements as of January 1965 were:

	Milk	Fat	Classification	Proportion of registered daughters past 4 years old
	(pounds)		(score)	(percent)
Approved Sire	8,500	425	84	70
Superior Sire	9,500	475	84	50
Senior Superior Sire	10,500	525	85	50
Century Sire	10,000	500	84	

PREDICTED DIFFERENCES AND PERCENTAGE REPEATABILITY

The USDA developed a rapid electronic method of analyzing many dairy production records. Analyses compared average production of a sire's progeny with that of herdmates calving in the same year and season. Partial lactations were projected to estimated completion. Production of daughters was compared with that of their herdmates, and a predicted difference in milk yield was computed. Repeatability of anticipated future daughters was estimated on a percentage basis, adjusted for the number of herds and daughters represented. The average of available type classification scores was appended in a sire's analysis when ten or more daughters had been classified. The Club published such analyses for active Jersey bulls three times a year.

The previous sire recognition system of Gold and Silver Medals, Superior and Senior Superior, Century, and Approved Sires was applied through 1967 and terminated. A new recognition system initiated in January 1968 accepted the latest (thrice yearly) USDA Sire Summary. The last available type score in 1967 and all subsequent scores were averaged for each later sire analysis. The Club published such analyses for active Jersey bulls thrice yearly. Awards prior to 1968 are designated as GM(ROM) or SM(ROM) to distinguish them from 1968 medal recognitions.

Medal recognitions in 1968 were at three levels of predicted differences in milk production of daughters over herdmates. The requirements were:

	Predicted difference over herdmates	Minimum repeatability	Minimum average classification
	(milk pounds)	(percent)	
Gold Medal	+ 400 or more	20	83
Silver Medal	+ 250–399	20	83
Bronze Medal	+ 100–249	20	83

STAR BULLS AND SELECTIVE REGISTRATION

A Star Bull program was instituted in 1939 as an attempt at pedigree evaluation. Scoring of qualifications was based on production and type classification of the dam, and upon average production and type scores of progeny of the sire and grandparents. The first standard provided that the pedigree of a young bull might rate up to Four Stars. Reliability of the plan was tested by surveying the background of 238 Tested Sires after rating their pedigrees by the Star method, as follows:

Number	Star rating	Production of daughters, mature basis	
		365 days	305 days
		(fat, pounds)	
139	0	540	470
45	1	576	501
29	2	595	518
19	3	621	541
6	4	687	598

The Star rating program attempted to analyze transmitting potential of an individual, based on ancestry. The scores were amended in 1943 to allow up to seven-Star ratings.

Selective registration of bulls was authorized by the Directors in 1939, to begin in 1942. The dams of 26 percent of male calves being registered had either Register of Merit or HIR records. Six percent were by Tested Sires. Others may have had either DHIA or private records, or the sires proved thereby. The proposed program provided two classes of registration certificates: (a) Selective Registration based on production and classification of ancestry, and (b) qualified registration certificates based on purity of descent only.

To merit Selective Registration, a male calf met one of the following requirements:

1. The sire must be proved in DHIA with daughters whose production averaged 400 pounds of butterfat in 305 days.
2A. The sire must be a Tested Sire under the Club's programs, his daughters' production averaging 400 pounds of butterfat in 305 days mature equivalent.
2B. His dam must be a Tested Dam with production as above.
3. His sire must be a Star Bull, and the pedigree of his dam equivalent to a One Star rating if a male.
4. His dam must have an approved record of 400 pounds of butterfat, mature equivalent.

Constructive Breeders Registry

The Constructive Breeders Registry was established in 1936 to recognize breeders who have developed superior herds through good management and consistent use of the breed programs. The award was based on status of the herd on the last day of each HIR testing year. The requirements were:

1. The applicant shall have bred at least 50 percent of the cows in the herd that have calved, and owned 65 percent of them at least 4 years.
2. The herd must average 400 pounds of butterfat in HIR on a 305-day $2\times$ mature equivalent basis and be entered on test for the ensuing year.
3. At least 60 percent of the females that have calved shall have been type classified and have a minimum average score of 82 percent.
4. The herd must be accredited as free from tuberculosis and be in a modified accredited area.
5. The herd must be free from brucellosis, or on an eradication program under state supervision.
6. The applicant must be a member of the state Jersey Club, or of the local or county Club.

Since 1952 the owner must have bred 75 percent of the herd and the type score be 83 or higher. The certificate applies for that year and is applied for annually.

The requirements were revised in 1964. A minimum herd of ten cows must exceed the breed average by 500 pounds of milk, and at least 60 percent of cows that have calved shall average 85 percent for type classification.

RELATION OF TYPE TO PRODUCTION AMONG JERSEYS

Relation of body conformation to production has been debated by biometricians and by advocates of the respective breeds. Lynn Copeland analyzed 10,363 Jersey type classification ratings and official production records in 1941, and stated:

> . . . on the average, the higher producing cows also receive the higher classification ratings. . . . It should be observed that a much greater percentage of the higher rating animals have records than the lower rating animals. . . . It appears that production and type can be combined and that there is a correlation between the conformation of a cow and her producing ability. . . . Since it is realized that there are sometimes pronounced variations in the producing ability of animals of similar con-

formation, such differences can only be determined through production testing. It does appear, however, that as Jersey breeders improve the conformation of their herds, the production will be increased.

This analysis included many cows under selective testing programs.

An analysis of 14,143 ratings in the respective classification groups up to March 31, 1948, found that the average production on a 305-day 2× mature equivalent basis increased from 420 pounds of butterfat with the Fair group to 483 pounds for Excellent cows. Copeland indicated that cows in the upper levels of type were not necessarily high producers, and vice versa. This analysis is presented in Table 15.4.

TABLE 15.4
THE RELATION BETWEEN AVERAGE TYPE CLASSIFICATION AND BUTTERFAT
PRODUCTION OF JERSEY COWS

Classification	Number of cows	Butterfat production[a] (lbs.)
Excellent	801	483
Very Good	4,213	460
Good Plus	6,060	448
Good	2,700	434
Fair	369	420

a. Butterfat records were computed to 305-day 2× milking, on a mature equivalent basis for comparison.

Copeland searched the Jersey Performance Register for records completed during 1943–50 by type-classified heifers with at least 9,000 pounds of milk on HIR test in first lactations. A total of 654 heifers qualified. These animals were traced through the following volumes to 1963, seeking possible average relationships between type ratings and useful longevity. The numbers of lactations by cows with the higher type ratings increased steadily. He concluded that ". . . heifers of desirable conformation have a better chance of achieving high lifetime totals than do the poorer type heifers, even though the poorer type heifers did produce well in their first lactations." A check of ownerships found that a large proportion of these Jerseys were owned in farmer-breeder herds where production might be the main reason for retaining them.

SPECIAL JERSEY MILK

The American Jersey Cattle Club established a registered trademark for "Jersey Creamline" milk in 1930. Licenses to use the trademark were granted to producers or to processors who contracted qualified producers. The contract specified conditions of production, handling, and quality of the product under the trademark. The Club aided in organizing Jersey Creamline, Incorporated, which supervised the milk program. Over 60,000,000 units of Jersey Creamline products were distributed by 295 licensees in 1939. World War II interrupted the program in some areas, and the organization disbanded in 1944. Its duties were re-assumed by the Club.

An "All-Jersey" trademark for dairy products from Jersey milk originated with Ed Jackson in the state of Washington. The Frederick J. Baker advertising firm developed promotional material for the Jersey producers in the Seattle area and copyrighted the name. The Oregon and Washington Jersey Cattle Clubs purchased exclusive rights to use the name, and granted franchises to Jersey organizations in California, Idaho, Montana, Nevada, Pennsylvania, Utah, and Wyoming. The American Jersey Cattle Club arranged an agreement with the Oregon and Washington Clubs in 1954 to franchise the All-Jersey trademark in other than the nine states. The Club's Jersey Creamline program was continued in established areas. National All-Jersey was organized in 1958 to promote the increased production and sale of Jersey milk and milk products. The purpose is to merchandize market milk of high food value. This promotes the breed indirectly.

COMPOSITION OF JERSEY MILK

Some 203,127 Jersey HIR records in 1949–58 averaged 5.33 percent of butterfat. Some lactations averaged as low as 3.5 percent and others well over 7.0 percent butterfat. There appears to have been little tendency for change, based on selection of herd sires. Some breeders have expressed interest in higher milk yields regardless of butterfat percentages.

Some 829 lactations in 13 Jersey herds in four states ranged in average composition from 5.19 to 5.69 percent fat, 3.77 to 4.32 percent protein, 9.27 to 9.75 percent solids-not-fat, with averages of 5.42, 3.85, and 9.41 percent, respectively. Milk from individual cows varies widely around these averages within herds.

EXTENSION OR FIELD SERVICE

The Club established an Extension Service in 1916 with Wallace MacMonnies as chief. Publicity was given to production records and to breed activities. Cooperation was extended to state and other organizations promoting purebred dairy bulls, junior dairy clubs, Jersey displays at fairs, and contacts between buyers and sellers of Jersey cattle.

Jersey fieldmen were financed later through increased fees for registrations and transfers of animals. Breeders organized local and state Jersey clubs. Their activities included exhibits, dairy tours and picnics, consignment sales, production testing, health programs, group type classifications, junior activities, and meetings. State organizations helped to sponsor statewide dairy projects and participated in unified programs of the Purebred Dairy Cattle Association.

The United States was divided into districts, with a Jersey fieldman representing the Club in each district. They rendered many services to state organizations, breeders, and prospective buyers of Jersey cattle.

Three national youth activities are sponsored. The All-American Junior Jersey Show is held each October in Columbus, Ohio. The Youth Production Contest award is made on the highest total solids produced by a project cow. The Achievement Contest award is based on all activities and achievements since the member first owned a Jersey bull or heifer. A national heifer sale during the annual meeting of the Club contributes toward support of the youth projects.

JERSEY BREED PUBLICITY

The staff prepared pamphlets, booklets, and artistic brochures from time to time. R. M. Gow, an employee for 54 years, wrote *The Jersey,* published by the Club in 1936. Guy M. Crews compiled *History of the American Jersey Cattle Club 1868–1968* as a reminder of the activities and programs that have improved the Jersey cow during the first century in America.

The Jersey Bulletin was the first magazine published exclusively as a dairy breed paper. D. H. Jenkins developed the magazine as a private enterprise on October 1, 1883, in Indianapolis, Indiana. Royer H. Brown became co-owner in 1912. The Club contributed news of the breed and annual Club reports and bought advertising space periodically. Interest in ownership of the breed magazine culminated in establishing the *Jersey Journal* by action of the Board of Directors in 1953. Negotiations soon concluded purchase of *The Jersey Bulletin*, its files and good will. The *Jersey Journal* is published from the Club office. News releases of local and special interest are distributed to local, state, and national papers by the Club.

The American Jersey Cattle Club moved from an office in New York City to 1521 East Broad Street in Columbus, Ohio 43205. The location was selected for its centrality in the Jersey cattle population and activities.

IMPACT OF ARTIFICIAL BREEDING

Jersey bulls were in demand for upgrading dairy cattle in many small herds. One male was registered for each 2.77 females between 1868 and 1931. Cooperative artificial breeding of dairy cattle began in 1938 in New Jersey and Missouri, and the practice spread. The numbers of dairy farms decreased slightly and size of herds increased. Some 2,074 male and 35,945 female Jerseys were registered in 1969, or a ratio of 1 male to 17.3 females. The proportion of registered Jerseys conceived by artificial insemination amounted to 50.8 percent in 1969. Males retained for breeding purposes have been selected more critically.

A number of farms are believed to have added some registered females because of accessibility of good quality Jersey bulls in artificial service. Average milk yields of cows have increased from the combined effects of improved feeding, better management, and more capable cows than previously. Some 173 Jersey bulls were in A.I. studs in 1969.

BREEDING JERSEYS

Some hereditary traits are known concerning Jersey cattle. Prior to the first breed association in America, it had been found that a cow sired by a Jersey bull and out of a common cow, produced richer milk even though it might be less in amount. The solid coat color was a dominant character, while broken color was recessive. The small size of Jerseys in their native home was due partly to environment, since they grew to larger sizes when exported to England or America before reaching mature growth.

Polled Jerseys long have existed. James Wilson, Professor of Agriculture at the Royal College of Science for Ireland, believed that the dun or fawn color traced to cattle of the Norsemen. He also attributed the polled character of some British breeds to the same source, since they developed along the coastal areas once occupied by Norsemen. However, polled individuals that breed true for the character also have been born from horned parents. Polled Jerseys were recognized by the prefix "X" before the registration number, later replaced with the prefix "P."

Early shows on the Island of Jersey helped to establish selections for the characteristic level rumps soon after 1833, except in some large and high-producing Jerseys developed in England and America. These included the Stoke Pogis, St. Lambert, and Tormentor strains as well as others in Pacific coast states.

A correlation of 0.18 was found by Harvey and Lush between transmitting ability for type and for production in 2,786 daughter-dam pairs of Jerseys in 226 herds during 1943–47. Selection on type alone would bring slower improvement in production than selection for production alone. Lynn Copeland found a small positive relationship between conformation of 4,587 Jersey cows and their production. He stated, however, that a high conformation score in an individual animal did not necessarily guarantee superior producing capacity.

Hereditary dairy temperament (light natural fleshing, in part) was demonstrated in a meat-cutting demonstration by John Gosling of Kansas City at the Iowa State College Farm and Home Week in 1915. He contrasted the muscle covering which an Aberdeen Angus and a Jersey calf possessed at birth, as shown in Figure 15.5.

Occurrence of some anatomical faults of hereditary character helped to confirm some small relationship between dairy breeds. Ear notch was carried from the Jersey to the Ayrshire in the formative years of the latter breed. Wry-tail occurred in Guernseys and Jerseys, and less frequently among Ayrshire cattle. Desirable characteristics may be intensified in uniformity of transmission by inbreeding or close linebreeding. Such breeding practices, unfortunately, also can bring out recessive hereditary faults that may be carried unseen in the germ plasm of both parents. Some recessive

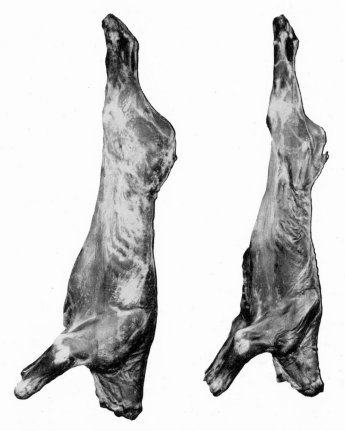

FIG. 15.5. John Gosling demonstrated that a beef calf (Aberdeen Angus at left) was born with good muscles into which fat would be deposited as the animal developed. Conversely, the Jersey calf (at right) was born with light natural fleshing or "dairy temperament." Photograph by F. E. Coburn, Iowa State College in 1915.

characters are weak udder attachment, buff nose, wry-tail, and similar simple characters.

A nonlethal type of bulldog head or prognathism was reported by Charles Darwin to have occurred among cattle of European origin in Argentina about 1760. The same condition occurred in a grade Jersey herd in Florida in 1949. The character is controlled by a single recessive gene. The recessive character crampy or progressive posterior paralysis (spastic syndrome) develops in adult cattle of the major dairy breeds past 2 years old. It may be chronic or critical. Some Jerseys lack hereditary resistance against the two types of lump jaw which affect the bones or the soft tissues.

Blood antigen combinations have been studied at the Ohio, Wisconsin, and California stations. All bulls in artificial breeding under contract with the Purebred Dairy Cattle Association are required to be blood typed. Their blood antigen patterns are retained by the breed associations and are used when needed to establish and confirm the parentage of progeny in cases of doubt or question between two possible sires of known antigen makeup. Positive identification of parentage by this means was recognized by a Canadian court.

Conformation and ideal size of Jersey cows have changed gradually from refined animals like Imported Lady Viola, to cows with capacity for greater producing ability. Type classification, as revised in 1968, gave greater emphasis to stature and udder character than formerly. These included a long cleancut head with strong jaw, about 50 inches in height at the withers, and corresponding skeletal structure. Strength of udder attachments were indicated by some definition between the halves for a strong middle suspensory ligament rather than a flat udder floor. A soft pliable udder that milked out well denoted much secretory tissue. Strong legs, pasterns, feet not toed outward, and a straightforward stride showed their ruggedness and wearing ability. Six hereditary defects were recorded when observed at type classifications.

JOINT OCCUPANCY

Karl B. Musser suggested in 1947 that the five dairy breed organizations build one building to house their offices at an accessible midwestern location, and do routine clerical work in cooperation. The Board of Directors of the American Jersey Cattle Club considered the project. A Guernsey-Jersey joint committee met in Peterborough, New Hampshire, in July 1960. The Jersey committee visited the Holstein-Friesian office in Brattleboro, Vermont. The latter organization would consider joint occupancy only if it were at their home office, which proposal did not fulfill joint occupancy goals.

The joint Guernsey-Jersey committee concluded that (a) joint occupancy could result in operational economies to both clubs; (b) moving to Peterborough could have only a short-time advantage; and (c) needs for additional space precluded moving into the Jersey office at Columbus, Ohio. Because of evident advantages, both Boards of Directors resolved in October 1964 " . . . that the Joint Occupancy Committees . . . be given appropriate authority to accomplish joint occupancy of their national headquarter facilities, including but not limited to, proposals for disposition of existing buildings, and for obtaining facilities in a new midwest location."

Modernization occurred with conversion to electronic data processing in the Jersey office over 3 years. The first registration certificates were processed thus in 1967. Registration and transfer fees were increased in line with advancing costs.

Executive Secretary J. F. Cavanaugh is located with the American Jersey Cattle Club office at 1521 East Broad Street, Columbus, Ohio 43205.

REFERENCES

Anonymous. *Memoirs of the Philadelphia Society for Promoting Agriculture* 4:155; 5:47.

Arms, Florence. 1941. One hundred years before Pope 652. *Jersey Bull.* 60: 1177, 1198–99.

Becker, R. B., and P. T. Dix Arnold. 1949. "Bulldog head" cattle. *J. Hered.* 90:282–86.

Brooks, John. 1867. *Herd Record of the Association of Breeders of Thoroughbred Neat Stock. Jerseys.* Vol. 1.

———. *American Jersey Herd Book.* Vol. 2.

Brown, Campbell, Thomas A. Malone, William J. Webster, and M. M. Gardner. 1884. *Butter tests of Jerseys.*

Copeland, Lynn. 1926. *Development of the Jersey breed.* Edward Bros., Ann Arbor, Mich.

————. 1932. Raising the breed level of production. 64th ann. meeting, American Jersey Cattle Club.

————. 1941. The relation between type and production. *J. Dairy Sci.* 24: 297–304.

————. 1965. Type and useful lifetime production. *Jersey J.* 12(8):3–31.

Crews, G. M. 1969. *History of the American Jersey Cattle Club, 1868–1968.* Columbus, Ohio.

Fuller, Valency E. 1893. *The Jersey herd at the World's Columbian Exposition.*

Gow, R. M. 1907. *About Jersey cattle.* (American Jersey Cattle Club pamphlet.)

————. 1936. *The Jersey.* American Jersey Cattle Club, New York.

Gowen, J. W. 1938. On the genetic constitution of Jersey cattle as influenced by heredity and environment. *Genetics* 18:415–40.

Gridley, S. R., and W. Barnes. 1869. *Bristol Jersey Herd Book.* Bristol, Conn.

Harvey, W. R., and J. L. Lush. 1952. Genetic correlation between type and production in Jersey cattle. *J. Dairy Sci.* 35:199–213.

Hazard, Willis P. 1872. *The Jersey, Alderney and Guernsey cow.* Porter & Coates, Philadelphia.

Linsley, John S. 1885. *Jersey cattle in America.* Burr Printing House, New York.

McKittrick, J. W., and A. D. Meyer. 1969. The type classification program of the American Jersey Cattle Club. *Jersey J.* 16(8):67-70.

Rotch, F. W. 1861. Select breeds of cattle. *Rep. Comm. Patents. Agriculture,* p. 465.

American Jersey Cattle Club Publications

1868–. Annual reports.

1871–1931. *Herd Register.* Vols. 1–117.

1883–1953. *The Jersey Bulletin.* Vols. 1-72.

1891. Butter tests of registered Jersey cows.

1909. Jersey sires with their tested daughters.

1946. *The Jersey Review.* No. 3.

1953. *Jersey judging made easy.*

1953–. *Jersey Journal.* Vols. 1–.

1954–. *Jersey Performance Register.* Vol. 1–.

1965. Joint Occupancy. *Jersey J.* 12(2):41, 49, 54.

1968–. Official Jersey Sire Summary List. Vols. 1–.

The Jersey Bulletin

1923. Some "ancient history" of the A.J.C. Club. 42:1991–97, 2074–75.

1932. Type classification plans for Jerseys. 51:891, 909.

1939. Super Bull and Selective Registration plan adopted. 58:872, 915.

1942. Star Bull program. 61:602–11.

1942. November AJCC Board meeting. New regulations of testing and classifying Star Bull changes. 61:1626, 1660.

1943. The requisites for Starring Bulls. 62:1324.

1943. Judge Adams' proposals anent directors. 62:1507.

1945. Directors' meeting makes history. 64:1831.

1948. Selective and Qualifying Registrations. 67:1249.

1948. Constructive Breeders Registry. 67:1347.

1949. Herd classification—what it does. 68:253.

Jersey Journal

1955. Jersey Handbook Issue. 2:82–90.

1956. All-Jersey milk . . . cream. 3:21–22, 62.

1956. All-Jersey milk program. 3:23–24.

1964. New Sire Award Program adopted by AJCC Board. 11:20–21, 23.

1968. Jersey type classification rules and regulations, effective January 1, 1968.

1968. The new Jersey Sire Award program. 15(8):26–27, 31, 97.

1968. A change in age conversion factors for standardizing lactation to a mature basis. 15(8):43-45.

1968. Special plan for evaluating type of a Jersey bull's progeny. 15(8):73.

1968. The American Jersey Cattle Club, 1868–1968. 15(19):21–29.

CHAPTER 16

DAIRY SHORTHORNS IN THE BRITISH ISLES

SHORTHORN CATTLE were developed in the drainage basin of the Tees River, a fertile level part of Yorkshire called the Holderness district, and contiguous areas near the "German Ocean" (North Sea) in northeastern England. Some soils in the area were moist clay loams which were relatively fertile. Nutritious forages grown on them permitted cattle to grow large, attracting attention to that kind.

The County or Bishopric of Durham borders the north bank of the Tees River. The North Riding of Yorkshire lies to the southward, extending toward the east coast, down to the Humber River. The West Riding occupies the southwestern part of Yorkshire. Lincolnshire is southward across the Humber. The counties of Northumberland, Cumberland, and Westmoreland join Durham and

Headpiece: Shorthorn vignette.

Yorkshire on the west and north. The accompanying map (Fig. 16.1) shows these areas.

The climate is cool and moist near the North Sea. Some freezing weather occurs during the winter months. The high latitude, around 54° to 55° North, is offset by the proximity of the Gulf Stream to

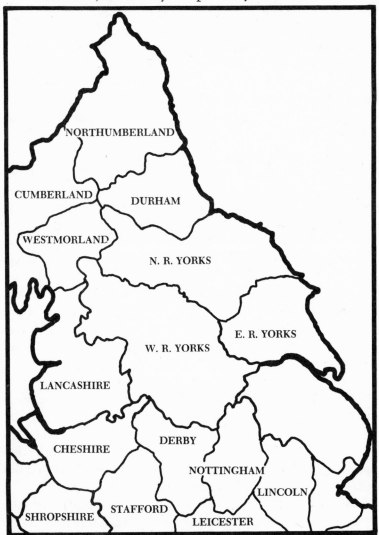

FIG. 16.1. Shorthorn cattle were developed in the bishopric of Durham, the Holderness district in the River Tees drainage basin and adjacent areas.

the northern inlet of the North Sea. Altitudes extend from sea level to the hills and low mountains in the west.

PREHISTORIC AND EARLY CATTLE

B. primigenius Bojanus was distributed widely over England and Scotland in the Pleistocene and Recent eras. These wild cattle were hunted for food. They became extinct, and participated little if any in later development of cattle in Britain. Neolithic man brought *B. longifrons* Owen (short-horned) in migrations westward over a land connection with the continent around 2000 B.C., before the first lake dwellers on the islands. *B. longifrons* remains were found with human remains in the cromlechs, as well as on sites of the later lake dwellers.

When the Carthaginian navigator Himilco landed on the south coast of England about 360 B.C., he found native Britons in possession of domesticated cattle. They stretched cattle hides over wooden frames to form boats with which to navigate the sea. Wilson assumed largely from circumstantial evidence that many of these cattle were black.

The natives drove the cattle inland when Romans invaded the island, hoping to deprive them of food. McKenny Hughes found that cattle brought by the Romans were a large lighter-colored kind. Roman influence extended over most of England, up to the Grampian Hills in Scotland, including Durham, Yorkshire, and the east coast area. A few skulls of early Roman cattle were polled, although most of them had horns.

Angles and Anglo-Saxons came as raiders and later as settlers with their families and cattle from across the North Sea. Norsemen plundered the coastal areas at intervals from A.D. 787 to 832, when they landed a large force along the Thames. Wilson claimed that settlers brought red cattle during the next century, some with polled heads. Most of the polled cattle of the British Isles developed near coastal settlement areas of the Norsemen.

Cadwallader John Bates cited the antiquity of short-horned cattle in the North Riding of Yorkshire from a will of John Percy, of Haram, near Helmsley in 1400, as follows: "To my son John I bequeath two stots with short horns; to John Webster a small horned stot; to John Belby, a cow with a white leske; to my son John, a heifer with a white head; to Thomas Peke, a heifer called Meg, and to Margaret Percy, another heifer."

James Sinclair assumed that this special mention of short horns signified that they were not common, nor were solid colors widely prevalent.

The figure of a cow was carved in the northwest corner tower of the eastern chapel of Durham Cathedral about A.D. 1300. When restored about 1778 from the Gothic original, the cow portrayed was of Shorthorn type.

END OF FEUDALISM

During the period of feudal rule, land was owned in large units mainly under control of lords, the church, or other large holders. Dependents, tenants, and free men attached to the large estates had little opportunity to plan their smaller operations separately. R. E. Prothero (Lord Ernle) remarked of feudal times: "No open-field farmer could farm with spirit. Unless all moved together, no one could move hand or foot, and what was every man's business was no man's business, they could make use of no improved methods of cultivation, new crops, better live stock, or mechanical inventions."

The system of feudal tenure and use of commons, where animals on the estate ran at large or were herded, broke down gradually with the movement for separate enclosures of land. Daniel DeFoe journeyed over Great Britain during 1724 to 1726, and observed that the greatest need " . . . is for more enclos'd Pastures, by which the Farmer would keep Stock of Cattle well fodder'd in the Winter, and, which again, would not only furnish good Store of Butter, Cheese, and Beef to the Market, but would by the quantity of Dung, enrich their Soil according to the unanswerable Maxim in grazing, 'that Stock upon Land improves Land.' "

Many separate Enclosure Acts by Parliament authorized division of Commons among the individual users thereof. Seventy Acts were passed between 1700 and 1760, and over 1,000 Acts between 1760 and 1815. These involved 561,494 acres in Durham and Yorkshire alone. Some of these were in the district where the Shorthorn breed developed.

Prothero wrote concerning the early agricultural leaders Jethro Tull, Lord Townshend, Robert Bakewell, Arthur Young and Coke of Norfolk: "With their names are associated the chief characteristics in the farming progress of the period, which may be summed up in the adoption of improved methods of cultivation, the introduction of new crops, the reduction of stock-breeding to a science, the provision of increased facilities of communication and of transport, and the enterprise and outlay of capitalist landlords and tenant-farmers."

Markham (1683), Mortimer (1716), and Hale (1756) mentioned the spread of spotted Dutch cattle to other districts, including Yorkshire. Lewis F. Allen, American Shorthorn historian, stated that agricultural progress had been made by diking and reclaiming salt marshes of Lincoln, Cambridge, and other east coast counties of England. Little mention of cattle improvement was noted before about 1720. Hale assumed some attention had been given to cattle, since a breed could not spring forth complete in type, color, and condition from the unimproved cattle of previous centuries.

FOUNDING THE SHORTHORN BREED

The main foundation of any breed consisted largely of local cattle. Written and genetic records introduced cattle that also contributed to the foundation of the Shorthorn breed. The polled character traced to cattle around the early Norse settlements near the coast. The "pyde" coat color resembled that of cattle in the homeland of migrants from across the North Sea.

Gervaise Markham (1660) mentioned that the best cattle in England were bred in Yorkshire, Derbyshire, Lancashire, Staffordshire, and three other counties. Cattle in four of these counties were "generally all black." Also "those in Lincolnshire are for the most part 'pyde', with more white than the other colours; their horns little and crooked, of bodies exceedingly tall, long, and large, lean and thin-hided, strong-hoved, not apt to surbate, and are indeed fittest to labour and draught."

He recommended that cattle of the different colors not be mixed, but pyde and red races could be mixed.

Sinclair believed that the Shorthorn breed probably had its origin soon after the union of England and Scotland in 1707. Peace did not come fully, however, until the English forces routed 5,000 Highlanders under Prince Charles Edward in 1746 in the battle of Culloden.

Several families in Holderness and Durham improved their cattle between 1700 and 1750. These included John Bates (grandfather of Thomas Bates), Sir Edward Blackett of Newby, the Smithsons of Stanwick, and Sir James Pennyman. Michael Dobinson of Witton Castle and Sir William St. Quinton of Scampston were mentioned as importing Dutch cattle to improve their herds.

John Bailey viewed the introduction of clover as a new era in agriculture, improving the feed supply for livestock. Robert Colling, a leading improver, used clover for pasture, soiling, and hay in 1796.

Thomas Bates dictated to his tenant in 1845, from which his grand-nephew wrote *Thomas Bates and his Kirklevington Shorthorns*, in 1897. He stated: "Notwithstanding that the then unrepealed statute 18 Car II, had in 1666, prohibited the 'importation of all great cattle' as a 'common nuisance,' and that no official records have been found of cattle landing from Holland, the fact that they did so land hardly admits of doubt."

William Ellis cited Dutch or Flanders white cows imported before 1744. An outbreak of murrain in 1745 was believed to have been introduced into the east coast with cattle from Holland. Some believed that Dutch cattle had been brought into England when Dutch engineers were engaged in diking and draining the coastal salt marshes. Another author mentioned in 1756 "the fine Dutch breed, having large legs, short horns, and a full body." Also "they are to be had in Kent and Sussex and some other places, where they are still carefully kept without mixture in colour; they will yield two gallons at a milking but, in order to do this, they require great attendance and the best of feed." This was before the period when cattle of the northern Netherlands became predominantly black-and-white in color.

George Culley, who was a friend of Robert Bakewell, recognized several breeds in England in 1786, including:

> The short-horned breed of cattle, is the next to be described; and it is pretty evident that our forefathers have imported these from the Continent: First, because they are still in many places called the Dutch breed; 2dly, because we find these cattle nowhere in this island, except along the east coast, facing those parts of the Continent where the same kind of cattle are still bred; and reaching from Lincolnshire southwards to the borders of Scotland northwards. But, 3rdly, I remember a Gentleman of the county of Durham (A Mr. Michael Dobinson), who went in the early part of his life into Holland in order to buy bulls; and those he brought over, I have been told, did much service in improving the breed; and this Mr. Dobinson, and his neighbors even in my day, were noted for having the best breeds of short-horned cattle, and sold their bulls and heifers for very great prices.

Culley mentioned that some other people brought some bulls that proved less desirable. Culley and his brother were extensive breeders of Shorthorn cattle.

David Low, Professor of Agriculture in the University of Edinburgh, wrote in 1842: "The Dutch breed was especially established in the district of Holderness, on the north side of the estuary of the Humber, whence it extended northward through the plains of Yorkshire; and the cattle of Holderness still retain the distinct traces of the Dutch original, and were long regarded as the finest dairy cows of England."

OTHER BLOOD IN FOUNDATION

Charles Colling mated two red-polled Galloway cows to his herd sires for a neighbor, and took as payment the red-and-white brindled bull calf from one of them. This animal sired the bull Grandson of Bolingbroke (280), and he in turn sired the cow Lady, lot 7, dropped in 1796. Progeny from this strain were termed the "alloy" blood.

Sir William St. Quinton of Scampston imported Dutch cattle to improve his stock. "The stewart at Ormesby repeatedly assured Major Budd that Sir James told him his breed was a cross between the old shorthorn and the Alderneys." Other writers also cited this cross with Alderneys. The Alderneys possessed short horns that already had been acquired from the Dutch stock. John Lawrence believed that smaller size and earlier maturity may have been gained from the Alderney cross.

The Reverend Henry Berry (1851), the Reverend John Storer (after 1877), and three other writers pointed to some Wild White Park cattle in the Shorthorn foundation. The hereditary color pattern—a red tinge on tips of the ears—and some historical evidence supported their statements. John Bailey was appointed by the Board of Agriculture to survey Durham in 1796 and again in 1807–9. He observed that color of the cattle was red-and-white around 1740, and white with a little red about the neck, or roan.

DEVELOPING BEEF QUALITIES

Robert Bakewell (1725–94) had chosen longhorn cattle, Leicester sheep, and German coach horses for improvement. He selected the best animals obtainable in his travels; bred the best to the best irrespective of family relationship; leased out bulls and rams for a season; drew back the ones he preferred to use in his own herd; and culled severely. George Culley and his brother worked with shorthorn cattle in the same period.

The brothers Charles Colling of Ketton (1750–1836) and Robert Colling of Barmpton (1749–1820) visited Bakewell in 1783 and studied his methods. They selected the best individuals of shorthorn cattle available as foundation stock about 1784 and practiced in-and-in breeding of their better animals even more than Bakewell had done.

THE BULL HUBBACK (319)

A great transmitting bull changed the character of shorthorn cattle. A former tenant farmer, John Hunter, retained a small cow and her bull calf when he became a bricklayer. The animals went to public market and were bought by a Mr. Basnett. Calves sired by this bull caused Robert Colling and a neighbor to buy and use him in their herds. Charles Colling later used him for four years. A Mr. Hubback took the bull to Northumberland when the bull was 10 years old. Thomas Bates saw this bull and his calves when the bull was about 13 years old. This bull, known later as Hubback (319) after establishment of the herdbook, became regarded as an important foundation animal of the improved beef-type Shorthorn breed. Bailey wrote of Hubback and his dam: "This bull and cow selected with so much judgement, are the original stock from which the celebrated *Durham Ox,* and the justly superior breeds in the possession of Mr. Charles Colling, Mr. Robert Colling and Mr. Christopher Mason are descended."

These men let bulls out on a year's lease. Also they kept full pedigrees of their animals as a guarantee of pure breeding. The fashion of keeping pedigrees of cattle by owners of the better herds began soon after 1730, some 90 years before formation of the breed herdbook.

Foundation of Dairy Shorthorns

Dutch breeding apparently contributed to the milking quality of the local cattle. Thomas Bates was an early proponent of milking ability in Shorthorn cattle. His grandfather, John Bates, admired a short horned cow, property of a Mr. Dobinson, at the Yarm Fair in 1730. Learning that she came of a tribe brought from Holland by Michael Dobinson, he bought six cows and a white bull of the same tribe from Mr. Dobinson's brother. He bred this strain until his death in 1777. George Bates, father of Thomas, bought a heifer calf from a heavy milking cow that in turn yielded up to 24 quarts in a day. She was fattened off when 17 years old. Her daughters in turn were quite good milkers.

Thomas Bates (1776–1849) leased the extensive Halton Castle estate and fattened kyloe cattle for market. His father sent him several excellent Shorthorn cows. He visited Colling Brothers, where *Durham Ox* and *White Heifer That Travelled* (got by Favorite) interested him so that he determined to breed the best Shorthorns possible. A legacy from an aunt enabled him to select some of the best animals, including some from the Collings in 1800. A Duchess cow was bought in 1804 and Young Duchess (renamed Duchess 1st) was obtained at the dispersal. Red Rose and Wild Eyes were bought elsewhere. The latter tribe descended from Richard Dobinson stock obtained in Holland over a century earlier. He regarded the Duchess family highly, and inbred his cows with bulls of that family. Belvedere (1706) was added in 1832. Bates seldom exhibited but was highly successful at the first Royal Show at Oxford in 1839. His Duchess and Oxford families became famous. He preferred milking ability over large carcasses. Many of his cattle were red-and-white, with some roans. Breeders interested in selling cows to dairies in the London area followed Bates's leadership in maintaining milking qualities in the breed.

Cattle tracing to Bates's herd became popular in America after his death in 1849. In 1873, 109 head brought an average price of $3,504 at the New York Mills Sale. Those were boom prices, and a depression followed.

BEEF AND DAIRY TYPE DISTINGUISHED

Bailey distinguished between beef and dairy strains within the Shorthorn breed in 1840:

> It has been already stated that the short-horned cattle were great milkers, this cannot be said of the variety, which has an aptitude to fatten, for though they give a great quantity for some months after calving, they decline considerably afterwards, but the variety of great milkers is yet to be found wherever the dairy is the chief object, and this variety is as carefully preserved and pursued, as the graziers do that of the fattening tribe. It is very common for cows of this breed to give thirty quarts a day.

Robert Brown (of Rennie, Brown, and Shirrell) who surveyed the West Riding of Yorkshire in the 1790s, noted, "In the southern part of the Vale of York breeding of cattle is not so much attended to as in the northern part: the object of cattle there being for the dairy, for the making of butter and old milk cheese; and consequently the milk alone is attended to. . . . This breed is generally coarse about the hips and rump, but rather shorter legged than in the northern part of the Vale."

The Colling Brothers bred mainly for the graziers, while Thomas Bates emphasized the milking ability of the earlier strains. David Low deprecated the loss of milking ability in Durham cows, but agreed in 1842 that many cows of the modern Holderness variety still ranked first among dairy cows of the country.

Youatt (third edition, 1851) observed concerning two classes of Shorthorns that:

> the Yorkshire cow was brought to the present state of perfection retaining with little diminution, the milking properties of the Holderness, and the grazing ones of the improved shorthorn. . . . The old and comparatively unimproved breed is still indeed found in the possession of most of the dairy farmers of this part (North Riding) of the country, for they are prejudiced against the improved short-horns, that their milking properties have been sacrificed to the accumulation of fat, still widely prevails.

Shorthorn colors in 1887 were white, red-white mixtures, or roan. Many breeders crossed the types, trying for a middle ground but losing the extreme of either beef or milk. As population of the British Isles increased, the Board of Agriculture estimated in 1908 that farms derived greater income from milk, butter, and cheese than from beef and veal.

AGRICULTURAL SOCIETIES AND THE SHOWS

The Agricultural Society for the County of Durham was organized in 1783. They held the first show for bulls at Durham in March 1784 and for cows at Darlington in September 1784. Robert and Charles Colling and George Coates, first editor of the herdbook, were successful exhibitors at these shows. The *Durham Ox* and the *White Heifer That Travelled,* from the Colling herd, drew wide attention to the large size of Shorthorns.

The Smithfield Cattle and Sheep Society (later the Smithfield Club) organized in London in 1798; the Highland and Agricultural Society (Scotland) formed in 1822. The Bath and West of England Society, and the Royal Agricultural Society of England all sponsored prominent shows. The Royal was organized in 1839. All except the Smithfield Club entered beef and dairy breeds of cattle as well as other livestock. Agricultural editor Arthur Young was the first secretary and treasurer of the Royal. Thomas Bates bred and exhibited Duke of Northumberland (1840) and Oxford Premium Cow that won first prizes at the first Royal show at Oxford in 1839. His cow Red Rose 3d led at the Royal show in 1840.

An experimental society organized in Rushyford in 1803, limited to 21 members, experimented with "seeds, sowing and soils, fencing, draining, working, cleaning land, manuring in its various forms, stock for different situations, food, feeding, etc." Robert Colling used the first threshing machine in Durham, and invented a horse-rake. Charles Colling introduced green manuring and experimented with roots and fodder crops.

The early shows helped to establish standards of quality, measured comparative breeding achievement, and brought improved animals to public attention. The Highland and Agricultural Society laid down principles for future guidance, since "exhibitions of the Society were exerting increasing influence on the character of the stock bred in the country." They recognized the Shorthorned, West Highland, Ayrshire, and the Polled breed of Galloway and the northern (Aberdeen) district as the only ones to distinguish as classes of stock. Others might be entered in their show under the general class of "any breed, pure or cross." This policy put an end to the Fifeshire as a dairy breed around Edinburgh and the Horned Aberdeenshire beef cattle in the northeast.

A Metropolitan Dairy Show was held at Islington in 1876, at which time the British Dairy Farmers' Association was organized. The Association sponsored later exhibitions at the London Dairy Show. All cows were exhibited in milk since 1920, and participated in the 2-day milking trials since 1922. Classes for bulls were held until 1939.

Records at the London Dairy Show now include the birthdate, live weight, calving date, days since last calving, milk yielded in three consecutive milkings during 24 hours, and the percentages of butterfat, total solids and solids-not-fat. Production points are allotted for pounds of milk in 24 hours, pounds of butterfat × 20 and solids-not-fat × 4. Twenty days are deducted from the time since calving, and one point is added for each additional 10 days. The leading cow for conformation is allowed 60 points on inspection, and others are given correspondingly lower scores. The total scores rate the order of the awards.

A contest based on production-lactation-inspection began in 1950. Each cow must have three complete lactations, based on minimum yields for the respective breeds. Production was included during 1,400 days from the base calving date. The scores were computed, according to breed, as follows:

Qualifying minimum milk yield	25 points
For each 1,000 pounds additional	1 point
Weighted average fat percent, minimum	25 points
For each additional 1 percent	20 points
Calving date, a basic 1,200 days, for each 30 days	
less, or more, add or subtract	1 point
Inspection score, for top cow (others proportionately less)	60 points

Scores were computed to the third decimal place, and awards were based on total score.

The Bledisloe trophy has been awarded at the London Dairy Show since 1922 for interbreed competition by six-cow teams. Dairy Shorthorns won the trophy in 1934, and provided the reserve team in 1931, 1933, 1935, 1947, and 1957. A Dairy Shorthorn cow was Supreme Champion at the London Dairy Show in 1958.

Nonpedigree Shorthorns "Typical of the majority of dairy cows in the United Kingdom" were regarded as superior milkers to most cows in the *Coates's Herd Book* during the early years of the London Dairy Show. They competed along with other breeds.

SHOW CLASSES FOR DAIRY SHORTHORNS

The Shorthorn Society of Great Britain and Ireland provided prizes in 1893 for Shorthorn cows in milk. They were judged as beef animals. Richard Stratton, president of the Society, and two friends solicited subscriptions and awarded prizes for pedigree Shorthorns on the milking characteristics. The Shorthorn Society eventually sponsored these prizes. After 1907 these prizes were awarded only if the cow yielded 25 pounds of milk in a day, or 20 pounds if fresh over 3 months. One-day milk weights were discontinued in 1933, and requirements were changed to lactation yields.

Shows have provided classes for Dairy Shorthorn bulls since 1918. They must have been entered in the Dairy Shorthorn Year Book in order to be eligible, the qualifications being: The dam and paternal granddam must have an official record in 305 days of (a) at least

7,000 pounds of milk if calved at 3 years or younger; (b) 8,000 pounds if calved between 3 and 4 years old; or (c) 10,000 pounds of milk in 305 days if calved at 4 years or older. These yields were for twice-daily milking, and were increased by 15 or 25 percent if milked three or four times daily after the first 30 days from date of calving.

Coates's Herd Book AND THE SHORTHORN SOCIETY

Accurate records of pedigree began to be kept in England when certain thoroughbred horses were recognized as sires of speed. Private records to assure honesty of pedigree culminated in publication of the Thoroughbred Stud Book. James Weatherby of London assembled the first volume in 1791, entitled *An Introduction to a General Stud Book*. It was the first herdbook published for any breed of farm animals.

Soon after 1730, leading breeders of short-horn cattle began to keep private records of pedigrees of their better animals. Purity of descent came to be considered important with improved animals. Studley Bull 626 calved in 1734 was the earliest named animal of the breed. At least 25 prominent *named* cattle born before 1780 were mentioned in herdbook pedigrees. Bailey wrote that Robert and Charles Colling and Christopher Mason kept books concerning the full pedigrees of their cattle, so that any person could see how they were descended.

Sir Henry Vane Tempest urged need of a trustworthy herdbook at a Wynyard annual gathering in 1812, attended by 18 prominent guests. Those present included Charles and Robert Colling, Thomas Booth and two sons, Thomas Bates, George Coates, John Whitaker, and others. Robert Colling and Mr. Paley assisted George Coates and his son to solicit and compile pedigrees of 710 bulls in Volume 1 of the *Coates's Herd Book* in 1822. Mr. Coates's son succeeded him with the *Herd Book* until 1846 when the enterprise became the property of Henry Stafford, assisted by Mr. Thornton.

Rumors of inaccurate pedigrees in 1872 incited breeders to form the Shorthorn Society. This organization purchased the herdbook and continued publication under the original name. The Shorthorn Society of Great Britain and Ireland was incorporated in 1875.

THE SHORTHORN DINNER CLUB

Lord Dunmore inspired organization of the Shorthorn Dinner Club in 1871 for an annual fellowship meeting and dinner among breeders interested in the breeding of Shorthorns.

DECLINE AND RISE OF DAIRY SHORTHORNS

The Improved Shorthorns became widely popular. Hall W. Keary wrote of the Holderness district in 1848: "The Holderness breed . . . of the Short-horn class, strangely resembles the old Yorkshire, except that it is larger and rather lighter of bone, and altogether a better fleshed animal. The cows of this breed are profuse milkers, and used to be eagerly sought after by the London dairymen. They have, however, lately been much crossed by the Durham bulls; and the old-fashioned Holderness Cow is far more rarely to be met with than formerly."

Increased demands and higher prices for beef bulls followed the opening of extensive grazing areas in the United States west of the Mississippi River and in Argentina. An agricultural depression about 1889 reduced the prices of milking-strain Shorthorns even lower. Many breeders discontinued registrations, or crossed their cows with beef-type bulls. "Deep milking qualities" of the breed were in danger of being lost. James Sinclair wrote:

> Teeswater and Yorkshire Shorthorns were wont to be the cows most prized for town dairies. Their descendants, the big, roomy, large-uddered, red, white, and roan Shorthorn without registered pedigrees, but carrying their descent unmistakably in their appearance, are still the favorites in town and country dairies . . .
> It may be further observed that since the establishment of the important dairy shows the Shorthorn cows, both those that are purebred and those not eligible for the Coates's Herd Book, have been very successful in carrying off the leading prizes. The Shorthorn is in fact clearly established as the general-purpose breed, combining . . . the dual purpose of early maturity and deep milking.

Dairy farmers avoided using pedigreed bulls which were largely beef type. A few leading breeders who maintained registration of the milking strains included Charles Adeane, of Barbraham; Lord

Henry Bentinck, Underly; Robert Hobbs, Kelmscott; J. C. Robinson, Iford; Lord Rothschild, Tring Park; George Taylor of Cranford; and a few others.

DAIRY SHORTHORN ASSOCIATION

The few breeders maintaining registrations of milking-strain Shorthorns organized the Dairy Shorthorn Association within the Shorthorn Society (*Coates's Herd Book*) in June 1905. To encourage breeding of the dairy strain, they offered prizes at the Royal and other leading shows for pedigreed Shorthorns of good milking type. This Association amalgamated in 1936 with the Shorthorn Society of Great Britain and Ireland. The latest development was publication of Volume 106 of *Coates's Herd Book* in two sections—Beef and Dairy—entering calves born in 1959.

PRODUCTION OF DAIRY SHORTHORNS

Some cows owned by Robert Colling yielded up to 28 to 36 quarts of milk in a day. One of Bates' Duchess cows was credited with 294 ounces of butter churned from her milk in a week. A cow owned by Mr. Larkin, of Pewyke, Worcestershire, produced an average of 1,050 imperial gallons of milk for 15 consecutive years.

The London Dairy Show added a class for Shorthorns in 1893, where every cow participated in the 2-day milking trials. The average yield of 106 pedigree and nonpedigree Shorthorns during 1895–1900 was 48 pounds 10 ounces of milk, with 1 pound 11 ounces of butter. After butterfat replaced churned butter, Shorthorn cows at the show averaged 43 pounds of milk, 3.73 percent fat, and 12.87 percent total solids in their milk. Twenty-five Dairy Shorthorns at the show in 1953 averaged 60.9 pounds of milk on three milkings daily.

The Dairy Shorthorn Association established minimum milk yields for cows to be eligible for prizes at shows. Records were published in the *Dairy Shorthorn Year Book* from 1907 to 1936, when the Shorthorn Society took it over. The Association established a scheme of milk records on farms in 1912, requiring occasional supervision by a local representative. The plan soon was superceded by official milk recording under the Ministry of Agriculture, and is now supervised by the farmers' cooperative Milk Marketing Board of England and Wales.

MILK RECORDING IN ENGLAND AND WALES

Several agricultural colleges and county councils organized local milk recording circuits soon after 1900 to obtain production records of individual cows. The Ministry of Agriculture sponsored improvement of livestock. Their milk recording scheme for all dairy breeds began under uniform methods early in 1914, but was hampered during World War I. In 1920 delegates from recording societies organized a Central Council as an advisory to the Ministry on policies. Milk recording expanded in October 1933 to include butterfat tests by the Gerber method at the owner's option.

The Ministry relinquished supervision to the Central Council in 1943 during World War II. The Milk Marketing Board of England and Wales assumed supervision in 1943 under the name of National Milk Records. A supervisory board represented the Central Council of Milk Recording Societies, breed registry associations, Milk Marketing Board, Ministry of Agriculture, and the National Farmers' Union.

The Milk Marketing Board adopted a single plan of official milk recording in 1957 after the government withdrew support. Members with herds on test bore over 80 percent of the costs, supplemented by the Board. The latter justified assistance, since production records were used in analyzing sires in the artificial insemination branch and in research applicable to the entire industry. Tests for total solids in milk and for solids-not-fat by difference were begun in 1955 with daughters of Friesian bulls, and with daughters of Ayrshires and Dairy Shorthorns in 1956. The Milko-Tester replaced the Gerber fat analysis in 1968, and tests for protein were begun.

The Milk Marketing Board formerly priced milk on the butterfat content, but also included solids-not-fat in October 1962. Chairman Richard Trehane of the Board reasoned that fluid milk rather than cheese dominated use of milk in English markets.

REGISTER OF MERIT FOR BULLS

In 1929 the Dairy Shorthorn Association established a Register of Merit which a bull entered when ten daughters qualified with milk yields according to age. An Improved Register of Merit was adopted in 1944. Requirements now are that at least six daughters must have yielded as follows:

	Milk	Butterfat
	(pounds)	
Calving at 3 years or younger	7,500	300
Calving at 3 to 4 years	8,500	330
Calving at 4 years or older	10,000	375

The average butterfat must be at least 3.5 percent for the lactation. At least 60 percent of his registered daughters of milking age must have qualified in production, and a sufficient proportion approved for body conformation.

ADVANCED REGISTRY FOR COWS

An Advanced Registry was established in 1947 for cows with production in 305 days; cows were listed according to their sires. The average milk and fat production of cows in the several branches of the Shorthorn breed in England and Wales was reported by the Production Division of the Milk Marketing Board of England and Wales for the year ending March 1968 as given in Table 16.1.

TABLE 16.1
AVERAGE PRODUCTION OF DAIRY SHORTHORNS IN ENGLAND DURING 1968/69
UNDER MILK CONTROL AND OPTIONAL BUTTERFAT TESTS

	Number of herds	Milk (lbs.)	Test (%)	Butterfat (lbs.)
Dairy Shorthorns	362	8,684	3.62	314
Lincoln Red Shorthorns	2	7,722	3.65	282
Northern Dairy Shorthorns	11	8,020	3.59	288

One of the older herds of Dairy Shorthorns in England was founded at Kelmscott in 1876. Eight homebred cows in this herd, owned by Mr. R. W. Hobbs, averaged 72.5 pounds of milk in a day when in full lactation. They are shown in Figure 16.2. Animals exhibited from this herd have won the Breeders Challenge Cup (bull and two homebred females) 13 times.

Registration of males as Dairy Shorthorn bulls in the *Coates's Herd Book* began in 1920. A cow qualified as a "bull breeder" when she produced in the first 315 days of lactation:

Heifer, 3 years 3 months or younger	5,500 pounds milk
Cow, up to 4 years 3 months	6,500 pounds milk
Cow, 4 years 3 months or older	8,000 pounds milk

The requirements were increased 15 percent if milked three times daily, or 25 percent when milked four times a day. Records were supervised in a recognized Milk Recording Society in England, Scotland, and Irish Free State or North Ireland. Both the dam and sire's dam qualified to enter Dairy Shorthorn bulls for the Register of Qualified Dairy Bulls.

GRADING-UP REGISTER

A plan to bring nonpedigree Shorthorns of milking strain into *Coates's Herd Book* was adopted in 1917. The cow's birthdate had to be known and an identifying tattoo placed in her right ear. She must have been approved by the Society inspector as having desirable Shorthorn characteristics.

A Foundation Cow must have given at least 6,000 pounds of milk with first calf, 7,000 pounds with the second calf, or 8,000 pounds in later lactations. Such a cow was admitted to Class A. Her female progeny by a registered Dairy Shorthorn bull were admitted to Class B of the Grading-up Register if tattooed and registered within 30 days of birth. Likewise, females qualified for Class C and Class D in turn. Female progeny of Class D cows, bearing four crosses of pedigreed Dairy Shorthorn sires, were eligible for full registration in *Coates's Herd Book*. All animals had to bear an approved ear tattoo. More than 300,000 females had entered this register up to 1959.

LICENSING BULLS

The Improvement of Live Stock Act passed by Parliament in 1931 required that bulls born after October 1933 be licensed before being used for breeding. Applications were made for inspection by a Ministry officer before a male calf was 9 months old. The ear of an approved animal was tattooed with a crown and serial number, or with an *R* in a rejected animal. Licenses were transferred to the next owners. The dam and sire's dam must have produced 6,000, 7,000, or 8,000 pounds of milk according to her age for a bull to be granted a "dairy" license in 1945. A minimum butterfat percentage of 3.5 also was required.

Artificial Breeding

Organized artificial breeding of cattle increased from one center in England inseminating 2,599 cows in 1944–45, to 1,783,583 cows in England and Wales in 1968–69. Twenty-three centers are operated by the Milk Marketing Board, and six are associated centers. Bulls of 18 breeds are used. Dairy Shorthorns accounted for 0.6 percent of services. Six Dairy Shorthorns were among 1,196 bulls used by the Board in 1969–70.

Consulting panels, officers, and others examine and obtain mature bulls satisfactory in natural service. Some young bulls are bought from breeders. Health inspections are made before entering A.I. service and periodically thereafter. Selected cows are mated under contract to leading bulls and their male calves reared at the Board's farm at Chippenham. Training for A.I. use begins at 10 months old. One-third of replacements came from the rearing unit in 1959–60.

FIG. 16.2. These eight cows yielded an average of 72.5 pounds of milk in a day. They were bred and owned by R. W. Hobbs, Kelmscott, Locklade, Gloucestershire, England.

Since 1955 young bulls have been used for 500 first services to cows in herds under milk recording, changing to "500 cows-in-calf per bull" in 1962. They are withheld from much further use, pending analysis by herdmate comparisons of daughters for production and conformation. Production of daughters in the first 180 days of lactation provides preliminary proof of some promising bulls. Progeny proved bulls then are used widely. Joseph Edwards concluded of the "first 500" plan: "Frankly, we have found this system to be far superior to earlier methods of sire proving, which included daughter-dam comparisons and straight daughter averages."

Thirty of the progeny-analyzed bulls were Dairy Shorthorns. About 85 of the top bulls of several breeds were in the nomination scheme for selective matings, their frozen semen being available through all centers. Their semen is prepared at the Northampton Deep Freeze Unit. Bulls in the contract mating program were blood typed in a laboratory at Copenhagen, Denmark. The Cattle Blood Typing Service was established in Edinburgh in September 1966 to serve all of Britain.

Five Dairy Shorthorn sires were progeny tested during 1958 at the British Oil and Cake Mills' Barley Farm, Selby, Yorkshire, England. The 10 or 12 daughters of each sire were leased from the owners for the trial; accustomed to the environment in advance of calving; then continued under uniform management for an average of 270 days in lactation. They received 14 pounds of hay and 6 pounds of dried beet pulp daily for maintenance. Concentrates (21 percent crude protein and 4.5 percent fat) were offered at 3.5 pounds per imperial gallon of milk with 3.7 percent of butterfat, and an increase of one-fourth pound for each increase of 0.2 percent butterfat above the base standard.

Records were kept of milk yield, butterfat, and solids-not-fat percentages. Speed of milking was measured during the peak of lactation. Breeders were able to observe conformation transmitted by the respective sires. Results of the trial in average production per cow are given in Table 16.2

Numbers of cattle of the dairy breeds have increased in England and Wales between 1908 and 1955 from 2 to 57 percent; beef cattle increased from 15 to 16 percent; and dual-purpose animals decreased from 83 to 27 percent of the cattle population. Influence of artificial breeding on demands for bulls for natural and artificial use is suggested from changes in the proportion of males and females registered in the herdbooks, as given in Table 16.3.

The average letdown of milk was measured in pounds per minute and pounds of strippings. Lactation curves were graphed for each cow to show maximum daily yield and persistency of production. Feed conversion was reported for each in terms of digestible protein and starch equivalent, considering nutrient values of feeds consumed, requirements for body maintenance, for gains in live weight, and the nutrients remaining for each 10 pounds of milk produced.

Twenty unselected daughters of a Dairy Shorthorn bull tested at Darlington Hall, Devon, averaged 9,031 pounds of milk, 3.80 percent fat and 9.04 percent of solids-not-fat.

TABLE 16.2

AVERAGE PRODUCTION OF DAUGHTERS OF DAIRY SHORTHORN BULLS THAT WERE UNDER PROGENY TEST FOR AN AVERAGE OF 270 DAYS IN LACTATION

Bull	Number of daughters	Milk (lbs.)	Butterfat (%)	Solids-not-fat (%)	Body weight at close (lbs.)	Milk in 3 minutes (lbs.)
Histon Dairy Farmer	12	7,963	3.76	9.02	1,213	12.1
Yewdon Rubio Musicism	12	7,788	3.84	8.90	1,224	13.8
Churchill Brilliant Boy 8th	11	7,870	3.57	8.97	1,202	11.8
Whatcote Lord York 13th	12	7,115	3.71	8.76	1,279	8.4
Avoncourt Lord Barrington 2nd	10	7,464	3.84	8.98	1,291	18.5
Average		7,640	3.74	8.92	1,241	

TABLE 16.3

EFFECT OF ARTIFICIAL BREEDING UPON THE PROPORTION OF DAIRY SHORTHORN BULLS REGISTERED, 1935 AND 1957

Herdbook	Year	Male	Female	Grading-up female	Ratio
Shorthorn, Coates's	1935	6,509	10,068		1:1.55
Shorthorn, Coates's	1957	2,961	18,958	9,185	1:9.51
Lincoln Red Shorthorn	1957	450	1,717	485	1:4.89
Northern Dairy Shorthorn	1957	402	3,531	505	1:10.04

364 DAIRY CATTLE BREEDS

DAIRY SHORTHORN JOURNAL

The Dairy Shorthorn Association published the *Dairy Shorthorn Journal* monthly from 1932 to 1936; since then the *Journal* has been published by the Shorthorn Society. This journal contains general news and advertising of the breed, and announcements from the Society. It is the only breed Society periodical published monthly in Great Britain.

REFERENCES

Allen, Lewis T. 1878. *History of the Short–horn cattle: Their origin, progress and present condition.* Buffalo, N.Y.

Ashton, E. D. 1949. The use of records in breeding dairy cattle in Denmark and the Netherlands. *Agriculture* 56:255–59.

Bailey, John. 1810. *General view of the agriculture of the County of Durham with observations on the means of its improvement.*

Bailey, John, and G. Culley. 1805. *General view of the agriculture of the County of Northumberland.* London.

Bates, Cadwallader John. 1897. *Thomas Bates and the Kirklevington Short-horns—A contribution to the history of the pure Durham cattle.*

Berry, Rev. Henry. 1851. The Short–horn. In W. Youatt, *Cattle.* 3rd ed.

Bradley, R. 1757. *A general treatise of agriculture both philosophical and practical.* London.

Brown, Robert (Rennie, Brown, and Shirreff). 1793. *General view of the agriculture of the West Riding of Yorkshire.* London.

Bull, Fred J. 1956. The London Dairy Show, 1876–1955. *Dairy Sci. Abstr.* 18:795–808.

Burrows, George T. 1945. How a few Englishmen saved the Milking Shorthorn. *Milking Shorthorn J.* 26(8):3–4.

———. 1950. *History of Shorthorn cattle.* Vinton & Co., London.

Buxton, Gerard J. 1924. The development of the Dairy Shorthorn in England and the influence of the breed on English agriculture. *Proc. World's Dairy Congr.* 2:1349–53.

Coleman, John. 1887. *The cattle, sheep and pigs of Great Britain.* Horace Cox, London.

Culley, George. 1786. *Observations on live stock.* G. G. J. & J. Robinson, London.

Dawkins, William Boyd. 1866. On the fossil British oxen. Part 1. Bos urus Caesar. *Quart. J. Geolog. Soc. London. Proc.* 22:391–402.

DeFoe, Daniel. 1778. *A tour thro' the whole Island of Great Britain.* 2 vols.

Edwards, Joseph. 1957. D.H.I.A. costs are shared in Britain. *Hoard's Dairyman* 103(24):1229, 1248.

Ellis, William. 1744. *The modern husbandman.* London.

Evans, John. 1923–24. Lincoln Red Shorthorn cattle. *Agriculture* 30:17–20.

Garner, Frank H. 1930. Milk recording and its sequel—Grading-up. *The Dairy Shorthorn* (pamphlet).

Griffiths, Col. J. 1934. The Improvement of Live Stock (Licensing of Bulls) Act, 1931. *Dairy Shorthorn J.* 3(7):197–98.

Hale, Thomas. 1756. *A compleat body of husbandry.* Vol. 2. London.

Housman, William. 1880. The management of a Shorthorn herd. *J. Roy. Agr. Soc. Engl.* 16(ser. 2):381–435.

Hughes, T. McKenny. 1896. *On the more important breeds of cattle which have been recognized in the British Isles.*

Keary, Hall W. 1848. Management of cattle. *J. Roy. Agr. Soc. Engl.* 9:424–52.

Lawrence, John. 1726. *A new system of agriculture.* London.

———. 1805. *A general treatise on cattle, the ox, the sheep and the swine.* C. Wittingham, London.

Leatham, Isaac. 1794. *General view of the agriculture of the East Riding of Yorkshire and the Ainstry of the City of York, with observations on the means of its improvement.* London.

Low, David. 1842. *On the domesticated animals of the British Isles.* Longmans, Green, & Co., London.

Lydekker, R., et al. 1901. *The new natural history.* Vol. 2. Mammals. Merrill & Baker, New York.

Markham, Gervaise. 1660. *The way to get wealth.* 10th ed. I. Harison, London.

———. 1683. *The English house–wife.* London.

Marshall, W. 1796. *The rural economy of Yorkshire.* Vol. 2.

———. 1805. *General treatise on cattle.*

Massingham, H. J. 1926. *Downland man.* London.

Moorhouse, Sydney. 1958. Britons adopt bull progeny test program. *Dairyman's League News* 42(21):14.

Mortimer, John. 1721. *The whole art of husbandry.* 5th ed. Dublin.

Plumb, C. S. 1920. Status of the Dairy Shorthorn in England. *Milking Shorthorn Year Book* 5:3–7.

Prothero, R. E. (Lord Ernle). 1908. Landmarks of British farming. (Cited by James Sinclair in *History of Shorthorn cattle.*) P. 11.

———. 1912. *English farming, past and present.* Longmans, Green & Co., London.

Ramsay, Alexander. 1879. *History of the Highland and Agricultural Society of Scotland.* W. Blackwood & Sons, Edinburgh.

Rowlin, Joshua. 1794. *The complete cow doctor.* Glasgow.

Rudd, Bartholemew. 1821. *An account of some of the stock of Short Horned cattle of Charles and Robert Colling.*

Sinclair, James. 1908. *History of Shorthorn cattle.* Vinton & Co., London.

Storer, Rev. John. 1875. *The Wild White Cattle of Great Britain.* Cassell, Petter, & Galpin, London.

Thornton, John. 1869. *Thornton's circular. A record of Shorthorn transactions.* 1:162.

Tuke, J., Jr. 1794. *General view of the agriculture of the North Riding of Yorkshire.*

Wallace, Robert, and J. A. Scott Watson. 1923. *Live stock of Great Britain.* 5th ed. Oliver & Boyd, Edinburgh.

Wilson, James. 1909. *The evolution of British cattle and the fashioning of breeds.* Vinton & Co., London.

Wright, John. 1846. On Short-horn cattle. *J. Roy. Agr. Soc. Engl.* 7:201–10.

Miscellaneous Breed Publications

1955. Lincoln Red Shorthorns. Lincoln Red Shorthorn Society. Lincoln.

1956. National Milk Records. Milk Marketing Board.

1965. Dairy herd census Milk Marketing Board. Breeding and Production Organization, Thames Ditton, Surrey.

The Dairy Shorthorn Year Book. Shorthorn Society. London.

CHAPTER 17

MILKING SHORTHORNS IN AMERICA

Sₕₒᵣₜₕₒᵣₙₛ were the first improved breed of cattle introduced to America. Messrs. Gough and Miller brought cattle of Durham extraction to Baltimore in 1783, and a second importation was made in 1795. Half-blood descendants were taken from Virginia to Clark County, Kentucky, in 1785 by Matthew Patton's sons. The elder Mr. Patton took a bull and a cow descended from the original "milk" breed. Lewis F. Allen wrote of the animals: "It was not at all uncommon for cows of this breed to give thirty-two quarts of milk daily. The Short-horn bull, red in color, with white face, rather heavy horns, yet smooth and round in form was called Mars. . . . The cow was called Venus, white in color, with red ears, small, short horns, turning down." Descendants were known as Patton stock.

Other importations came into Kentucky, Maryland, Massachusetts, New York, and Pennsylvania between 1815 and 1830, including cattle from Thomas Bates. Breeding records were kept, and animals were registered in the (English) *Coates's Herd Book*. A bull and three cows came from the Bates herd in 1849 and 1850. They included Red Rose 2nd that produced 49 pounds of churned butter in 25 days in second lactation. Allen listed 30 importations of 113 bulls and 229 females between 1850 and 1857. Many early Shorthorns were of the dual-purpose strains.

After the War Between the States, agricultural expansion west of the Mississippi River created interest in beef cattle. Importations of beef-type Shorthorns predominated, with animals representing Colling, Booth, and Cruickshank breeding. Many milking herds were crossed with popular "Scotch" bulls. A new herdbook rule barred registration of imported Shorthorn bulls that traced to the Appendix Register of *Coates's Herd Book*. This barred most Dairy Shorthorns. A change in organization of the Shorthorn Breeders' Association eventually eliminated this rule.

Frank S. Peer, of Ithaca, New York, imported 72 choice Milking Shorthorns which auctioned in 1916 for an average of $753 per head. James J. Hill, president of the Great Northern Railway, imported 2 bulls and 25 cows to his Minnesota farm in 1912. W. J. Hardy, later fieldman and Secretary of the American Milking Shorthorn Society, brought these cattle. Still other choice animals were imported.

HERDBOOK ASSOCIATIONS

Kentucky breeders appointed a committee to establish a herdbook about 1840. Meetings were held and pedigrees submitted but the matter was postponed. American-bred Shorthorns were registered meanwhile in Volumes 2 to 13 of *Coates's Herd Book* in England.

Lewis F. Allen of Buffalo, New York, assembled pedigrees of Shorthorns privately in 1845 and published Volume 1 of the *American Herd Book* in 1846. This volume attracted considerable attention. Volume 2 followed in 1855, and others later.

To help preserve purity of the stock, the Association of Breeders of Thorough-Bred Neat Stock published one volume of a herdbook for "Short Horns" in 1863 in Connecticut. Four of the cows had produced 14 to 17 pounds of butter in a week, and up to 59 pounds of milk in a day.

The American Short-horn Record was assembled in Kentucky by Major Humphrey Evans and published by A. J. Alexander. Ten volumes appeared. The Ohio Shorthorn Breeders Association organized in 1876 and published three volumes of a herdbook between 1878 and 1882.

The American Shorthorn Breeders' Association was organized in Indianapolis, Indiana, in 1872. They adopted a resolution that Shorthorn cattle trace wholly to imported animals or to those correctly recorded in the *American Herd Book*. A cow's owner at time of service was recognized as breeder of the progeny. Allen followed this resolution in later registrations.

J. H. Pickerall, secretary of the American Shorthorn Breeders' Association, at Chicago, Illinois, aided in consolidating the three American herdbooks into the *American Shorthorn Herd Book*, new series.

The American Polled Durham Breeders' Association was organized in 1889; changed its name to Shorthorn in 1919; and was absorbed by the larger organization in 1922.

The *Canadian Shorthorn Herdbook* appeared in 1867, and the *British American Herdbook* in 1881. These united as the *Dominion Shorthorn Herd Book* in Canada in 1886.

AMERICAN ORIGIN OF POLLED DURHAMS

Polled Shorthorn cattle were known in England, but selection favored horned strains. Horned animals were brought to America. James Sinclair mentioned that three females born in America, tracing wholly to horned cattle from England, were the source of Double Standard Shorthorns. Also, pure Shorthorn bulls, mated with native cows in America, resulted in Single Standard Polled Durhams. The Polled Shorthorn Association voted to join the American Shorthorn Breeders' Association when Double Standard animals were recognized with an X prefix in the herdbook. Single Standard animals were dropped from registry. The American Milking Shorthorn Society designated polled cattle with a P.

DECLINE AND RISE OF MILKING SHORTHORNS

Shorthorn cattle in America before 1860 were largely the dual-purpose strain, some being from Thomas Bates's herd. After Bates died in 1850, prices of Bates-bred cattle reached speculative levels. Colonel Powell's herd of 109 Bates-bred animals averaged $3,504 at public auction in 1873. Rising popularity of beef-type Shorthorns eclipsed expansion of dual-purpose cattle for two decades. A trend toward beef production favored importing improved beef-type animals. Many dual-purpose herds were crossed with Scotch (beef) Shorthorn bulls. Joe Anderson, David Barnard, and George Taylor of Massachusetts, Frank Holland of Iowa, Arthur Simpson of Vermont, and a few other breeders in the Northeast and Midwest avoided use of beef-type bulls.

Amos Pendergast bought six cows from Henry Clay in Kentucky in 1848 and drove them to his New York farm. L. D. May bought two of their descendants for his brother-in-law J. K. Innes at Glenside Farm. These were Kitty Clay 3rd and Kitty Clay 4th. The turning point in Shorthorn interest occurred partly with these cows in a public milking contest at the World's Columbian Exposition at Chicago in 1893, as will be mentioned later. This contest drew considerable attention to milking ability.

MILKING SHORTHORN ORGANIZATIONS

L. D. May and W. Arthur Simpson assembled Milking Shorthorn breeders during the Vermont State Fair in 1910 and organized the American Dairy Shorthorn Association. They adopted the constitution and bylaws of the American Shorthorn Breeders' Association in 1912. Thirty-nine breeders had joined. The first *Milking Shorthorn Year Book* was published in 1915.

A separate midwestern group led by James J. Hill organized the American Milking Shorthorn Breeders' Association during the Minnesota State Fair in 1915. They adopted a uniform plan for classes at fairs; approved a list of Milking Shorthorn judges, and started the *Milking Shorthorn Journal* in 1919. The two organizations united as the Milking Shorthorn Society in 1920 by a mail vote of 201 to 2. Glenn A. Cobb of Independence, Iowa, was elected secretary and editor, succeeded after his early death by Roy A. Cook. Registrations were continued in the *American Shorthorn Herd Book*. L. D. May started a "Voluntary Fund" in 1920 with contributions to promote Milking Shorthorn cattle.

Milking Shorthorn breeders increased to 27 percent of the American Shorthorn Breeders' Association membership by 1935. There were 41.99 percent of Milking Shorthorn members by August 1942; 46.76 percent beef Shorthorn and 11.25 percent Polled Shorthorn members. Differing views on policies led to separation on February 28, 1948. The American Milking Shorthorn Society built a headquarters at Springfield, Missouri.

SOCIETY OFFICERS

The Society conducts business at an annual meeting of delegates and members. Members paying annual dues are eligible to nominate and vote for delegates; the ballots are mailed from the Society headquarters. Sixty-five delegates were elected for 1968. There were 2,015 active members reported in 1961. The officers appointed the secretary, who served also as managing editor of the *Milking Shorthorn Journal*.

GRADING-UP PLAN

Nonpedigree or grade Milking Shorthorn cows in the United States may be entered in a grading-up plan. The foundation cow was owned by a member, classified Good Plus or higher, had Milking Shorthorn characteristics, and was tattooed. She must qualify in production for the Record of Merit. Such a cow entered Class A of the Grading-up Plan, and was served by a registered Milking Shorthorn bull. Female descendants by registered Milking Shorthorn bulls entered Class B, Class C, and Class D in successive generations. A heifer calf bearing four top crosses of registered Milking Shorthorn sires was eligible for full registration. Membership, entry, and transfer fees were involved for such transactions. New Grading-up rules in 1966 granted a Class A foundation certificate to first-cross females meeting requirements for type, production or gains. Daughters of a registered sire became Class B animals, and their daughters in turn were eligible for full registration.

An attempt was adopted at the 1968 annual meeting, to improve milking potential of Milking Shorthorns through use of red-carrier Holstein blood. An approved Red-and-White or a red-carrier Holstein male may be mated with a Shorthorn female scoring 82 points or above. Male progeny would be known as an EXP #1 sire. When such an EXP #1 bull was mated with a Milking Shorthorn female, the female progeny must score 82 points or higher, and yield 13,000 pounds of milk and 455 pounds of butterfat to continue under the plan. A birth record EXP #2 male, mated with a Milking Shorthorn with the above qualifications might have progeny that classified 82 points and become fully registered. This involved the parental and three succeeding generations of grading up toward full registry. No animal with an EXP prefix in a three-generation record certificate would be accepted for a full registration certificate. The next generation was eligible for full registration.

The five-member Grade-up Committee gave unanimous approval to proposed matings of cows producing in the top 2 percent of the breed, with six Red-and-White or red-carrier Holstein bulls whose semen was available, as sires for the 50-50 generation. The committee's approval was required *before* the A.I. breeding.

PUBLIC MILKING TRIALS

A turning point in popularity occurred in 1893 when 25 Shorthorn cows competed with Guernseys and Jerseys in milking trials at the World's Columbian Exposition. Kitty Clay 4th won third place in the 30-day butter test with 1,592.9 pounds of milk and 51.98 pounds of butterfat. The Shorthorn cow Nora was second in the 90-day trial with 3,678.8 pounds of milk, 134.5 pounds of butterfat. Twenty-three Shorthorns averaged 2,881 pounds of milk in 90 days and gained nearly 123 pounds in weight per cow. The Kitty Clays are shown in Figure 17.1. The Babcock test to determine butterfat percentage in milk was used publicly first in this contest (see Fig. 13.5).

Five Shorthorn cows yielded an average of 261.5 pounds of churned butter per cow in 5 months during the Pan-American Exposition at Buffalo, New York, in 1901. Twenty-four Shorthorn cows averaged 4,152 pounds of milk with 153 pounds of fat in 120 days during the Louisiana Purchase Exposition at St. Louis, Missouri, in 1904. Rowena 2d was the champion dual-purpose cow, making 210 pounds of "butter" in 120 days. Many people saw the cows or read reports about them in farm papers.

SHOWS AND FAIRS

An imported bull was exhibited by Colonel L. G. Morris in 1848 at the New York State Fair. The judging standard published in *The Northern Farmer* in September 1854 favored beef conformation. Forty points were allowed for tracing direct descent in the English (Coates's) herdbook. The animal's crops must be full; back, loin, and hips broad and wide; hips well covered; rump with plenty of flesh; twist well filled out; quarters well developed downward; and the carcass round. Of 100 points for perfection, one point was assigned for udder. This disregarded dual-purpose type.

Arthur Simpson wrote of Milking Shorthorn cattle at the Vermont State Fair in 1909: "This fair was the first in America to provide full classes for our (Milking Shorthorn) cattle." Show ring competitions were independent of public milking trials.

FIG. 17.1. Kitty Clay 4th and Kitty Clay 3d produced 1,592.9 pounds of milk, 51.98 pounds of butterfat and 1,230.6 pounds of milk and 41.87 pounds of butterfat, respectively, in 30 days during the World's Columbian Exposition in 1893. An artist's drawing.

Few state fairs offered a classification for dual-purpose Milking Shorthorns by 1916. One breeder going to midwestern state fairs carried a tent to accommodate his cattle. By 1919 a Milking Shorthorn classification was offered by the larger fairs—in California, Eastern States Exposition, Illinois, Iowa, Minnesota, New England Fair, New York, and the Pacific International Exposition at Portland, Oregon.

The National Dairy Show entered only dairy breeds. However, the breed society and California officials cooperated in 1939 so that 19 herds with 168 Milking Shorthorns competed when the National Dairy Show was held at San Francisco. Four cows in milk that won the Dairy Herd class in their breed are shown in Figure 17.2.

The Dairy Cattle Congress at Waterloo, Iowa, has entered Milking Shorthorn cattle since 1941, assisting the Society to build their own barn on the grounds. A class for "Steer, Spayed or Martin Heifer" was added. Judges considered conformation, coverage, and quality of the live animals. Federal carcass grade, dressing percentage, loin, eye area, and carcass value were obtained after slaughter at Waterloo. Nineteen steers entered in the 1958 show averaged 1,059 pounds live weight, 495 days of age, 2.14 pounds daily gain, and dressed 62.27 percent.

FIG. 17.2. First prize Milking Shorthorn dairy herd at the National Dairy Show, San Francisco in 1939. Duallyn Juniper, Grand Champion cow, stands at the right.

Sires of steers entered in the 1959 show were required to be out of Record of Merit dams.

A "Parish Herd" contest was initiated in Iowa in 1936 whereby three or more herds in a district contributed animals that were shown as a group. Minnesota breeders adopted a similar plan in 1937. This plan allowed small breeders with worthy cattle to participate in major shows.

Milking Shorthorns competed at the International Fat Stock Show in Chicago.

A Futurity class was instituted at the National Milking Shorthorn Show at Waterloo, the first show in 1958. Owners nominated heifer calves two years in advance of the show, paying nomination fees of $3, $5, $10, and $20 at 6-month periods prior to the show. Prizes from this purse were paid at 15, 12, 10, 8, and 6 percent to winners of the first five places, and lesser amounts for others. Any balance reverted to the purse for future futurity shows. The eighth Futurity showing of two-year-old cows was held in 1965.

MILKING SHORTHORN MODELS

Seven prominent breeders and Secretary W. J. Hardy studied photographs of 20 leading show cows of England, Canada, and the United States. An artist retouched the outlines of pictures of an imported cow and of a former Grand Champion Milking Shorthorn bull at the Eastern States Exposition, to meet combined ideas of the committee. These model type pictures were approved by the Society in 1933. A revised red Milking Shorthorn model cow, either polled or horned, was developed of durable plastic in 1963.

TYPE CLASSIFICATION

While the Directors were considering type classification, James W. Linn of Kansas State University discussed the subject in the *Milking Shorthorn Journal* (November 1942), based on his experience as an Ayrshire breeder, type classifier, and Extension Dairyman. He stated that the foremost objectives with purebred livestock were to maintain purity of breeding and to improve and promote better animals. Show rings set standards for breed type and were the show windows for better animals, yet few breeders exhibited. A prize ribbon represented only selected animals and affected few individuals from the larger herds. He described type classification with three dairy breeds, measuring progress in entire herds, similar to use of official production records.

Herd classification was adopted in May 1943 with approved judges serving as classifiers. Bulls over 15 months old and all cows in a herd were submitted for inspection. The ratings were:

	Points
Excellent	90 or more
Very Good	85–89
Good Plus	80–84
Good	70–79
Fair	60–69
Poor	under 60

A cow could be rated Excellent after the second calving and after it scored in detail and taped at least 73 inches at the heart girth, equivalent to 1,100 pounds live weight. Small size was insufficient reason for a Fair or Poor rating. A bull scored 90 points and had at least two Good Plus daughters, to be classified Excellent. Only females could be registered from Fair parents. Registration certificates were cancelled of Poor animals. Some 939 cows and 67 bulls were classified in 1969. Average scores for bulls and cows were 87.0 and 85.4 points, respectively.

A spot sample analysis of Milking Shorthorns classified in Canada and the United States before May 1951 was made, seeking some relationship between type and production. It was thought herds possessing a higher proportion of high scoring cows had better management. This analysis is shown in Table 17.1.

TABLE 17.1

TYPE CLASSIFICATION OF MILKING SHORTHORN COWS PRIOR TO 1951 AND THE AVERAGE PRODUCTION OF A SPOT SAMPLE OF THEM

Classification rating	Cows classified		Number of cows with records	Average production		
	(No.)	(%)		Milk (lbs.)	Test (%)	Butterfat (lbs.)
Excellent	945	5.65	794	10,743	3.99	428
Very Good	5,380	32.15	180	10,067	3.95	398
Good Plus	6,535	39.06	250	10,150	3.90	396
Good	3,323	19.86	150	9,663	3.93	380
Fair	514	3.07	50	8,358	4.00	336
Poor	36	.22	1	Not Representative		

The scale of points was revised in 1967 to allow Fair, 65 to under 75 points; Poor, fewer than 65 points. Animals with dark pigmented noses were not rated above Fair. Scores and a letter describing character of the animal, as adopted then, were as follows:

Descriptive Terms for Milking Shorthorn Cows

General appearance and breed character   50 points

Head and neck
  A. Clean cut, well proportioned, with character and strength
  B. Strong head but lacking character
  C. Short head and/or neck
  D. Plain and/or coarse head and neck
  E. Weak, narrow head, especially in muzzle

Balance and carriage
  A. Well balanced with straight topline and style
  B. Acceptable balance but lacking style
  C. Lacking balance with weak loin and/or back
  D. Lacking balance and coarse
  E. Lacking balance and weak or frail

Hips, rump, and thighs
  A. Rump long, wide, and nearly level
  B. Rump medium in length, width, and levelness
  C. High tail head
  D. Narrow rump at thurls and/or pins
  E. Sloping rump

Feet and legs
  A. Clean, flat bone with strong pastern and deep heel
  B. Acceptable with no serious fault
  C. Bone too light or refined
  D. Sickled and/or close at hocks
  E. Spread toes and/or shallow heel

Withers and shoulders
  A. Smoothly blended
  B. Intermediate but not compact
  C. Crops weak
  D. Open or loose

Body capacity   20 points

Stature
  A. Upstanding with stretch and scale
  B. Intermediate but not compact
  C. Low set and/or compact
Chest and barrel
  A. Long barrel with deep open rib and wide chest
  B. Average barrel and chest
  C. Barrel lacking depth and/or narrow chest and heart girth
  D. Short barrel

Mammary system 30 points
Fore udder
    A. Moderate length, firmly attached
    B. Moderate length, slightly bulgy
    C. Short
    D. Broken or very bulgy
Rear udder
    A. Firmly attached high and wide with strong suspensory ligament
    B. Intermediate in height and width but with well-defined halving
    C. Narrow and pinched
    D. Lacking defined halving
    E. Loosely attached and/or broken
Teats and udder quality
    A. Teats plumb, desirable length and size, and squarely placed
    B. Teats not uniformly placed and/or strutting
    C. Teats of undesirable size and/or shape
    D. Udder lacks quality (if determined by handling)
Remarks: A. Patchy; B. Small for age; C. Crampy; D. Wing shouldered; E. Dark muzzle.

Final scorer   points

Descriptive Terms for Milking Shorthorn Bulls

General appearance and breed character 70 points
Head
    A. Clean cut, well proportioned, with character and strength
    B. Strong, lacking character
    C. Short
    D. Plain and/or coarse
    E. Weak, narrow head especially at muzzle
Neck, withers, and shoulders
    A. Smoothly blended
    B. Too thick and/or chine not prominent
    C. Crops weak
    D. Open or loose
Topline and carriage
    A. Straight topline, full crops, strong, wide loin, and moving with style
    B. Strong topline, maybe slightly uneven but lacking style
    C. Weak topline and/or weak, narrow loin
    D. Roaches back when walking
Hips, rump, and thighs
    A. Rump long, wide, and nearly level
    B. Rump medium in width, length, or levelness
    C. High tail head
    D. Rump narrow, especially at thurls and/or pins
    E. Sloping rump
Feet and legs
    A. Clean, flat bone with strong pasterns and deep heels
    B. Acceptable with no serious faults
    C. Bone too light or refined
    D. Sickled and/or close at hocks
    E. Spread toes and/or shallow heel

Balance and smoothness
   A. Well balanced and smoothly fleshed
   B. Lacking balance, coarse and/or patchy
   C. Lacking balance and weak
Stretch and scale
   A. Long bodied and upstanding with proportionate depth
   B. Intermediate in length and/or height
   C. Short coupled and/or low set
Body capacity   30 points
Substance and size
   A. Well muscled and large for age
   B. Average in substance and size
   C. Frail in muscling
   D. Small for age
Heart Girth
   A. Wide chest and full back of shoulders
   B. Intermediate in chest width
   C. Narrow chest and/or pinched back at shoulders
Barrel
   A. Long barrel with deep, open rib
   B. Average barrel
   C. Barrel lacking depth or spring of rib
   D. Short barrel

Final score   points

The Board of Directors appointed the Herd Classification Committee to administer the program. The secretary supplied blank forms, received applications, arranged for inspections, and issued final certificates showing the score of each eligible animal classified in a herd.

A rule adopted in 1968 allowed a Milking Shorthorn to be classified Excellent once when under 5 years old; 2 Ex from 5 to 9 years; 3 Ex from 9 to 12, and Ex at 12 years or older.

PRODUCTION OF MILKING SHORTHORNS

Shorthorns imported early into the United States were popular because of size and milking ability. Private records were reported up to 55 pounds of milk in a day, and 15 to 17 pounds of churned butter in a week. Public milking trials in 1893, 1901, and 1904, mentioned previously, called attention to the milking ability of selected cows. Rose of Glenside, owned by L. D. May at Granville Center, Pennsylvania, was credited with a private record of 18,075 pounds of milk with 675 pounds of butterfat in 365 days.

RECORD OF MERIT

The Milking Shorthorn Society established a Record of Merit in 1915 to obtain milk and/or butterfat records. Cows qualified for entry upon yielding at least a minimum of either milk or butterfat in 365 days. Records were supervised by representatives of the Colleges of Agriculture. Original requirements were:

Class	Age	Milk	or	Butterfat
			(pounds)	
A. Official	30 months	5,250		210.0
Increase per day older at calving		3		0.1
	5 years	8,000		300.0
B. Semiofficial	Under DHIA supervision			
C. Private records	Subject to supervision as determined by the Society. Daily milk weights reported monthly over owner's affidavit.			

CT. Under DHIA supervision. Daily milk weights not required.

The American Milking Shorthorn Society adopted the Unified Rules for Official Testing for supervision as approved by the American Dairy Science Association and the Purebred Dairy Cattle Association. Under a new standard, a cow qualified for the Record of Merit by producing 8,000 pounds of milk or 300 pounds of butterfat in 305 days, computed to a $2\times$ mature equivalent basis with the USDA Age Conversion Factors. Daily milk weights were required, and monthly butterfat tests were supervised by the State Superintendent of Official Testing. Class C Private records were not accepted.

Cows that met production requirements and dropped a living calf within 15 months of previous calving qualified for the corresponding Double Letter class—AA, BB, or CCT.

The rules required in 1959 that the entire herd be entered on test. Daily milk weights were required of all cows in Class A, with a preliminary dry milking and a 1-day test each month. In Class B (Herd Improvement Registry), daily milk weights were not required nor was the preliminary dry milking. The entire herd could start on test in any month in Class B. Cows could be omitted from test in Class B and CY with previous records if they (a) were past 12 years old; (b) were used temporarily as nurse cows; or (c) had lost two or more quarters of the udder.

Class CT records (DHIA or DHIR) were similar to Class B. The owner obtained forms from the Society on which the DHIA supervisor reported the lactation records of every cow and signed the report. Some 2,497 records accepted during 1968–69 averaged 10,120 pounds of milk, 3.74 percent and 378 pounds of butterfat in 305 days on a mature equivalent basis. Owners could enter individual cows in Class D under any of the methods supervised. Such individual records were not used in proving sires or for other recognition awards.

Advanced Record of Production

Superior animals were recognized on meeting requirements if the owner applied for recognition and paid the fee. A cow that classified Excellent or Very Good and produced 20,000 pounds of milk or 800 pounds of butterfat in 28 consecutive months on a $2 \times$ mature equivalent basis qualified for the Advanced Record of Production. Any Record of Merit cow also became eligible when two or more daughters qualified for the Advanced Record of Production. The production requirements were increased in 1962 to 24,000 pounds of milk and 906 pounds of butterfat in 26 months.

Bulls were eligible for the Advanced Record of Production on meeting two requirements: *Production*—Five or more daughters (50 percent or more of daughters recorded 40 or more months previously) shall equal or exceed the Record of Production of their dams in 305 days $2 \times$ mature equivalent, and 15 percent above the requirements. *Type*—At least five classified daughters (50 percent or more, as above) shall average at least 83 points. They must be the same daughters used for production requirements.

The Record of Merit was replaced in June 1962 by the Record of Production, and the requirements were raised for a mature cow to produce 10,000 pounds of milk and 400 pounds of butterfat in 305 days. Animals then on test competed under the earlier plan.

Medal Awards

A Bronze, Silver, or Gold Medal was awarded when a cow produced 2,000, 3,000, or 4,000 pounds of butterfat, respectively, in her lifetime on a $2 \times$ mature equivalent basis.

Meritorious lifetime production was recognized with the W. J. Hardy Memorial Award to any cow with a total production of 100,000 pounds of milk or 4,000 pounds of butterfat in the Record of Merit. Several cows have produced over 135,000 pounds of milk or 5,000 pounds of butterfat in 10 to 17 lactations.

STAR RATING

A young bull or a heifer rated from one to five Stars based on the number of Advanced Record of Merit sires and dams in three generations of ancestors.

PROGRESSIVE BREEDER AWARD

A Progressive Breeder Award plaque or an additional bar for the plaque was earned when a breeder met seven requirements in a year: (a) All eligible breeding animals over 1 year old are registered; (b) at least 50 percent of the females (at least ten) were bred by the owner; (c) the herd averaged 10,000 pounds of milk or 400 pounds of butterfat in 305 days 2× mature equivalent basis for the last completed year; (d) at least two-thirds of eligible animals classified with an average of 83 points or higher; (e) the owner shall have shown at least five Milking Shorthorns at a local or other show within 2 years; (f) the owner shall be a member of the state organization and the American Milking Shorthorn Society; and (g) the owner must make application with proof of the qualifications. Twelve breeders qualified in 1967. Production requirements were increased to 11,000 pounds of milk and 445 pounds of butterfat in 1968.

In April 1969 the Directors voted the statement unanimously: "Milking Shorthorn breed is a dairy breed." Steer Classes also were dropped from the Standard Show Classification.

USDA Sire Summaries of active bulls in artificial breeding, with daughter-herdmate comparisons, repeatability percentages, and predicted differences for daughters' production were published first in January 1968. Average production of cows tested from 1922 through 1969 is given in Table 17.2.

The average butterfat test in individual lactations in 1915–25, 1950, and 1957 ranged between 2.81 and 5.27 percent, with the modes at 3.80 to 3.89 percent fat. Averaged tests of records published in the Year Book from 1915 to the present year varied between 3.74 and 4.01 percent. There was little indication of selection of breeding animals to change richness of their milk.

GAIN REGISTRY

Bulls and steer calves or grade steers born since January 1962 may be entered in the Gain Registry program. Entries were made of all calves born in quarters of the calendar year, calves were

TABLE 17.2
AVERAGE PRODUCTION OF MILKING SHORTHORNS

Year	Number of records	Milk (lbs.)	Test (%)	Butterfat (lbs.)
1922	264	8,408	3.96	332.9
1932	495			331.0
1942	919	8,312	3.92	325.3
1952	2,702	7,995	3.95	315.8
1962	2,072	9,895	3.86	381.0 [a]
1969	2,497	10,120	3.74	378.0 [a]

a. Computed to a 305-day 2× mature equivalent basis.

tattooed, and purebreds were registered. Male calves born within a quarter were weighed at 205 days plus or minus 45 days after birth, under supervision of the Agricultural Extension Service or Vocational Agricultural Instructor. Recognition was given when "the computed weight gain over and above an arbitrary 70 pound birth weight must equal or exceed 2.0 pounds a day for all bull calves and 1.8 pounds of steer calves." Heifer calves were added later, to gain 1.7 pounds a day. Weights were adjusted to 205 days. The symbol WR plus two digits represented the Weaning Recognition and average daily gains to this age. A testing period of 365 and 550 days was approved in 1965. During 1968–69, 75 bulls gained an average of 2.4 pounds, 22 steers 2.2, and 94 heifers an average of 2.0 pounds per day.

At least 10 young males, or 15 males and females, met the Gain Registry requirements to qualify their sire for the Progeny Weaning Recognition award. Retnuh Choice, bred and developed by Joe Hunter, Retnuh Farms, Genesco, Kansas, was the first sire to qualify. At least two progeny met the Weight Recognition requirements in the Gain Registry to qualify their dam as a Progeny Weaning Recognition Cow. Two bulls and 46 cows earned these awards in 1969. The owner pays a $2 recording fee for such a dam.

Secretary Ray Schooley viewed dual purpose qualities from early observations in seven states, as follows: "We have characteristics in our breed that are unique in Milking Shorthorns. These are (1) easy fleshing, (2) rapid growth, and (3) lean attractive tasty butcher cuts. However, to compete in this modern economy, we must produce cows that will milk without sacrificing our other good points."

ARTIFICIAL BREEDING

The Purebred Dairy Cattle Association required that bulls used in artificial insemination be blood typed for antigen pattern and a copy of this record be filed with the breed association. A filing fee was required.

The American Milking Shorthorn Society established three requirements on December 1, 1959, under any of which the purebred progeny of a Milking Shorthorn bull in an artificial breeding center were eligible for registration. These standards were: (a) The bull's dam must have classified Very Good or Excellent. Her Record of Merit records must average 10,000 pounds of milk or 400 pounds of butterfat within 365 days on a $2\times$ mature equivalent basis. All records of the three nearest dams must meet this standard. (b) The dam must qualify in production and type as above, and five of the seven nearest dams must average 10,000 pounds of milk or 400 pounds of butterfat on a $2\times$ mature equivalent basis, and each have qualified for the Record of Merit. (c) The bull must have earned the Advanced Record of Merit.

Restrictions on eligibility of bulls for use in artificial service were removed in October 1962. Requirements of the Purebred Dairy Cattle Association were continued. The bulls must be blood typed for antigens at a fee of $25, at the Serology Laboratory, School of Veterinary Medicine, University of California at Davis, before the semen is collected. The blood typing report must be filed with the Society with a $10 recording fee, even for within-herd artificial use. All registered Milking Shorthorn cows must be identified from the registration certificate at insemination, and a complete breeding receipt must accompany application for registration of the calf. Some 19.4 percent of animals registered in 1966 were conceived thus. Fifteen Milking Shorthorn bulls were in six A.I. studs in the United States during 1969. Frozen semen was available by transfers of title between studs in addition to custom freezing service with privately owned blood-typed bulls.

One Milking Shorthorn bull was registered per 4.7 females, plus 25 steers to qualify them for shows in the year. Some 1,066 active members paid annual dues for 1966, with 1,413 junior members.

Milking Shorthorn Year Book AND Journal

The *Milking Shorthorn Year Book* has been published annually since 1915. The *Year Books* contain production records, show winnings, USDA proved (analyzed) sire averages on production, type classification, gain of any bull, and records of five or more progeny during the year. Five volumes of the *Milking Shorthorn Herd Book* appeared in 1948–54.

The *Milking Shorthorn Journal* was edited bimonthly by Roy A. Cook from March 1919. W. J. Hardy became secretary and editor in December 1940 and moved to the American Shorthorn Breeders' Association office in Chicago. The *Journal* increased in importance as membership increased. The Society occupied a new headquarters at 313 Glenstone Avenue, Springfield, Missouri 65804. W. E. Dixon became secretary in 1955. Ray Schooley was secretary during 1961–67. Harry Clampitt became executive secretary in 1968.

REFERENCES

Allen, Lewis F. 1846–82. *American Short-horn Herd Book*. Vols. 1–24.
———. 1872. *History of the Short-horn cattle. Their origin, progress, and present condition*. Buffalo, N.Y.
Anonymous. 1854. Pedigree of the Short Horn cow. *Northern Farmer* 1(9): 417–18.
Anonymous. 1863. *Shorthorns*. The Association of Breeders of Thoroughbred Neat Stock. Hartford, Conn.
Anonymous. 1893. (Report of milking trials, World's Columbia Exposition.) *Breeder's Gazette* 24:277–78.
Burrows, G. T. 1947. How dual-purpose Shorthorns reached America. *Milking Shorthorn J.* 28(3):3.

Cande, Donald H. 1955. The Milking Shorthorn then—now—tomorrow. *Milking Shorthorn J.* 36(2):2.

Clevenger, C. L. 1942. Shorthorns in America. In E. P. Prentice, *American dairy cattle. Their past and future*. Harper, New York.

Cook, Roy A. 1919–. *Milking Shorthorn J.* Vols. 1–.

Linn, James W. 1942. Herd classification. *Milking Shorthorn J.* 23(11):3–5.

Plumb, C. S. 1920. *Types and breeds of farm animals*. Rev. ed. Ginn, Boston. Pp. 235–37.

Sanders, A. H. 1916. *A history of Short-horn cattle*. Chicago.

Simpson, W. Arthur. 1922. *Milking Shorthorn J.* 4(3):1–2.

Sparkman, John. 1951. The proven merits of classification. *Milking Shorthorn J.* 32(6):41.

Wood, Cliff. 1948. Polled Milking Shorthorn history. *Milking Shorthorn J.* 29(4):45–47.

Milking Shorthorn J.

    1937. Mrs. L. D. May passes away at her Glenside home. 18(9):3.

    1943. Editorial. Progress. 24(1):28–29.

    1945. Grading up plan. 26(1):26–27.

    1949. Division of herdbook society. 30(5):6.

    1965. Rules governing artificial insemination. 46(5):8.

    1967. A history of Polled Milking Shorthorns. 48(3):4–5.

    1967. Production rules and associated pedigree symbols. 48(3):12.

    1968. The USDA Sire Summaries. 49(3):6.

    1968. 20th annual meeting and Board of Directors report. Special notice of Grade Up Committee. 49(6):2–5, 28–30.

    1969. 21st annual meeting and Board of Directors report. 50(6):4–5, 28–29.

Other Breed Publications

    1916–. *Milking Shorthorn Year Book*. Vols. 1–. Chicago; Springfield, Mo.

    1948–54. *The American Milking Shorthorn Herd Book*. Vols. 1–5. Springfield, Mo.

CHAPTER 18

RED DANISH IN DENMARK

Denmark is situated east of the North Sea and in the western part of the Baltic, between 54° 33′ and 57° 45′ North latitude, 8° 5′ and 12° 57′ East of Greenwich, exclusive of Bornholm Island in the Baltic Sea. The main islands are Zeeland (Sjaelland), Funen (Fyn), Moen, Falster, and Lolland. The Faroe Islands and Greenland are in the North Atlantic. Jutland comprises 11,441 square miles and the islands 5,171 square miles. The population numbered 4,448,401 in 1958. The land is mainly a rolling low plain up to 564 feet above sea level. Annual rainfall ranges from 21 to 27 inches and mean temperature from 61° in July to 32° F. in January. Glacial boulder clay soil is distributed generally, with sandy soil in the west and organic (peat) soils in beds of old glacial lakes. The soils are adapted for agricultural and forage production. The main income is from agriculture, industry, commerce, and fishing. The agricultural census of 1958 showed 7,699,000 acres (73 percent of the country) divided into 3,438,000 acres of cereal grains (wheat, rye, barley, and oats); 1,453,000 acres of potatoes, sugar beets, beets, turnips, and carrots; 2,570,000 acres of hay crops and grasslands; 11,000 acres of fallow lands; and 227,000 acres of other crops.

The 3,400,000 cattle in 1959 included the Black-and-White Danish in Jutland, the Red Danish Milk Breed predominating on the islands, with 15 percent of Jerseys and 8 percent of Shorthorns. About 5,500,000 swine, largely of the White Danish (Landrace) breed, were produced. Poultry were important, and horses were used in farming. The main agricultural exports were bacon, butter, eggs, beef, and veal.

EARLY DANISH SETTLEMENT

Denmark was covered by icecaps three times. Neanderthal man hunted fallow deer and split their bones for marrow during the last warm interglacial period (50,000 B.C.). The earliest men known in the area were of the Noerre-Lynby culture; they used coarsely flaked flints and reindeer horn axes.

The Maglemose culture followed; it was named for the Danish site recognized by bone implements and weapons. Hunters of the Hamburgian culture lived in southern Jutland in the late Glacial period. They used flint arrows, scrapers, and tools of antlers and bone. Charcoal from the Bromme settlement was dated by radioactive carbon-14 to about 10,000 B.C.

The most widely known early people were of the Litterina period, when oak trees spread across the landscape. These people lived along the coast, ate some fishes and molluscs, shells of which accumulated in *Koekkenmoeddinger* ("kitchen middens") or shell mounds. Excavation of the shell mounds along the Kattegat and elsewhere yielded bones of many wildfowls and of hunted animals. E. Cecil Curwen wrote:

> The bones of animals found in Mesolithic settlement sites of the Mullerup culture (say 6000 B.C.) show that man was dependent for his living on wild animals such as the red deer, roedeer, elk, urus, wild boar, beaver, badger, pinemartin, wild cat, fox, and hedgehog, and on such birds as the grey duck, grebe, shoveller, coot, crane, cormorant, sea-eagle, and black

stork. The only domesticated animal at this period is the dog—
a large and a small. Later on, in the Ertebolle middens, dog-
bones are common, and the bones of other animals show marks
of gnawing by dogs.

The cultivation of corn (cereals) and the domestication of
animals appear in Denmark at the beginning of the Neolithic
period, about 3000 B.C. . . . both appear simultaneously in
northwestern Europe. . . . The domestic ox, pig, sheep, and
goat first appear in Denmark at the beginning of the Neolithic
period, simultaneously with wheat and barley, and must have
been introduced from the south. There was thus no purely pastoral stage preceding cultivation.

. . . From impressions of seeds found on early pottery, the
following facts were obtained:

Neolithic		Number of discoveries	Number of impressions
Einkorn wheat	(*T. monococcum*)	2	21
Emmer wheat	(*T. dicoccum*)	40	288
Dwarf or common wheat (*T. compactum* or			
	vulgare)	24	46
Naked barley	(*Hordeum coelesti*)	26	40
Husked barley	(*Hordeum vulgare*)	7	14
Bistort	(*Polygenum*)	3	3
Apple		6	13

From this, wheat was more common than barley, and the
husked barley to which modern varieties are related, forms
only a quarter of the total barley. In the late Bronze Age
(800–400 B.C.) the importance of wheat and barley appears
reversed, with Emmer still predominately among the wheat.
The probable ultimate source of wheat is the Near East . . . in
Turkestan, Afghanistan and northwest India, emmer in Abyssinia. Barley from Abyssinia, in the Himalayas and the neighboring parts of China. Oats came in during the Bronze Age;
rye at the beginning of the Iron Age.

EARLY CATTLE IN DENMARK

B. primigenius Bojanus was the major wild ox of northwestern
Europe, with *B. frontosus* Nilsson (a small polled species) in
the Scandinavian region to the north. Neolithic man brought
domesticated cattle of the *B. longifrons* Owen species and cereal
grains in migrations from western Asia, up the Danube valley, and
northward of the Alps. As people migrated, some domesticated

cattle crossed with the wild cattle, producing the Celtic Red (Werner).

Evidence was seen in the Danish National Museum, Copenhagen, that the wild ox (*B. primigenius*) was among early hunted animals. Small flint microliths are embedded in two ribs of an almost complete skeleton dug from a peat bog at Vig, on Sjaelland. See Figure 2.5. Their bones were found also in kitchen middens.

DEVELOPMENT OF DANISH CATTLE

History of early agriculture was scanty, with tribal groups or small kingdoms in different parts. Livestock appeared of minor importance. The peasants paid rents to the landowners with grain.

Christianity was introduced in A.D. 826 by the French prelate Ansgar, "the apostle of the North," beginning a new era. King Harold Bluetooth was baptized soon after, followed rapidly by conversion of his people and development of natural resources.

Bullocks from Jutland were sold for grazing in the Netherlands around A.D. 1200. Considerable demand developed late in the seventeenth century for fat cattle into Germany, southwestern Europe, and Copenhagen. An import duty levied against Danish cattle in 1724 reduced this trade. Rinderpest caused losses in 1745–52 estimated at 2 million head. Exports of cereal grains, butter, and cheese began to replace this loss of income.

Cattle in northern Jutland were mainly black-and-white. No distinct breed existed on the Danish islands then. The cattle were small, largely horned, and generally spotted—red and white, brindle and white, but seldom black. Hay and straw were fed in winter. Svendsen noted that during the spring, crews of men went to the cow stables each morning on many large farms to help up weak cows. A mural painting at the National Agricultural Exposition in 1938, celebrating 150 years after relief from serfdom, depicted this practice of "lifting days."

FREEDOM FROM SERFDOM

Count Christian Ditlev (1748–1827) and his brother Johan Ludwig Reventlow saw from English agricultural practices that abolition of villeinage would benefit the Danish people. Crown Prince Frederick decreed emancipation from serfdom on June 20, 1788.

ENCLOSURE OF LANDS

Separate strips of land tilled by tenants and small freeholders were exchanged for consolidated units replacing the "Community Plan" and open field system. Farmers were free to plan cropping systems and manage their own livestock. Farmsteads were built on many small farms. These changes led to growing more cereal grains. Wheat production increased in the Argentine, Australia, and the midwestern United States. Competition in European markets caused a crisis in Danish agriculture, turning many farmers to livestock production.

AGRICULTURAL ORGANIZATIONS

The village early governing body was the By-lag, representing landowners, tenants, and crofters without land. Rules to govern the community were agreed upon, read aloud, and recorded. These rules, confirmed by the King, became the district laws. Village fields were divided into strips requiring simultaneous working by the laborers. Government by the By-lag was lessened by the powers which estate owners exercised through the Middle Ages and during the absolute reign by Kings. Village fields became neglected when peasants were obliged to work the estates. This situation and wars impoverished the Danish nation, particularly the farm population subject to military service.

The Royal Agricultural Society of Denmark was organized in 1769. Members included the nobility, estate owners, and civil servants in towns but few actual farmers. The Society favored forming local agricultural societies, beginning with one on the Island of Bornholm in 1805. Local societies increased to 137 by 1835, dealing with every phase of agriculture. Bylaws of the Royal Agricultural Society stated as an objective " . . . to encourage by prices and prizes the agriculturist, the artist and the merchant within the domain and lands of His Royal Majesty."

Agricultural advisers or "Konsulents" were employed by societies to serve their members. T. R. Segelcke was appointed as dairy adviser by the Royal Society in 1860. The societies sponsored shows, production competitions between herds, and investigations of bulls' progeny; they also arranged for local experiments, demonstration fields with crops, and other research.

Federations of societies coordinated their activities, starting in 1872. The Federation of Danish Agricultural Societies was established. The first breeding society was organized in 1884 for joint ownership of bulls. By 1933 some 1,200 societies owned 1,443 bulls used by about 25,000 members. A small grant was available for each breeding society since 1887. The government granted a half million kroners by 1940 to sponsored cattle shows, breeding societies, etc., supplementing the membership fees. Cooperative ownership of bulls for natural service decreased to four animals by 1958. However, 98 cattle breeding societies owned 1,212 bulls that inseminated 1,536,946 cows and heifers in 172,000 herds. Some bulls were owned privately.

DEVELOPMENT OF THE RED DANISH MILK BREED

Svendsen described native cattle of Lolland-Falster as small, slender, angular, and often long legged. They were light fleshed with comparatively good udders, and some had narrow rumps and were polled. Colors ranged from light red to brownish black; many were red spotted, and black spotted or dun.

Cattle on Funen were shades of red or spotted. Some were *ryggede* (a broad white stripe along the back) with white speckles in the adjoining ground color. They were generally light fleshed, with large barrels, often cow-hocked, with long upturned horns. The cows had reasonable dairy conformation and fair udders. According to Prosch, native cattle on Sjaelland were red, brown, black, or spotted. They were angular, with prominent hips. Cattle on the large farms were improved first with introduced animals. Poor feeding and management restricted milk production.

Appel stated that during the best season on Lolland and Falster, milk production might average 5 to 6 Potter daily (1 Pot nearly equaled 1 kilogram); 10 Potter on Funen. A few cows in the Frederiksberg district yielded 12 to 14 Pots. Butter production was estimated at 1 Fjerding (61 3/4 pounds) yearly per cow. The Aalstrup herd averaged 45 to 50 pounds of butter yearly per cow during

1832–37 without grain in winter. When oats were added in the winter feed (1837), 68 cows averaged 77 pounds of butter. With more grain during 1838, 70 cows averaged 90 pounds of butter.

CATTLE DRIVES

Cattle from Slesvig were introduced onto Lolland and Falster by 1750. Cattle were driven overland during 1800 to 1840 for feeding on Funen and Sjaelland, and to supply milk for Copenhagen. Cows from the Angel peninsula in South Slesvig were favored but later were considered small. Larger cattle from the marsh district near Ballum and around Tonder in southwestern Slesvig were driven across Funen to Sjaelland in 1840 to 1863. This province was lost to Prussia in 1864, and restored by plebiscite on July 10, 1920. The popular Kristoffer bull line descended from these cattle. Anthony wrote: "As these droves of cattle passed thru Funen and the other islands, the drovers had stopping places along the way to rest and feed them. This . . . resulted in the buyer . . . selling an unusually good cow to some friendly farmer." Newborn calves were left along the route. Farmers became dissatisfied with poorer native cattle. These districts are shown in Figure 18.1.

OURUPGAARD FARM

E. L. Anthony studied development of the Red Danish Milk Breed. E. Tesdorpf took over Ourupgaard Farm on South Falster island in 1840, where 175 to 200 milking cows were maintained. Some 20 to 40 young Angler heifers or cows replaced native cows yearly. Tesdorpf kept permanent milk records of each cow, and earmarks permitted tracing the descent after 1850 to 1860. Gradually the replacements were homegrown, except for later herd sires.

P. A. Morkeberg, state konsulent in cattle breeding, improved the marking system for Tesdorpf, and started the "Family Herd Book" in 1891 based on cows. The herd exhibited and participated in "Breeding Center" production competitions. Development was based on purchased Angler stock along with native cows. Later bulls were from the Ryslinge farm and elsewhere on Funen, with cattle descended from Ballum and Slesvig foundations. Changes in average production per cow are seen in Table 18.1. Breeding stock sold from the herd influenced many Red Danish cattle. Forty-five of the first 77 bulls of the *Lolland-Falster Herd Book* traced to this herd, and 48 of the 253 cows in the *Cow Herd Book*. Some 866 bulls and 369 females were sold as breeding animals from 1875 to 1921.

Other herds important in development of the Red Danish Milk

FIG. 18.1. The black-and-white Jutland breed originated on the mainland. The Red Danish Milk breed was developed on the islands from crosses of native cattle with animals from the Angel peninsula, large Ballum cattle from the Tonder district, and elsewhere. The boundary between Jutland and Germany has been changed by wars and a plebiscite.

Breed, cited by Anthony, included the Ryslinge herd founded in 1840, the Braenderupgaard herd (1865–1915), Birkelund, and Lemberg herd, all on Funen. Prominent foundation herds or centers on Sjaelland included Kolle Kolle (1892–   ), a group of small breeders at the Tjustrup Center, and the Holsinge district herds. Some estates supplied breeding animals. Small farmers gave more attention to individual cattle, feeding, and management. Some authorities believe that cooperative use of bulls led small herds to more rapid

TABLE 18.1

AVERAGE PRODUCTION OF COWS IN THE OURUPGAARD HERD, 1841–1921

Years	Average number of cows	Milk (lbs.)	Fat tests (%)
1841–51	202	3,213	
1851–61	209	3,852	
1861–71	219	5,082	
1871–81	230	5,914	
1881–91	249	5,596	
1891–1901	294	5,573	
1901–11	515	6,190	3.34
1911–14		6,630	3.45
1914–18	545	5,685	3.53
1918–21		6,730	3.57

a. Nejsemhed and Skevlykkegaard farms were purchased in 1901.

advances than on large estates. More feed crops were grown; cows were fed more liberally; and management was improved. In addition, interest in production records increased. These measures speeded breed improvement. Few Angler cattle were brought in after 1890.

Improved feeding and management allowed cows to show their ability in proportion to hereditary capacity. Governmental and cooperative agencies worked together to improve the breed.

DANISH BUTTER EXPORTS

Consul Ryder believed that increased feeding around 1870 influenced milk yields and yearly butterfat production and butter exports, which were as follows:

1866	67,305 cwt. butter	1880	300,157 cwt. butter
1870	127,013 cwt. butter	1883	353,584 cwt. butter
1875	206,171 cwt. butter		

He believed Danish cattle had developed through improvement of Angler cattle to weigh 900 to 1,050 pounds at maturity. Young bulls weighed 1,200 to 1,400 pounds. Ryder estimated that cows yielded 6,500 pounds of milk with many exceeding 8,000 and some over 10,000 pounds of milk yearly.

FARMERS' ASSEMBLIES AT AGRICULTURAL SHOWS

Probably the first cattle show in Denmark was held in 1810. They were more systematical after 1845 with classes and premiums for different kinds of cattle. The *Landmandsforsamling* or general assembly for farmers was held at Randers, Jutland, with the show in 1845. Many of the 94 cattle competing were auctioned later.

General farmers' assemblies were held during the shows at Odense in Funen in 1846; at Aarhus, Jutland, in 1847, and Copenhagen, Sjaelland, in 1852. Discussions at these assemblies considered improving native cattle with foreign blood, and other subjects.

A Shorthorn bull came to Ejderstadt in Slesvig in 1843. Island farmers, however, decided on developing a red milk breed. The largest movements of red Angler cattle were from 1841 until the Slesvig war in 1863. Red cattle from northern Slesvig were sold at markets to small farmers.

The show at Aarhus in 1866 had eight classes for pure Jutland cattle (black-and-white), four for Anglers, two mixed classes, one class each for Slesvig, Ayrshire, Spanish, mixed Slesvig, and mixed Ayrshire-Angler cattle. By 1873 cattle at many shows on the islands were mainly of red color.

RECOGNITION OF RED DANISH CATTLE

The name "Angler or Red Danish Cattle of Pure Race" was applied at the 14th Farmers' Assembly and at the Svendberg show on Funen in 1878. Later shows had classes for Red Danish Milk Cattle. Shows favored rapid improvement in type.

Anthony regarded 1845 to 1885 as the great molding period in development of the Red Danish Milk Breed on a native foundation. He wrote: "The three prominent breeds which really left an influence was first of all the Angler, in sections of Lolland-Falster and Sjaelland, and to a certain extent in South Funen, with the Ballum cattle having the greater influence in North Funen and the northern part of Sjaelland, and the North Slesvig . . . closely akin to the first two, having a prominent part also in Funen and Sjaelland and to a lesser extent in Lolland-Falster."

Although Ayrshire, Breitenberger, Brown Swiss, Shorthorn, and Tyrol cattle were used to a limited extent, their influence was largely lost. Red Danish Milk Cattle were recognized generally by 1885.

UNITED STATES CONSULAR REPORTS

United States Consul Ryder reported to the State Department in 1883: "Denmark possesses two breeds of cattle, namely the Red Danish and the Black Spotted Jutland. The first named constitute the cattle herds of the islands, as also those of a few districts in the southern part of Jutland, whilst the Black Spotted are to be found throughout all the Jutland districts."

Some Shorthorn cattle survived and spread in southwestern Jutland.

Angler cattle developed in Schleswig (Slesvig). The cows were a red-brown color and weighed 750 to 900 pounds. They yielded 2,200 to 3,000 quarts of milk yearly. Selected cows in Saxony during 1877 to 1881 yielded an average of 2,939 quarts yearly. Consul Ryder wrote that Angler cattle were smaller than those on Funen. Importance of Angler cattle was shown from the census of bulls.

	Island provinces		Jutland	
Year	Angler bulls	Native bulls	Angler bulls	Native bulls
1866	1,981	10,894	266	4,833
1871	1,961	7,907	266	4,208
1876	2,388	7,091	300	4,376

DANISH CATTLE SHOWS

L. Hansen Larsen stated: "The cattle shows may be characterized as our oldest existing measure for the furtherance of cattle breeding." Cattle shows originated in the British Isles but were modified in Denmark. The Agricultural Society of Randers, Jutland, awarded prizes on conformation to bulls at its first show in 1810. The first government Act of Breeding of Domestic Animals in 1852 appropriated $4,000 as premiums for livestock at shows. These efforts encouraged local enterprise with premium moneys in proportion to local contributions. Improvement of livestock was encouraged through educational features at the shows.

The scale of points for judging cattle at the Soree County Farmers Economic Society show in 1852 was as follows:

Guenon milk mirror, size and fineness	16 points
Conformation, hide, hair, lay, head, horns, eyes, and size of animal	8 points
Total	24 points

The proprietor of the Ourupgaard herd went before the Farmers' Assembly in 1869 to object to so much emphasis on the escutcheon.

The government appropriated $7,300 for prizes to bulls past 3 years old at district agricultural shows in 1887. To avoid inbreeding, bulls often had been discarded before their daughters were of milking age. Only 371 bulls competed at district shows in 1887, as compared with 1,200 bulls at state shows in 1908.

The first show for young stock was held in 1892 to encourage greater uniformity in animals. Young stock shows became important for purchase and sale of breeding animals. Provincial shows began in 1892. Before 1900, production records from milk recording societies replaced opinions concerning udder, teats, and milk veins as indications of milking capacity.

GOVERNMENT PROVISIONS FOR SHOWS

The Law of 1902 increased government grants for prizes at State Shows to $18,280. If prize money was accepted, the owner was obligated to retain the bull in service until the following May. Bulls 5 years old or over were judged upon the character of daughters and sons examined at the owners' farms before the show. The number of progeny varied with age of the bull, as follows:

Age of bull	Length of service in district	Progeny over 1 year old shown
3 years		
5 years	2½ years	10
6 years	3 years	12*
7 years	4 years	14**

*Three must be yearlings.
**Four must be yearlings.

Awards to old bulls constituted a progeny show of transmitting ability. A committee appointed by the Minister of Agriculture managed progeny shows, cooperating with the Agricultural Societies.

Competitions between entire herds and Milk Recording Societies began in 1895. Production records have been required of all cows exhibited since 1906.

The Law of 1912 specified:

> After the expiration of two years from the enactment of this Law, no grant shall be given for prizes for bulls of dairy breeds unless reliable information be given of the yields of milk of their dams by quantity and percentage of fat.
>
> After five years . . . no grant shall be given for prizes of cows of dairy breeds unless reliable information be given of their yield of milk by quantity and by percentage of fat.

This law was a minimum standard for the grant. Cows met minimum standards for milk yield and butterfat percentage. The government withdrew from supervision in 1927. Over 100 shows were held in Denmark in 1930. Government grants provided about half the prize moneys, supplemented by the societies. By 1954 about 2,000 bulls and 13,000 cows and heifers were awarded prizes annually at shows.

Bulls under 5 years old were awarded prizes as first, second, third, or fourth class bulls for conformation and for "pedigree." The pedigree points were given for conformation scores of parents and production records of ancestry in the first four generations, with more points given to sire and dam than to earlier generations.

Bulls were divided into two groups. A bull 4 to 5 years old might earn prizes for conformation and for progeny. Yields of daughters were considered with bulls past 6 years old. Rules differed for bulls in natural use, and those in artificial service.

Herds of 6 to 8 cows or more might exhibit an individual cow; 3 cows represented herds up to 30 cows; 4 cows for herds of 31 to 50 cows; etc. Separate prizes were given for conformation and for production. Prizes for heifers were not based on size of the herds represented.

EVALUATION OF PRODUCTION RECORDS

Average production records of each cow shown formerly were computed to "butter" by formula, based on milk yield, the fat percentage minus 0.15, and the amount divided by 86. A cow must have produced at least 150 kilograms of butter, with 3.7 percent butterfat in her milk. More points were earned with higher yields and butterfat percentages, thus:

Butterfat	Fat	Points
(kg.)	(percent)	
170	3.9	1
180	4.0	2
190	4.1	3
200	4.2	4

More butterfat and a higher fat test earned additional points at the same increasing rate. An age conversion factor of 20 kilograms of butterfat was added to a record begun under 2 years 9 months of age, and 10 kilograms in the second lactation. If the first lactation began after 2 years 9 months, 10 kilograms of fat were added only once. Cows with 4, 5, or 6 lactations received 1, 2, or 3 points in recognition of longevity. Points for all cows shown were added and prize groups divided as follows: A 12.0-point average rated First Prize class; 9- to 11.9-point average rated Second Prize class; 6- to 8.9-point average rated Third Prize class; and 3- to 5.9-point average rated Fourth Prize class.

Uniformity of animals as to size, conformation, breed type, and mammary development was considered important within groups. Separate prizes were awarded for conformation and for pedigree to each individual from a small herd, or to groups from larger herds. Bulls past 5 years old received a prize only according to the merits of their progeny.

The last government-sponsored national show was held in 1900. The federated local societies sponsored provincial shows. The federated local societies held an agricultural exposition at Bellahoj in 1938 celebrating 150 years of liberation from serfdom. Cattle were exhibited from every province. A Red Danish bull at this show is shown in Figure 18.2.

Larsen commented on the influence of shows: "The cattle shows have contributed very materially to the improvement of the conformation of Danish cattle. . . . The highly prized bull, born 1886, is tall and coarse of forequarters, soft of back and light of hindquarters, while the highly prized bull, born 1932, has a far better back line, and deeper and stronger hindquarters."

PRODUCTION RECORDS AND MILK RECORDING SOCIETIES

An estate-owner offered prizes in 1833 for herds on Soroe or Sjaelland that produced milk especially high in butterfat. Milk was recorded from individual cows on Ourupgaard Farm since 1841. Other large herds kept production records later. N. Jacobsen, a Jutland farmer, devised a churn test in 1877, as did N. J. Fjord in 1879. The Gerber butterfat test superceded these in butter factories in the 1890s.

Several farmers discussed producing ability of cows in the folk school teacher's home in Askov. His wife, Mrs. Annine Hansen, suggested that they employ a "Control Assistant" to record the yield and feed consumption of individual cows. Subsequently, 12 farmers owning about 300 cows organized the Vejen and Omega Society on January 24, 1895. They employed Emil Konradi as su-

FIG. 18.2. This typical Red Danish bull was exhibited in 1938 at the agricultural exposition at Bellahoj.

pervisor to begin work May 1. Milk samples were taken regularly and tested for butterfat by the method devised by Dr. Nicholas Gerber of Zurich, Switzerland, in 1888.

The first Control Society desired to develop cow families that would produce much milk with a high percentage of butterfat. By 1958, 1,721 milk control societies operated in Denmark. More cows were under test on Funen than in other parts of the country. Other countries patterned after this first society. King Christian X conferred the title "Knight of the Danish Flag" upon Emil Konradi for his pioneer leadership.

The Law of 1902 on Breeding of Domestic Animals granted over $32,000 toward milk control societies with this provision: "The society shall have for its aim to make dairy farming more profitable, by examining into the feeding of the individual cows and their yield of milk by quantity and quality and to help to form strains of dairy cattle producing a higher yield of butter."

The Law of 1912 set the maximum grant at $50 to a society of at least ten members with 200 cows. In return, each society submitted an annual report to the provincial Federation of Agricultural Societies listing all cows with their birth dates, sire and dam, milk yield, average fat percentage, feed consumption calculated to feed units, calving date, and identification of calf. Transmission of high butterfat test was important, as butter was the main dairy product exported.

The Red Danish bull Birk, three sons, and three grandsons illustrate the transmission of high butterfat percentage (see Table 18.2). Hermed and Thjalfe among the Birk bulls decreased the amount of milk from their daughters, but increased the fat percentage enough to increase the total butterfat yields.

Danish cows under milk recording in 1899 averaged 4,322 pounds of milk and 145 pounds of butterfat. Eskedal showed that there was appreciable improvement in average milk yields and butterfat percentages of cows under milk control. The change over 54 years is shown in Table 18.3.

Better feeding, management, and improved hereditary capacity resulted in a higher average production per cow. Limited feed supplies reduced production in World War II. Part of the recent increase in production is attributed to improvement of the cows by widespread use of proved bulls in artificial breeding.

RED DANISH MILK BREED

Red Danish cattle have horns of medium size, often directed outward and forward. The head is refined; skin thin and pliable; chest deep; and barrel capacious. Hips and rump are square and level,

TABLE 18.2

DAUGHTER–DAM COMPARISONS OF 7 BIRK BULLS OF THE RED DANISH
MILK BREED (BIRK, 3 SONS, AND 3 GRANDSONS)

Bull	Number	Daughters			Dams		
		Milk (lbs.)	Test (%)	Fat (lbs.)	Milk (lbs.)	Test (%)	Fat (lbs.)
Birk	59	8,424	3.94	332	7,962	3.70	295
Birkfus	10	9,214	3.91	360	9,000	3.46	311
Birk Nakke	30	9,053	3.83	347	8,639	3.38	292
Hermod	21	9,303	4.35	405	9,731	3.81	371
Jason	33	9,299	4.25	395	8,884	3.73	331
Kretheus	27	9,429	4.05	382	8,534	3.65	313
Thjalfe	34	8,582	3.91	336	8,794	3.56	313
Average		9,043	4.03	364	8,792	3.61	318

TABLE 18.3

INCREASE IN PRODUCTION BY DAIRY COWS UNDER MILK CONTROL IN DENMARK,
BETWEEN 1904 AND 1958

Island or province	1903–4			1957–58			
	Milk (lbs.)	Test (%)	Fat (lbs.)	Cows under control (%)	Milk (lbs.)	Test (%)	Fat (lbs.)
Bornholm	6,413	3.39	216	60	9,006	4.47	403
Funen	7,395	3.39	251	77	8,878	4.72	419
Lolland-Falster	6,243	3.36	210	57	9,176	4.41	405
Sjaelland	6,905	3.42	236	59	9,264	4.44	411
Jutland[a]	6,731	3.41	229	58	8,960	4.34	389
Entire country[b]				60	9,001	4.41	397

a. Part of the cattle on Jutland are Black-and-White Danish cows. Most of the island cattle were of the Red Danish breed. Some Jerseys and Shorthorns are in Denmark.
b. Red Danish cows at the bull testing stations in 1967 yielded an average of 11,469 pounds milk, 4.53 percent and 493.8 pounds of fat, 3.80 percent and 423.6 pounds of protein. Black-and-White Danish cows averaged 11,484 pounds milk, 4.32 percent and 496 pounds of fat, 3.65 percent and 419 pounds of protein.

sometimes with prominent sacrum and tailhead; thighs are fairly full. The udder is capacious but tends to be pendulous; teats from 3 to 4 inches long and sometimes irregular in shape. The color markings were described by Larsen thus:

> The colouring is most frequently a medium red, varying from a light red to a dark red. Markings are now of fairly rare occurrence, they are not desired, but are permissible even in herd book and show cattle, if small and found under the belly or around the udder. The muzzle must be of a dark slate colour. The sparsely haired skin around the natural openings is not infrequently of a light red or yellowish red colour.

Mature cows generally average around 1,150 pounds, but are about 220 pounds heavier in elite herds. The cattle are of a typical dairy build. Red Danish cows exhibited at the Jubilee Cattle Show in Copenhagen in 1930 weighed 1,389 pounds (1,168 to 1,647 pounds); at Odense, 1,360 pounds. Mature Red Danish bulls averaged 2,095 pounds.

The selected 670 cows in the *Cow Herd Book* (volume 39) averaged 12,637 pounds of milk, 4.53 percent and 573 pounds of butterfat. Selection of breeding animals is for at least 4.0 percent fat, with above 7,721 pounds of milk and 485 pounds of butterfat.

BLACK-AND-WHITE DANISH MILK BREED

In 1949 crossing two black-and-white breeds—Jutland and Friesian—formed a breed called the Black-and-White Danish Milk Breed, now entered in the same herdbook. By 1959 all but a few of the bulls were of Friesian pedigree. This new breed, which was improved in conformation over the Jutland, is distributed in all parts of the province. The Black-and-White Danish cows exceed Red Danish cattle slightly in size.

Cows admitted to the herdbook in 1959 measured:

	(centimeters)
Height at withers	133.5
Chest circumference	204.7
Depth of chest	75.8
Width of hips	59.7
Width at thurls	54.9

Average production of 94,196 Black-and-White Danish cows in control societies during 1957–58 was 9,524 pounds of milk, 4.07 percent and 388 pounds of fat. The selected cows registered in the herdbook during 1959 averaged 12,873 pounds of milk, 4.21 percent and 542 pounds of butterfat.

Larsen listed typical feeds given to a milk cow in 1933 as being about 40 kilograms of swedes (rutabagas) or mangolds, 4 kilograms of straw, and 1 or 2 kilograms of hay daily, with concentrates mainly as high protein oilcakes. The typical winter feed in 1959 consisted of about 50 kilograms of swedes or 35 of beets, 5 to 10 kilograms of beet-top silage, 4 kilograms of straw, and 1 or 2 kilograms of hay daily. Mainly oilcakes with some grain made up the concentrates.

Most herds are on fenced pastures in the summer, though some are tethered on clover-and-grass pastures. Elite herds graze upwards of 5 months in summer and receive some extra concentrates in proportion to milk yield. Young stock, and bulls that must be shown, are developed to good size, especially in elite herds.

BREEDING SOCIETIES

Before the late 1800s, most Danish farms were small, except for a few estates. Farmers often took turns in maintaining a "town bull," but excessive service often rendered these bulls impotent while young. Sentiment was against inbreeding. Older bulls sometimes got unruly, so few were retained until 3 years old. This was the situation in 1874 when the Taurus Cattle Breeding Society organized with 11 Jutland bulls on members' farms. The organization disbanded in 1878. The bylaws of a smaller society organized in 1881 pledged the members to use only bulls approved by a judging committee on cows from which the calves would be raised. This society succeeded and set the plan for later societies.

The Society for Improvement of Cattle Breeding was organized in 1884 in Boholte Parish on Sjaelland by J. M. Fries and Fr. Hvass. A superior bull was bought for use on cows from which to raise calves. One hundred such societies formed before 1887, when a Domestic Animals Law provided small grants for retaining bulls that had received prizes. Breeders' societies began to federate about 1880 to foster cooperation with agricultural organizations. The Law of 1912 provided that the society eligible for a breeding-bull grant must have at least ten members, the bull must have received a prize, and 50 percent of members' cows must have been in a Milk Control Society.

The agricultural societies joined with cattle breeders' societies and employed an agricultural adviser, or konsulent. Production records from the control societies came into use by the breeding societies. The konsulent worked for advancement of both societies under a common management. His duties included:

1. (a) Taking part in meetings and fairs; (b) assisting in selecting cows for breeding purposes for the breeders' associations; (c) giving lectures, as arranged with the society presidents; (d) attending state and other shows important to his work; (e) assisting in organizing milk control societies and in preparing their yearly reports; (f) assisting members as much as time permitted, and (g) making all entries in herdbooks for herds under the common management according to arrangements, particularly the cow family herdbooks.
2. Working under direction of the president, planning with him. The konsulent furthered the societies in every way; he received information of well-bred bulls to direct prospective buyers, with no financial interest in the animals.
3. Assisting the societies in arranging catalogs for fairs and similar work.
4. Preparing an annual report before May 1 of each year.

The first step in selective breeding was to locate cows to be bred to association bulls to raise progeny. Cows met conformation requirements to be eligible for grants, and (in later years) for production and type of dam and granddams. The sire and grandsires also were considered later.

Bulls receiving grants must have been exhibited at the local shows and, under some conditions, must have participated in larger shows. The grants and rules concerning age of bull, premiums won, number of progeny, and their quality contributed to improving cattle. Breeding societies received small grants also for developing good families through the cow family herdbooks.

About 1,200 breeding societies were operating by 1935. Most societies owned one bull; a few had three; and the largest—"Rudme" in Funen—owned eight bulls that served 870 cows in the herds of 114 members. Generally, 75 to 100 cows were owned per bull. Many small herds benefited from use of better bulls for longer periods than was possible under individual ownership. Cooperation between breeding societies and milk control societies contributed toward higher production per cow.

ARTIFICIAL BREEDING SOCIETIES

Konsulent Jens Gylling-Holm organized breeders of Red Danish cattle on the island of Samsoe into the Elite Breeding Society in 1936 with veterinarian K. A. F. Larsen as technician. About 277 farmers listed part of their cows, and bred 1,151 of them with semen largely from one of two bulls. Some 880 cows settled to the first, second, or third service—equal to efficiency of natural service. There were 98 societies in 1958, averaging nearly 12.4 bulls, with 1,268 cows per bull and 63 percent being Red Danish cows. About 27 percent of the bulls were five years or older. Some studs consolidated, giving more cows per bull and more efficient operation. From 85 to 90 percent of all cows in Denmark were bred artificially in 1958. Some 93.5 percent of cows inseminated that year were with calf, an increase of 2.3 percent above the record of 1950.

The scope of the National Association of Danish Cattle Breeding Societies and the breed distribution are shown in Table 18.4.

BULL TESTING STATIONS

Three testing stations were established by breeding societies in September 1945 to analyze bulls by testing their daughters under uniform conditions. In 1969–71, 20 stations operated. During 25 years, 1,024 groups of cows had been daughters of Red Danish bulls, 355 of Black-and-White Danish, 243 of Jerseys (two from American frozen semen), and six Shorthorn bulls. The societies realized the daughter-dam comparisons of production often were misinterpreted with cows under different environments. The stations were society-owned farms that could produce the roots, hay, silage, and pasture desired. Milk from the cows reimbursed the farmer for feed. A local manager was paid and supervised by a superintendent from the

University, independent of the local breeding society. The societies owned 17 of the stations in 1963.

Thirteen to 21 unselected heifers by each bull were under test and were delivered on lease to the farm by September 1. Preferably the heifers were 27 to 33 months old, free from tuberculosis and brucellosis, and due to calve between October 1 and November 15.

TABLE 18.4

Extent of Artificial Breeding, and Breed Distribution in Denmark, 1958

Province	Societies	Herds	Bulls[a]	Cows	Cows per bull
Funen	16	18,671	178	158,329	890
Jutland	57	124,358	802	1,117,795	1,394
Lolland–Falster	5	5,653	43	49,470	1,150
Sjaelland-Bornholm	20	24,014	189	211,442	1,119
Total	98	172,696	1,212	1,537,036	1,268 (ave.)
Breed		Breed Distribution			
Red Danish Milk Breed		87,085	754	968,188	
Black-and-White Danish		26,594	218	331,076	
Jersey		20,011	193	201,409	
Shorthorn		1,055	15	12,630	
Other breeds and crosses		37,951	32	23,643	

a. These numbers exclude some privately owned bulls in artificial insemination.

They were housed in stanchion barns. The test continued for about 304 days after calving. A concentrate mixture was provided to all testing stations for winter, and another for the pasture season. Feeding standards during the winter period provided:

	Digestible true protein (grams)	Feed units
Body maintenance	250	4.0
4 percent fat corrected milk, per kilogram	67	0.41
Weight increase, per kilogram	120	0.8

Some 449 Red Danish cows during the 1969–70 tests yielded an average of 11,676 pounds of milk, 4.49 percent and 524 pounds of butterfat, 3.84 percent and 448 pounds of protein in 304 days. Nineteen daughters of the leading Red Danish bull averaged 14,769

pounds of milk, 4.55 percent and 672 pounds of butterfat, 3.68 percent and 543 pounds of protein. Progeny tests disregarded conformation. All bulls under test, however, had passed their first examination at the cattle shows. Also, "Cattle shows and progeny tests are the primary breeding purposes and they are complementary to each other in the endeavors to improve our cattle stock in conformation as well as in producing ability."

Rate of milk letdown of each cow was determined twice during the winter at three consecutive milkings. Twenty-four groups of

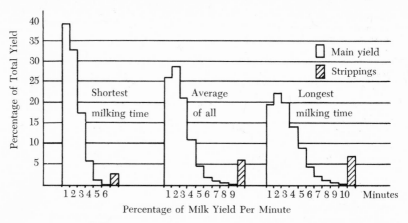

FIG. 18.3. Rate of milk let-down per minute by daughters of 67 bulls at co-operative bull testing stations in Denmark.

Red Danish cows ranged between 2.60 and 5.10 pounds of milk per minute, averaging 3.90 pounds. An example of the rate, milking time, and weight of strippings is shown in Figure 18.3. High producers tended to let down their milk faster than did low producers. Protein content of the milk was measured by Kjeldahl analysis since 1951. Milk from 449 Red Danish cows in 1969–70 ranged between 3.38 and 4.26, and averaged 3.84 percent of protein. Butterfat tests ranged from 3.64 to 5.41 percent and averaged 4.49 percent during the investigations. Fat and protein percentages varied independently, hence R. D. Politiek believed them to be inherited independently. However, with large numbers of records, the protein percentages tended to follow the fat irregularly and at a lower rate. Persistency of lactation, body size, and conformation were observed together with yields of individual quarters of the udder, ability to consume feeds, and other characteristics.

About 95 percent of the cows were pregnant by artificial insemination when returned to their owners. Conditions of management and individual feeding in proportion to production were believed to permit a relative measure of heritable characters of these daughters. Wide variations occurred between daughters of different sires and between sire groups. Pedigrees did not necessarily measure breeding value of a bull. Two groups of daughters were sired by three-quarter brothers of equal pedigree value. Daughters of Sire A averaged 10,700 pounds of 4.0 percent fat-corrected milk, while Sire B's daughters averaged 7,200 pounds. About half the increase in milk yield was believed to be sufficient to pay for the extra feed required to produce it. The breeding societies considered that results from the testing stations reduced the chances of using poor bulls widely in artificial service.

Records of daughters assembled at bull testing stations are a more random sample of the breed and under more uniform feeding and management than are the records of cows selected for entry in the herdbook. Cows at the stations tended toward an increase in butterfat percentage in milk of the breed.

The 21 groups of young cows in 1968–69 dropped male calves averaging 88 pounds and females that averaged 81 pounds at birth. The cows weighed 1,098 pounds after calving, and 1,235 pounds at conclusion of the tests. Most of them were pregnant by artificial insemination at the close, affecting final weights slightly.

A high-protein concentrate mixture (Mixture I) was given in almost constant amount daily, with Mixture II given in addition according to milk yield.

Mixture I		Mixture II	
	(percent)		(percent)
Cottonseed cake	40	Mixture I	35
Soybean meal	17	Barley	48
Sunflower seed cake	15	Wheat bran	15
Coconut cake	10	Dicalcium phosphate	2
Linseed cake	8		
Molasses	5		
Animal fat	3		
Dicalcium phosphate	2		

The average feed intake during the 1968–69 tests included Mixture I, 496 feed units; Mixture II, 882 feed units; swedes, 185; stock beets, 786; beet tops, 22; grass silage, 262; grass hay, 247; and straw, 40 feed units. These feeds provided a total of 684 pounds of digestible true protein and 2,920 feed units per cow on the average. The cows were on pasture an average of 85 days during that summer.

The 21 groups of Red Danish cows varied in average milk letdown rate from 5.10 to 2.60 pounds per minute, varying between groups of daughters and within groups by individual cows.

Year	Groups of daughters	Production in average of 304 days				
		Milk (pounds)	Fat (percent)	Butterfat (pounds)	Protein (percent)	Protein (pounds)
1945–46	6	9,150	4.29	399		
1951–52	33	9,396	4.40	414		
1964–65	49	11,069	4.52	500	3.77	417
1968–69	21	12,211	4.58	559	3.80	464

Bulls at stud are given health tests to eliminate such diseases as vibriosis.

PROGENY TESTING FOR MEAT PRODUCTION

Feed conversion was measured with groups of male calves by four Red Danish sires in a winter and a spring feeding trial from 14 days old to 200 kilograms (440 pounds) live weight. Calves were fed on a uniform schedule, with colostrum 1 to 4 days old, whole milk in reducing amounts, and skim milk in increasing amounts during the trial. Concentrates comprised of 20 percent linseed cake, 10 percent soybean meal, 10 percent linseed meal, 30 percent rolled barley, and 30 percent rolled oats. This supplemented equal amounts of rolled barley and rolled oats. Each calf received 500,000 international units (I.U.) of vitamin A and 100,000 I.U. of vitamin D on arrival. When whole milk was discontinued, each calf received 2,000 I.U. of vitamin A and 200 I.U. of vitamin D daily. Store calves got 30 grams of dicalcium phosphate daily, supplementing the stock beets fed. From birth to 27 days of age, calves received 100 grams of Aurofac antibiotic daily. Slaughter tests and carcass grades terminated each trial.

Calves in four groups in 1966–67 averaged 2.24 pounds of gain daily from 1.38 feed units per pound of gain. The carcasses dressed out at 53.7 percent, 70 percent of which was lean meat.

Cow Family Herdbooks

Since 1841 permanent individual milk records were kept of cows in the Ourupgaard Farm herd, owned by E. Tesdorpf. Peter A. Morkeberg, state konsulent in cattle breeding, analyzed the records and set up the first cow "Family Herdbook" from them in 1891. This served as a pattern. Family Herdbooks were not published, but they were recognized officially, since only the Konsulent made entries in them.

The family herdbook was compiled, based on prominent cows, when production had been recorded for three generations and type was satisfactory. The records included birthdate, sire and dam of each cow, description by the konsulent, prizes received, and milk yields and butterfat percentages under Control Society supervision, coupled with butterfat yields, progeny, transfers of ownership, and other desirable facts. New records were entered yearly by the konsulent or officials of the Danish Agricultural Society. Family herdbooks kept in 1941 with Friesian, Black-and-White Jutland (Danish), Jersey, Red Danish, and Shorthorn showed the following numbers of cattle: Bornholm, 70; Funen, 1,860; Jutland, 2,100; Lolland-Falster, 180; and Sjaelland, 835. The total for the five breeds was 5,045.

Only cows in family herdbooks were considered for entry in the *Cow Herd Book* of the Red Danish Milk Breed. The number of family herdbooks kept in 1958 with Black-and-White Danish, Jersey, Red Danish, and Shorthorns amounted to about 9,500.

Red Danish Herdbooks

The Red Danish Milk Breed was declared a distinct breed in 1881, although the breed had been used mainly as milk cows previously. A tenant farmer, Andersen, of Gunderslevholm, started a private herdbook in the 1860s. The Federation of Agricultural Societies in Jutland published the first public herdbook for the Black-and-White Jutland breed in 1881, divided into beef and dairy animals. This division was discontinued later, and only the dairy type was recognized.

Four herdbooks were initiated under the provincial agricultural societies—(a) in Sjaelland in 1885 by J. B. H. Andersen, a farmer, and published through the federated agricultural societies; (b) in Jutland for bulls in 1892, by Konsulent S. P. Pedersen; (c) in Funen in 1891 for bulls, and in 1905 for cows; and (d) in Lolland-Falster in 1896 by A. la Cour.

A central herdbook for Red Danish bulls was established in 1893 under the Federation of Danish Agricultural Societies; the first volume appeared in 1896. Between 1905 and 1919, herdbooks for Red Danish cows were maintained under control of the separate provincial agricultural societies.

The cow herdbooks also came under supervision of the Federated Danish Agricultural Societies. The consolidated herdbooks published since 1920 for the Red Danish Milk Breed include 63 volumes for bulls, 39 volumes for cows, and a young stock registry herdbook (up to 1959).

L. Hansen Larsen described the basis of the current herdbooks:

> First . . . the foundation for the public herdbook is the family herdbook kept by the Livestock Advisers of the Agricultural Societies. In the family herdbook the Advisor collects all information about the descent, yield, issue, prizes, measurements, weights of each animal, besides his own description of the animal. Before entering the various items of information, the Adviser checks the correctness of them, whereafter the book and its information are authorized by the Agricultural Societies; about 3500 prominent Danish elite herds are entered in family herdbooks.
>
> Secondly there is the peculiar feature of the Danish keeping of herdbooks that the public herdbook to a material extent is an elite herdbook, containing only elite animals from elite herds.

The *Registry Herdbook* was begun in 1921 under direction of Lars Frederiksen. It published short pedigrees, such as could be used for export purposes, and only *average* production records of the dams were entered. Four pamphlets or volumes were assembled. The work was given to P. A. Morkeberg in 1923 to reorganize. The first volume of the new *Register Herd Book* appeared in 1924.

Hansen Larsen stated:

> The *Register Herd Book* is a printed herdbook, comprising mainly young animals, the minimum age being 1 year. In order to be admitted, the bulls must have been awarded prizes at the cattle shows, and the cows and heifers must have come from herds entered in the Family Herdbooks. In the case of animals of dairy breeds the rule applies . . . that the cows in their descent must have been controlled by recognized Milk Recording Societies.

If a bull had not received at least a third prize at a local show, he could be entered if inspected by the Konsulent and if approved for body conformation. Females must have been at least 2 years old, have calved, and have passed the same type requirements as bulls.

Cow and Bull Herdbooks

The *Cow Herdbook* and *Bull Herdbook* were elite herdbooks functioning by a system of selective registration. Only cows from good family herdbooks were considered for entry in the *Cow Herdbook*. The dam and granddams must have been production recorded for at least 2 years, with at least 485 pounds of fat as an average of all recorded years. The annual butterfat yield in kilograms must have equaled the chest circumference in centimeters, and her milk must have averaged at least 4 percent of fat. At least one progeny must have met requirements for the *Register Herd Book,* and any daughters in milk must have yielded 441 pounds of fat in 1 year at mature equivalent age.

Herdbook standards were reviewed each year. Cows usually were 6 to 7 years old before qualifying for the *Cow Herdbook*. The degree of selection was great.

Strict requirements limited entry of bulls in the *Bull Herdbook* to superior animals at least 6 to 7 years old. Only 60 bulls qualified in 1959. At least the dam and both granddams must have been in Milk Control Societies. The bulls must have been awarded at least one prize at the shows, namely first class first grade, first class second grade, or second class first grade for conformation and for pedigree;

the bull also must have been entered in the *Register Herd Book*. He must have been awarded progeny prizes in one of three categories, as above. The production of all daughters in Milk Recording Societies was compared with that of their dams. Hansen Larsen wrote:

> This work was initiated by the Agricultural Society in Funen, in as much as they as early as about the year 1900 requested Mr. Sorensen, a veterinary surgeon in Odense, to undertake an investigation of the performance of the progeny of prize bulls. And in 1903 a report was issued, compiled by Mr. Jorgen Otto Pedersen, the agricultural adviser, on inquiries into the progeny of bulls. . . . This work was difficult, because the offices had to collect the material for his progeny performance tests. . . .
>
> The Government gives to the Provincial Federations of Agricultural Societies an annual grant of 15,000 Kroner toward this work and the reports.

Only 4,700 Red Danish bulls were entered in 63 volumes of the bull herdbook, only 712 being in the recent 10 years (1950–59). Entries contained the birthdate, show prizes, transfers of ownership, and progeny records. Improvement in the selected herdbook cows was shown by a comparison of their average production records.

Volume	Year	Number of cows	Milk (pounds)	Test (percent)	Butterfat (pounds)
1	1921	267	9,632	4.05	390
15	1935	549	11,762	4.32	508
39	1959	670	12,637	4.53	573

BREEDING RED DANISH CATTLE

The Island farmers decided to develop a breed of red dairy cattle. The dappled or mottled coat color may come from a different gene combination than the recessive red of some Friesian cattle. The horns are of medium length, inclining outward and forward from the head. Breeders of some elite herds favored retaining the horns to bear brands of identification and achievements. Some uniformity in size and shape of teats has been attained for convenience in machine milking. A gradual increase in butterfat percentage in the milk came from selecting herd sires from high-testing cow families.

Protein content of the milk has been determined on bulls' daughters at the testing stations since 1959.

Some hereditary defects are known. Jens Nielsen reported in 1950 that congenital paralysis of the hind legs of some Red Danish calves was reported first in 1924 in southeastern Sjaelland. Affected calves had incomplete paralysis of the hind legs at birth, and soon died. A recessive gene appeared responsible, as symptoms were not evident among animals of heterozygous constitution. He believed the trait traced to the bull Tjalfe Kristoffer, born in 1913. Højager, transmitting good production, was heterozygous for this gene. Attempts to avoid bulls known to transmit the defect is reducing its occurrence. Some cases of mummified fetuses have been observed.

Applying diagnostic methods with Red Danish bulls in A.I. service has contributed to reduce or eliminate trichomoniasis, vibriosis, and certain defects of the reproductive organs.

REFERENCES

1932. Det danske Landbrugs Historie. Vol. 3. Copenhagen.

Andersen, K. M. 1925. Bull associations in Denmark. *Hoard's Dairyman* 69(8):260.

———. 1946. *Haandbog in Kvaeghold.* Aksel Hansens Forlag. Copenhagen.

Anonymous. 1938. Denmark's anniversary. A century and a half of progress. *Hoard's Dairyman* 83(12):346.

Anthony, E. L. 1925. The history and development of the Red Danish race of cattle. Unpub. manuscript by permission.

Appel, A., and P. A. Morkeberg. 1896. *Kvaebrugets Udvikling i Danmark.* Copenhagen.

Ashton, E. D. 1949. The use of records in dairy cattle breeding in Denmark and the Netherlands. Agriculture. *J. Ministry Agr.* 56:255–59.

Blom, E. 1965. The history of artificial insemination in Danish cattle breeding with special regard to its influence on improved sexual health control. *Veterinarian* 3:243–48.

Bruce, William. 1905. Some features of dairy farming in Denmark. *Trans. Highland Agr. Soc. Scotland* 17(ser. 5):228–39.

Curwen, E. Cecil. 1938. Early agriculture in Denmark. *Antiquity* 12(46): 135–53. Taken from G. Hatt, *Landbrugi: Danemark old rid.* Copenhagen. 1937.

Dunne, John J. 1920. Cows that made Denmark famous. *Hoard's Dairyman* 60:528, 542–43.

Eskedal, E. Wenzel. 1935. Milk recording societies. In *Denmark. Agriculture.* Copenhagen. Pp. 277–82.

———. 1955. How the Danes are testing bulls for artificial insemination. (Notes prepared for Minnesota Valley Breeders' Assoc. Courtesy of Wallace Miller, Manager.)

Faber, H. 1921–22. Improvement of dairy cattle in Denmark. *J. Ministry Agr., Great Britain.* 28:598–607, 704–11.

———. 1931. *Cooperation in Danish agriculture.* Longmans, Green & Co., London.

Fisker J. 1910. *Kvaegavlarbejdets i Kobenhavns Amt.* Copenhagen.

Frederiksen, Lars. 1910. *Dairy cattle. From Denmark.* Danish Ministry of Foreign Affairs.

———. *Danish milk recording societies and their influence on cattle breeding.*

Goldschmidt, Herald. 1886. *Kvaebrugets Udvikling i Danmark.* Copenhagen.

Hansen, J., and A. Hermes. 1905. *Die Rinderzucht in In- und Ausland.* Vol. 1. R. G. Schmidt & Co., Leipzig. Pp. 532–49.

Houghton, F. L., and W. S. Moscrip. 1926. The system of judging cattle at Danish shows. *Holstein-Friesian World* 23(4):117, 132. (Cited a summary of Peter Aug. Morkeberg.)

Johannson, Ivar. 1960. *Genetic aspects of dairy cattle breeding.* Univ. of Illinois Press, Urbana.

Klindt-Jansen, Ola. 1957. *Denmark before the Vikings.* Praeger, New York.

Knudsen, A. F. 1935. The organization of Danish agriculture. In *Denmark. Agriculture.* Copenhagen. Pp. 37–48.

Larsen, L. H. 1933. *Die Rindviehzucht in Danemark inbesondere die Zucht der Roten danischen Milchrasse.*

———. 1935. Cattle breeding and cattle races. In *Denmark. Agriculture.* Copenhagen. Pp. 187–210.

Maarssee, Sv. Al. 1951. *Aarsberetning fra 1950.* Aaehus.

Morkeberg, P. A. 1886. *Kvaebrugets Udvikling i Danmark.* Copenhagen.

———. 1910. The Danish system of cattle breeding. *J. Board Agr.* 19:998–1005.

Nielsen, Ejner, and B. Vesth. 1951. Afkomsprover med tyre. *Beret. Forsogslaboratist* 15:387.

Nielsen, Jens. 1960. *Arvelig lamherd hos kalve.* (English summary.) Copenhagen.

Nielsen, K., and C. J. M. Hinks. 1967. Malkbarhed og ydelse undersogelse fra afkomsprovstationerns. *Beret. Forsogslaboratorist* 359.

Nielsen, M. 1931. Cow testing associations in Denmark. Trans. by Edwin C. Voorhies. *Hoard's Dairyman* 76:356.

Perry, E. J. 1939. *Among the Danish farmers.* Interstate, Danville, Ill.

Plum, Mogens. 1936. Cow testing in Denmark. *Hoard's Dairyman* 81:467, 480.

Prosche. 1925. Cited by E. L. Anthony. The history and development of the Red Danish race of cattle. Unpubl. manuscript.

Pullian, A. L. 1951. Denmark's new model cow. *Hoard's Dairyman* 96:612.

Rasmussen, Frederik. Cattle breeders' associations in Denmark. *USDA Bur. Animal Ind. Bull. 129.*

Robertson, Alan, and I. L. Mason. 1954. A genetic analysis of the Red Danish breed of cattle. *Acta Agr. Scand.* 4:257–65.

Rottensten, Knud, and Robert Foote. 1959. Development of A. I. and progeny test in Denmark. *A. I. Digest* 7(7):6–7.

Skovgaard, K., and Anton Pedersen. 1946. Survey of Danish agriculture. National Danish F. A. O. Committee. Copenhagen.

Svendsen, A. 1893. *Kvaegavl og Kvaegopdraet.* 1st ed. Copenhagen.

Touchberry, R. W., K. Rottensten, and H. Andersen. 1960. A comparison of dairy sire progeny tests made at special Danish testing stations with tests made in farmers' herds. *J. Dairy Sci.* 48:529–45.

Werner, Hugo. 1902. *Die Rinderzucht.* 2nd ed. Berlin.

Worsaae, J. J. A. 1849. *The primeval antiquities of Denmark.* Trans. by William J. Thomas. London.

References in U.S. *Consular* Repts.: 1887. Cattle and dairy farming. Vol. 2. GPO, Washington, D.C.

Brewer, M. S. Cattle in Schleswig-Holstein. The Angeln race. P. 395.

Ryder, Henry S. The Red Danish cattle. Pp. 490–95.

————. Butter exports of Denmark. Pp. 495–98.

Vogeler, F. Among the cattle of Germany. Pp. 410–12.

CHAPTER 19

RED DANISH IN AMERICA

THE REPUTATION of the Red Danish Milk Breed as milk producers preceded their importation by many years. Danish immigrants settled on many farms. United States consuls surveyed the dairy industry in Denmark in 1883–84 for the consular reports on *Cattle and Dairy Farming*. In 1905 Helmer Rabild organized a cooperative cow testing association in Newago County, Michigan (Fig. 19.1), built upon a plan from Denmark. Later he sponsored production testing of dairy herds in many states while in the Dairy Division in the USDA Bureau of Animal Industry. Cooperative bull societies operated in Jutland in 1874–78, and generally since 1881. One was organized in Michigan in 1908, and others spread from that example.

E. L. Anthony studied the origin and development of the Red Danish Milk Breed in Denmark during 1923–24. E. J. Perry observed the first artificial breeding cooperative on the island of

Samsoe, and organized a similar association in New Jersey in 1938. The Danish system of judging dairy cattle into blue, red, and white ribbon classes for conformation was adopted in junior dairy shows. Thus knowledge of Danish cattle and methods for improvement preceded introduction of Red Danish cattle into the United States.

IMPORTATION TO AMERICA

In 1915 Henry A. Wallace, Secretary of Agriculture, desired Red Danish cattle for breeding investigations, and Congress appropriated $10,000 for the project. E. L. Anthony was commissioned by

FIG. 19.1. Helmer Rabild organized the first cow testing association, patterned after those in Denmark, in Newago County, Michigan in 1905. Later, he sponsored similar organizations over the United States while with the Dairy Division, Bureau of Animal Industry in the United States Department of Agriculture.

the USDA to select the animals because of his acquaintance with the breed. Two young bulls and 20 bred heifers were selected from some 200 animals with excellent production background. The animals were quarantined 90 days in the Virgin Islands and another 90 days at Athenia, New Jersey, after which time they were placed at the Agricultural Research Center at Beltsville, Maryland. Three male calves were dropped to services of bulls in Denmark, but some calves were aborted, presumably acquired from swine at large dur-

ing the quarantine period on the Virgin Islands. The imported Red
Danish cattle developed into large animals with the solid red color
or the dappled red pattern common in the breed. The cattle were
used for research at Beltsville; some young bulls were leased out to
determine the milk and butterfat yields of their daughters. Four
Red Danish bulls of different bloodlines were imported from Sweden in 1939.

DISTRIBUTION IN AMERICA

Extension Dairyman A. C. Baltzer organized a Red Danish bull association in Sanilac County, Michigan, among ten dairymen with
240 grade dairy cows. Four Red Danish bulls were loaned for proving by the USDA on February 17, 1939. A second Michigan association formed in Alcoma County in June 1939 among 12 dairymen
owning 213 cows. Thirty herds with 230 cows in Oscoda County,
Michigan, formed four bull blocks in November 1941. Red Danish
bulls were placed in the Michigan Artificial Breeders Cooperative
at East Lansing in 1946. By 1948, 210 breeders in seven other Michigan counties agreed to follow rules for grading up their herds as
Red Danish cattle. Seven bulls were in artificial service in 1956, with
frozen semen from one bull for selective matings. Red Dane 555,
used in Michigan, is shown in Figure 19.2.

FIG. 19.2. Red Dane 555 was used in artificial service at the Michigan Artificial Breeding Cooperative as one of the pure foundation bulls. A recent proving credited him with 98 daughters whose 153 records in DHIA averaged
8,437 pounds, 4.1 percent and 348 pounds of butterfat.

Under the cooperative project, all cows and female progeny were tested for production in DHIA for at least one lactation to measure transmitting ability of USDA bulls on loan. Male calves were castrated. The original local cows were replaced by their half Red Dane heifers. Two Red Dane bulls were sent to the Alaska Agricultural Experiment Station at Matanuska in 1948 for artificial service. In 1956 breeders were located in 17 states, Alaska, and El Salvador.

New bloodlines were obtained when four Red Dane bulls were imported from Sweden. Some were leased to the Michigan Artificial Breeders Cooperative in 1959, replacing animals then in service. Four Red Dane bulls were in artificial service in 1967. The Beltsville herd was transferred to Purdue University Agricultural Experiment Station in 1948, and later moved to the Herbert Davis Forestry Farm for cross-breeding research. Frozen semen was available from the Michigan stud.

A HERDBOOK ASSOCIATION

Michigan farmers who cooperated in proving transmitting ability of the Red Dane bulls became interested in establishing a breed organization. After several meetings, rules were drafted for an open herdbook, similar to methods in European countries. The American Red Danish Association was organized on January 16, 1948, with Mrs. Harry Prowse, Route 3, Marlette, Michigan 48453 as secretary.

The rules required that grade or purebred cows of other breeds be recorded as foundation cows when the first calves by a Red Dane bull were birth reported. Each foundation cow must have a DHIA record for at least one lactation. This record became part of the Association's records, as did the production of all later females.

Its relatively recent arrival in America has allowed only a few years for the breed to appear before the public. Red Dane cattle have been displayed at field days on breeders' farms, at some local fairs, and at the Michigan State Fair in 1950 (Fig. 19.3), where they were judged as a dairy breed. Birth-reported calves were eligible for exhibition at shows.

PRODUCTION RECORDS

Red Dane cows in the Beltsville herd were reported by O. E. Reed to have produced creditable amounts of milk and butterfat, as listed in Table 19.1.

The Bureau of Dairy Industry required that transmitting ability of Red Dane bulls loaned to Michigan farmers be measured. Production records were supervised in Dairy Herd Improvement As-

TABLE 19.1

AVERAGE PRODUCTION OF RED DANISH COWS IN THE USDA HERD AT BELTSVILLE

Animals	Records	Average age at calving (years, months)	Milk (lbs.)	Test (%)	Butterfat (lbs.)
Imported cows	62	4, 2	7,633	4.21	318
12 daughters of D-501	38	4, 0	9,199	4.05	371
8 daughters of D-502	18	3, 2	9,307	4.03	374
Red Danish cows in DHIA in United States					
Year					
1954	417		8,744	3.96	386
1968	193		13,487	3.93	530

sociations for daughter-dam comparisons since 1949. Second- and third-cross heifers were obtained as bulls were exchanged between herds. Some 271 lactations in 1964 averaged 11,533 pounds of milk, 4 percent and 457 pounds of butterfat in 305 days 2× mature equivalent basis. Daughters of 12 Red Dane bulls were summarized in DHIA in 1965.

Successive generations of females by Red Dane bulls were known as first-cross, second-cross, etc. Third-cross females and fourth-cross males were considered purebred. No earlier male calves were recognized, so they were castrated.

BIRTH REPORTING

First- and second-cross heifer calves, Red Dane females and males eligible for full registration, were required to be birth reported before 90 days old. They could be reported at a higher penalty fee before 9 months of age. If not birth reported, the female could be recognized only as a potential foundation female. Birth report forms were purchased from the secretary.

Identifications were required by an approved ear tag, tattoo, ear notch, or a brand on the hip midway between the pinbone and the hipbone.

Rules for Registration

Red Dane *bulls* leased from the USDA were approved or accepted as foundation sires because they descended pure from the original importation. Bulls imported later must be approved for registration by the Board of Directors or executive committee. Males must be sired by a registered or approved Red Dane sire. The dam must be registered in the American Red Danish Cattle Association herdbook. The male calf must have been birth reported. Breeders were responsible for DHIA supervisors reporting 305-day production records of females to the Association when completed, on the secretary's form, or signed computer card.

Females were required to have 305-day DHIA records on the dam, maternal granddam, and great granddam, as defined under the rules for males. Any imported Red Danes or their progeny must be reported, approved, and registered in the American Red Danish Cattle Association herdbook before offspring became eligible for

FIG. 19.3. Buckner's Viola, a third-cross Red Dane, was Grand Champion at the Michigan State Fair in 1950. Her first two lactations averaged 9,261 pounds of milk, 3.64 percent and 337 pounds of butterfat. Bred and owned by Alfred C. Buckner.

registration. Females must have been birth reported before 9 months old. They must have 305-day production records duly reported. All females after the third cross also must have a fully registered sire and dam. Forms were obtained and registration fees paid to the Association.

TRANSFERS

Transfer forms were in triplicate. One copy went to the purchaser; the original and one copy were sent to the secretary with the registration certificate for inscription of the name of the buyer thereon. The fee was 50 cents.

MEMBERSHIP

Membership applications must be approved by the Board of Directors. The membership fee was $10.00.

There were 9,160 animals entered in the herdbook before 1959, including females in the grading-up process. Some 361 females and 204 males were fully registered. Bulls from the Red Danish herd at Beltsville were recognized as foundation sires. The foundation cows (grade or registered), as tabulated by Lester J. Cranek, Sr. in 1952, comprised 32.8 percent Guernseys, 29.0 Holsteins, 25.6 Milking Shorthorns, 9.2 Jerseys, 2.7 Brown Swiss, and 0.6 percent mixed breeding. He standardized production records between 270 and 365 days to a 305-day 2× mature equivalent basis. Some 1,358 foundation females in the cooperating herds averaged 8,536 pounds of milk, 354 pounds of butterfat. Three generations of crosses averaged 9,116 pounds of milk and 372 pounds of butterfat.

In 1956 N. R. Thompson and N. R. Ralston computed records of herdbook cows to a 305-day 2× mature equivalent basis and found:

	Lactations per cow	Average butterfat (pounds)
801 foundation females	2.10	351
703 first-cross cows	2.22	379
411 second-cross cows	1.99	377
123 third-cross cows	1.60	389

Daughters of four bulls in A.I. produced an average of 13,078 pounds of milk, 3.88 percent and 507 pounds of butterfat in DHIA in 305-day 2× mature equivalent basis during 1967.

BREEDING RED DANISH CATTLE

Four Red Dane bulls were in artificial service in Michigan Artificial Breeders' Cooperative during 1967. Average milk and butterfat yields were close to those of herdmates.

The solid red color of the Red Danish cattle appeared to be dominant over some other colors when bulls were crossed with cows of Guernsey, grade Holstein, and Milking Shorthorn characteristics. Color on sides of the body sometimes was "bloomed" or dappled with darker red on a light red background. White spots sometimes persisted on the underline of first-cross animals, disappearing largely with the second cross.

Hereditary light natural fleshing was less marked than with Channel Island breeds. Horns were light colored and of medium length, turning outward and forward from the head. Milk from Red Dane cows is similar in color to Ayrshire, Brown Swiss, and Milking Shorthorn milk. Maximum milk yield often was attained by cows past 6 years of age.

Some individuals carry the recessive lethal gene in heterozygous condition for posterior paralysis in newborn calves. It outcrops only when a calf acquires the gene from both carrier (heterozygous) parents. Affected calves die in a few days. Mummified fetuses sometimes occur as a recessive defect.

Mature cows in average condition weigh 1,300 to 1,500 pounds, and bulls 1,800 to 2,200 pounds.

Interest in Red Dane cattle arose mainly in herds whose owners wanted a grading-up program. The opportunity to locate and use bulls desirable as transmitters of high production was limited. Interest in the grading-up program decreased and the Association became dormant in the 1960s. Frozen semen still is available at Michigan Artificial Breeders' Cooperative from Red Dane bulls imported from Sweden. Frozen semen from leading bulls in Denmark could be imported under federal regulations. The breed possesses merit for constructive breeders, as evidenced by their status in their native country, Denmark.

BREED PUBLICATIONS

An Association pamphlet contained a brief history, facts, and rules for registration. A newsletter from the secretary's office carried reports about the breed.

REFERENCES

Anonymous. 1936. Import Red Danish cattle. *Hoard's Dairyman* 81(2):40.

Anthony, E. L. June 1955. Information to the author.

Cranek, L. J., Sr. 1952. Genetic and environmental influences affecting the Red Danish cattle in Michigan. Ph.D. Diss., Michigan State Univ., East Lansing.

Eckles, C. H., and E. L. Anthony. 1956. *Dairy cattle and milk production.* 5th ed. Macmillan, New York. Pp. 96–97.

Prowse, Mrs. Harry. 1956. Information concerning the Red Danish cattle. American Red Danish Cattle Assoc., Marlette, Mich.

Rabild, Helmer. 1911. Cow testing associations. *USDA Bur. Animal Ind. Bull.* 179.

Reed, O. E. 1946. Several generations of Red Danish cattle now tested. *Rept. Chief Bur. Dairy Ind., USDA Agr. Res. Admin.*

Sweetman, W. J., and H. J. Hodgson. 1954. Dairying in the land of the midnight sun. *Hoard's Dairyman* 99:68–69.

Thompson, N. R., and N. P. Ralston. 1956. Estimates of genetic progress in the development of the American Red Danish cattle. *J. Dairy Sci.* 39:931–32.

Winkjer, Joel G. 1916. Cooperative bull associations. *USDA Yearbook,* pp. 311–19. Also *USDA Farmers' Bull. 993.* 1918.

CHAPTER 20

RED-AND-WHITE DAIRY

The gene for recessive red coat color existed in Friesland and other provinces of the Netherlands prior to formation of herdbook societies. Red-and-white cattle were recognized in the breed and were registered under the same provisions as for black-and-white animals. The proportion of red-and-white Friesians registered was limited, as the gene for black color is dominant over that for red.

The first herdbook associations in the United States and Canada registered only black-and-white cattle; they barred red-and-white individuals from entry, despite their parentage. Thus presence of red color long was "swept under the rug." Solid-colored cattle and those with too great an extension of black on the legs, starting at the hoofs, or black in the switch, were barred likewise, hoping to avoid unethical entry of animals that might have some ancestor outside of the recognized breed. First-cross and later-generation grade Holsteins frequently had black hair in the switch or extending upward from the hoofs. Red-and-white calves seldom were retained among registered Holstein-Friesian herds.

A survey for ineligible color markings was conducted by the Holstein-Friesian Association of America among one-fourth of the active breeders in 1962 and 1963, 45 percent of whom replied. Their herds averaged 61.7 females per herd. At least one red-and-white calf was reported in 4.2 percent of the herds, or 0.12 percent of calves were red-and-white.

A few breeders did not discard red-and-white females of good producing ancestry. In the 1940s Larry Moore, of Suamico, Wisconsin, gathered desirable red-and-white heifers for potential milking purposes, along with registered Holsteins in his herd. Bulls in artificial breeding then were mated widely in commercial herds. This often afforded matings with grade cows carrying the red factor. When the red factor was identified, this information was indicated above the stalls, included in stud literature, and mentioned by technicians so that owners of registered Holstein-Friesian cows could avoid the bulls' use, if so desired.

About 1947 the brochure *Red & White Dairy Cattle Association* narrated discussion between the prominent mink breeder Larry Moore and the owner of an excellent Holstein-Friesian herd. Moore suggested developing a mutation breed of dairy cattle from red factor carriers. The breeder believed this could not be accomplished in his lifetime. A red-and-white double grandson of Posch Ormsby Fobes 14th 729449 was sent to the Moore farm and was named Larry Moore King 4710181. Moore assembled red-and-white Holstein females to build a herd. He recorded the breeding records in a private registry called *Colorstein*. Several Holstein bulls recognized as good transmitters of production and type also carried the gene for recessive red color.

An Appendix Registry for crossbred Milking Shorthorn × Red & White Holstein animals was established in the Milking Shorthorn herdbook in 1962. Directors of the American Milking Shorthorn Society voted to discontinue the Appendix Registry in the fall of 1963, suggesting a separate society.

Red and White Dairy Cattle Society

The Red and White Dairy Cattle Society was organized at a meeting in Springfield in February 1964 and incorporated under Missouri laws. *Association* replaced *Society* in the name in 1966.

An advertisement in *Hoard's Dairyman* (volume 111, page iii, Jan. 25, 1966) stated: "A great new dairy breed is being developed. It has the striking eye-appeal of rich cherry-red and white color and the uniformly fine type and production that comes from the best registered Holstein bloodlines of North America. Highest producer to date Strickland Reflection May 231—5 y. 2x, 348 d, 24,334 M. 879 F, with more production-bred Red & Whites coming along to give her competition. Crossbreeding potential with Guernseys, Ayrshires and Milking Shorthorns also being developed. Frozen semen available. Write: American Red & White Dairy Cattle Society, 6404 High Drive, Shawnee Mission, Kansas."

An advertisement in the *Holstein-Friesian World* (volume 63, page 1453, July 10, 1966) mentioned registering Red & White Holsteins.

Officers

Seven members are elected to the Board of Directors at annual meetings and serve 2-year terms. The Secretary and Treasurer need not be Board members. The directors named an Executive Committee.

Membership

The secretary granted membership in the Red & White Dairy Cattle Association when an owner registered an animal or purchased frozen semen through the organization for use after October 7, 1966. Membership discontinued on failure to register an animal in any 12-month period. The membership fee of $5 per year included the monthly *Newsletter*. Frozen semen from selected bulls was delivered from the Northern Illinois Breeding Cooperative, Hampshire, Illinois, when ordered through the secretary. The price of semen included a 15 percent service charge to finance the Association.

A person under 21 years old, one of whose parents is a member, or who is a member of a regularly constituted Calf Club, is entitled to record pedigrees and transfers, as is a Junior Member.

REGISTRATION

Foundation animals based on crossbreeding were recognized in the Appendix Registry of the American Milking Shorthorn Society as crosses between Milking Shorthorns and red-and-white Holsteins. Red-and-white animals of registered Holstein-Friesian parentage also were accepted in the Appendix Registry.

Red and White, or Red, White, or Roan animals of dairy type and with registered ancestry could be registered in the *Red & White Herd Book*, provided the owner became a member of the Association. If either parent of the animal was registered in another association, the name and registration number therein was required. If unregistered, breeding and bloodlines; (including name and registration number of registered ancestors) were required, along with satisfactory proof and identification of each animal in the pedigree, to establish that such sire or dam was entirely of registered breeding. Descendants of fully registered Holstein breeding were known as Certified F100s.

Symbols indicated breeding and parentage of crossbred foundation cattle as follows: F100 (100 percent Holstein of red-and-white color), F50MS (50 percent Holstein and 50 percent Milking Shorthorn), F50G (50 percent Holstein and 50 percent Guernsey), F50RD (50 percent Holstein and 50 percent Red Dane), F50A (50 percent Holstein and 50 percent Ayrshire). Second-generation descendants were designated similarly as F75A (75 percent Holstein and 25 percent Ayrshire), and so forth. The registration fee was $3 per animal.

Article V, Section 10 of the Bylaws stated that:

The Board of Directors may establish such conditions or restrictions on the recording of bulls with respect to sale of semen for artificial insemination purposes as it may deem necessary. Such conditions or restrictions may include, but are not limited to requiring membership in the corporation as a condition to recording of bulls; providing for cancellation of recording if bulls used for such purposes have not been sold or leased to the Association; and refusing to record artificially sired off-spring where any such conditions or restrictions have been violated.

A transfer fee of $5 applied within four months of sale, or $10 thereafter.

A MODEL COW

Larry Moore Betsy, daughter of the foundation bull Larry Moore King, was regarded as the model cow of the breed. She had two lactations with unofficial records of over 20,000 pounds of milk in

FIG. 20.1. Larry Moore Betsy, sired by the foundation bull Larry Moore King, is accepted as the model Red & White cow.

305 days, and a lifetime production of over 115,000 pounds of milk. Gene Hoy painted her picture (Fig. 20.1), used on the *Red & White Briefs* as the model cow. The breed paper *The Red Bloodlines* began in 1966.

SHOW DISPLAY

No classification existed in shows in the United States and Canada in 1966 for Red & White cattle. The American Royal Dairy Show at Kansas City in that year offered $350 toward transportation expenses for exhibited herds of Red & Whites. Thirty-seven animals were displayed voluntarily, one cow having a DHIA record of over 21,000 pounds of milk, 4.8 percent and 1022 pounds of butterfat in 307 days. The National Red & White Show at Waterloo, Iowa, in 1970 had 25 exhibitors from nine states and Ontario, Canada.

Type Classification

Through agreement with the Holstein-Friesian Association of America in 1967, official classification was conducted by Association classifiers during regularly scheduled classifications. Applications were submitted to the Red & White secretary 6 weeks in advance, with a $4.50 fee per animal. Jack Fairfield classified 41 animals in Larry Moore's herd at the fourth annual meeting and field day, July 1. 1967. He followed the descriptive classification for the Holstein breed and assigned a total score for each animal.

Production Records

Members were urged to submit production records on forms provided by the Association office and signed by the representative supervisor or officer. The records were published in *Red & White News Briefs*. The USDA was asked to recognize Red & Whites as a dairy breed. Such recognition would enable a system of official production records. It also would affect import duties on animals from Canada, where several prominent bulls transmitted the red gene to some progeny. It would also affect export regulations to Latin American countries where a demand exists.

Registered Red & White cattle have been exported to Brazil by a Canadian member. Other prospective buyers have inquired for available cattle of the breed.

Association Officers

C. L. Wasmer of Albuquerque, New Mexico, was elected president in 1970. Mrs. Susan J. Fortner is Executive Secretary and Treasurer, at P.O. Box 771, Elgin, Illinois 60120.

REFERENCES

Advertisement. 1966. A great new dairy breed. *Hoard's Dairyman* 111:111.
———. 1966. *Holstein-Friesian World* 63:1453.
Anonymous. 1964. Red and White Cattle Society formed. *Hoard's Dairyman* 109:426.
———. 1964. December Board meeting. *Holstein-Friesian World* 61:2851, 2868–69.
Cole, L. J., and S. V. H. Jones. 1920. The occurrence of red calves in black breeds of cattle. *Wisconsin Agr. Expt. Sta. Bull. 313.*
Scholl, M. 1968. The emergence of Red and White Holsteins. *Hoard's Dairyman* 113:743, 761.

Red & White Cattle Association publications:
   1966. Red & White Cattle Association brochure.
   1966. By-laws of Red & White Dairy Cattle Association.
   1967. *Red & White News Briefs.* Vol. 1.
   The Red Bloodlines. Vol. 5.
   1970. National Red & White Show.

CHAPTER 21

CONTRIBUTIONS TO BETTER DAIRYING

CATTLE (*Bos*) were not native to the western hemisphere. Early colonists appealed for cows to be brought from Europe, to be protected and increased in numbers. Columbus carried cattle on his second voyage from Spain to the West Indies in 1493; from this source they reached Florida, New Mexico, and California. Cattle from Portugal and France were taken to Sable Island and the St. Lawrence valley about 1518 and in 1541. English colonists brought animals to Virginia in 1611 and to Massachusetts colony in the 1620s. Captain John Mason imported large cattle from Denmark in 1622 and 1623 for colonists in New Hampshire. Large Dutch cattle reached New York from the Netherlands in 1626. Those from Sweden were introduced into Delaware in 1640. Intercolonial trade contributed to later distribution.

Some colonies prohibited slaughter of milking cows, to assure a supply of milk needed by the colonists. Cattle were herded, as on

Boston Commons, or ran at large. Rail fences were built to protect fields and farmsteads. Improved breeds had not been developed in Europe, but as farming in America was extended over larger areas and cattle were fenced into pastures, American owners controlled the breeding in their herds.

Agricultural societies took interest in improving farming practices and livestock. The Massachusetts Society for Promotion of Agriculture and the Philadelphia Society for Promoting Agriculture became prominent. Several presidents of the United States were members of the latter society. The Massachusetts Society imported animals of several improved breeds after 1800 and placed them with members to maintain purity and increase the numbers.

Perhaps the first livestock show in the country was held in 1793 at Waltham, Massachusetts, by an agricultural society. Elkanah Watson, a weaver of woolen goods, established an early agricultural fair. The movement spread, and Watson was in demand as a speaker in several states. These fairs were for display of livestock and farm products, not public sales. Many agricultural societies today sponsor local and state fairs and shows for livestock and farm products, with concessions to display improved machinery and supplies. Most fairs maintain household departments, and many educational contests and exhibits are features of the better fairs.

At the request of leading stockmen, the State Department circularized the Consular Services in Europe for information on improved cattle and management. The reports were published by the House of Representatives in two volumes entitled *Cattle and Dairy Farming* in 1887.

GOVERNMENTAL SPONSORSHIP OF AGRICULTURE

Some improvements in agriculture were promoted by a small Agricultural department in the U.S. Patent Office from 1836 until the Department of Agriculture was organized on July 1, 1862, under President Abraham Lincoln. The movement had been sponsored by the United States Agricultural Society and previously had been called to the attention of Presidents Harrison, Tyler, and Fillmore. The first Commissioner, Isaac Newton, continued the activities of the Agriculture office in obtaining statistics relating to agriculture;

collecting and distributing seeds and plants for trial; experimentation; and preparation of publications dealing with farm subjects.

A few states established schools for agricultural instruction. Land grant colleges and universities were established under provisions of the Morrill Act, passed by Congress in 1862. Beginning in 1840 several older state agricultural institutions had experimental farms.

Federal support for agricultural research had been sought by farm leaders since 1871. The Hatch Act, which became a law under President Grover Cleveland on March 2, 1887, granted $15,000 yearly to each state to investigate largely "practical" problems. The Adams Act was passed and signed by President Theodore Roosevelt on March 16, 1906. Secretary James Wilson directed the Adams Act funds of $15,000 annually to be used wholly for fundamental research.

The Purnell Act in 1925 provided for investigations in agricultural economics and home economics. Purnell funds also were applied toward problems of long-time duration, including value of feeds and livestock requirements for mineral nutrients. Some investigations increased knowledge of vitamins and protein quality. These discoveries also applied to human nutrition.

The Research and Marketing Act in 1943 pointed toward regional projects between states and with the Department of Agriculture. Improved varieties of vegetables, regional pasture investigations, animal diseases, swine breeding, and problems in reproduction were among those investigated.

The several federal acts were consolidated as the Hatch Act in 1955. A large proportion of research in soils, feed crops, feeding value of crops and products, breeding, dairy products, disease control, management, nutrition, and marketing had a direct or indirect effect on the dairy industry.

Agricultural Education

Farmers' Institutes, short courses, and other provisions carried results of experience and research to the adult farming population. The state Agricultural Extension Services were expanded in 1914 with funds under the Smith-Lever Extension Act matched with state funds for spread of agricultural knowledge. The United States

Department of Agriculture had pioneered in this field with agents under Seaman Knapp, in counties in South Carolina and Texas. The County Agent service expanded gradually on local county levels with a man in agriculture and a woman to work with homemakers.

Agricultural instruction in public schools gained greater impetus through assistance under the Smith-Hughes Act in 1916. Each of these acts by Congress supplemented state appropriations for agricultural teaching, research, and/or extension services. Improved methods were developed for handling dairy products, and their nutritional qualities were evaluated. The combined results influenced the dairy industry in many ways.

PRODUCTION RECORDS

Solomon Hoxie envisioned permanent records of production and conformation of purebred dairy animals as part of the herdbooks on which to base "a science of cattle culture." Dairy cattle were registered on purity of pedigree. They were advanced to the Main or Advanced Register later upon milk production, supplemented when possible with a score for conformation and a description of type.

Public churn butter tests were supervised at fairs and shows sponsored by agricultural societies. Dr. S. M. Babcock devised a rapid practical test for butterfat in milk in 1890 at the Wisconsin Agricultural Experiment Station, published in Bulletin 24. See Figure 13.5.

Dr. Nicholas Gerber devised the Gerber butterfat test in Switzerland in 1888 (published in 1891). Other methods for butterfat analysis were the Houberg test used in France and Roumania, and the Lindstrom method in Norway, Sweden, and Finland. Butterfat tests were used in evaluating quality of milk, computing butterfat yields, and investigating nutrient requirements for economical feeding in proportion to production. Butterfat content of milk has been an important hereditary character in selecting breeding animals. Attention now points toward the percentages of protein, lactose, and solids-not-fat in addition to butterfat content of milk.

A rapid Milko-Tester method for fat analysis in fresh milk was developed in Denmark. Three models of the complex tester are manufactured by the Foss Electric Company. Milk samples are placed on a conveyor which moves them into position at 20-second intervals. It mixes each sample. A 2-ml. sample is pipetted and heated to 60° C.; a Versene diluter is flushed in to dissolve the protein. The mixed sample is homogenized under pressure and is passed into a micro curvette where a photocell measures the optical interference, based on light interceptance by the emulsified fat globules. The light interference is converted into an adjusted fat percentage reading. The results can be printed on appropriate typed or punchcard forms within a minute. Even though the cost of the machine is high, it is economical for large numbers of analyses in central laboratories.

COOPERATIVE PRODUCTION TESTS

Helmer Rabild organized a cow testing association in 1905 at Fremont, Michigan, patterned after milk control societies begun in Denmark in 1895. He was called to the Dairy Division of the USDA to establish similar organizations through cooperation with Extension Dairymen in many states. These organizations now are called Dairy Herd Improvement Associations. Supervisors obtained records of milk and butterfat yields, feed consumption, feed costs, and reproduction with grade and registered cows. Owners used the records to guide the feeding and management of their dairy herds. Percentages of solids-not-fat and/or protein in milk were added officially in 1962.

Selective testing was begun with the Dutch-Friesian cattle in 1879, then with Holstein-Friesians on consolidation of the association in 1885, and later with all dairy breeds. The Ayrshire Breeders' Association began to record production of every cow in the herd under the official Herd Test plan. Other dairy breeds adopted the practice. The state superintendents of official testing, who supervise the production testing programs, are staff members of the College of Agriculture or Extension Service in their respective states.

Two lower-cost plans of obtaining records are the owner-sampler and weigh-a-day-a-month plans. Owners weigh and sample the milk from each cow for the DHIA supervisor or central laboratory to test for butterfat under the owner-sampler plan. The second plan, begun in September 1956, involves milk weights and a feed statement by the herd owner on the middle day of each month. Milk yields and costs computed from either form of record keeping are not authenticated since milk weights and samples were not taken by an impartial supervisor. Extension Dairymen of the states train the DHIA supervisors and direct the three forms of production testing.

The average production of cows under DHIA test has increased

TABLE 21.1
AVERAGE PRODUCTION OF COWS ON DHIA TEST IN THE UNITED STATES[a]

Year	Number of cows on DHIA test	Average cows per herd	Milk (lbs.)	Test (%)	Butterfat (lbs.)
1906	239		5,300	4.06	215
1926	327,653	16.8	7,331	3.96	290
1946	627,878	27.0	8,635	4.04	349
1956	1,406,306	34.4	9,713	3.94	383
1966	2,058,592	52.9	12,127	3.81	462
1970	2,122,011	61.9	12,750	3.79	483

a. Taken from *Dairy-Herd-Improvement Letter*, 47 (4), April 1971.

because of (a) improved feeding and management, (b) removal of lower-producing cows, and (c) wider use of good transmitting animals. The average production per cow has been reported through cooperation of the state dairy extension services and the USDA, as given in Table 21.1.

Including owner-sampler and weigh-a-day-a-month plans, 3,235,-552 cows were reported on production test in 63,084 herds in January 1970. This was 25.5 percent of the dairy cow population. Herds on standard DHIA test averaged 61.9 cows per herd.

Committees of the Purebred Dairy Cattle Association and the American Dairy Science Association modified the methods of supervising DHIA records so that dairy breed associations now accept them under the name of Dairy Herd Improvement Registry (DHIR) records when DHIA herds enroll in both. Revised rules, accepted in 1967 and earlier, recognize only DHIA and DHIR records for breed programs.

DHIAs in Wisconsin were encouraged to combine into county or district units in 1940 to utilize facilities and services efficiently. Extension Dairyman Lyman Rich of Utah began central machine computation of DHIA records in 1950. States cooperated later in establishing regional computing laboratories, of which 13 centers were operating in 1968. The centers provided a printed report with individual cow and herd records to each owner promptly. This service afforded current attention of owners to breeding and herd management.

Farmer members obtained a voice at national policy levels by organizing the National Dairy Herd Improvement Association. It— along with the Extension Section of the American Dairy Science Association, National Association of Animal Breeders, Purebred Dairy Cattle Association, and USDA Livestock Research Division— are represented on the National DHIA Coordinating Committee that makes rules for DHIA operation. The National DHIA group was formed during the National Dairy Cattle Congress at Waterloo, Iowa, in October 1965 under a Wisconsin charter. The purpose was for herd-owner representation in joint planning and administration of rules and related programs. Twenty-two states were represented in 1970, and the movement is expanding. Revised rules, accepted in 1967, recognize only DHIA and DHIR records for breed programs.

MILKABILITY

The cow's udder normally consists of four functional quarters. E. Lewis Sturtevant, of the New York (Geneva) Agricultural Experiment Station, perhaps was the first person before 1885 to measure their separate milk yields. At the Missouri Station, C. W. Turner used a four-quarter milking machine designed by H. C. Beckman of the DeLaval Separator Company, during 74 lactations by Ayrshire, Guernsey, Holstein, and Jersey cows. Each front quarter yielded slightly more than 20 percent of the dairy milk production and each rear quarter slightly less than 30 percent of the daily milk production. Similar milking machines are used in Denmark, the Netherlands, and Switzerland to determine average rate of letdown, milking time, and strippings with groups of daughters by separate sires or cow families. Reaction to machine milking also is observed.

Copeland analyzed bulls with daughters at different levels of production. Young bulls selected by A.I. studs averaged nearly as well as bulls "proved" when acquired. Daughter averages, daughter-dam, and herdmate comparisons gave progressively more dependable information. Copeland considered type classification averages important in breeding cattle for a long useful lifespan.

How Good Cows Differ

Trials conducted in Iowa with four "native" and three grade calves on uniform feed showed practically equal ability to digest various feed nutrients. C. H. Eckles and O. E. Reed found at the Missouri

TABLE 21.2

The Results of Grading–up with Purebred Dairy Bulls on "Native" Cows at the Iowa Station

Generation	Lactations	Milk (lbs.)	Test (%)	Butterfat (lbs.)
Native parents	36	4,110	4.67	192
First cross	42	5,815	4.59	267
Second cross	53	8,056	4.51	363
Third cross (Holsteins)	9	12,735	3.29	419

station that the maintenance requirements and ability of cows to digest feeds were quite similar. The good cow, however, utilized more nutrients above her maintenance to make milk than did the poorer cow.

Improving Dairy Cattle Through Breeding

The Massachusetts Society for the Promotion of Agriculture reported in 1848 on the consistent production and transmitting ability of pure Ayrshire cattle as compared with famous "native" cows that left no progeny equal to themselves in producing ability. Colonel Zadock Pratt stated in 1861 that cows could be improved by a single cross to a pure breed.

The Iowa station began to grade up a dairy herd in 1907 with 14 cows and heifers purchased in a region where no purebred bulls had been used. Those common cows were bred to Guernsey, Holstein, and Jersey bulls, grading up within each breed in succeeding generations. Results of the grading-up investigation, directed by H. H. Kildee, are given in Table 21.2.

The Oklahoma, South Dakota, and other stations obtained similar results with different foundation females. Use of purebred bulls to grade up dairy cows soon was practiced by dairy farmers, and influenced the entire dairy industry.

PROVED BULLS AND COW FAMILIES

In 1935 the Bureau of Dairy Industry began to "prove" (analyze) dairy bulls having five or more daughters with DHIA and/or HIR records. Individual records were computed to a 305-day 2× mature equivalent basis before averaging them for all of the daughters. Persons interested in such bulls were advised to visit the herds to study feeding and management, as well as conformation of the animals. Many valuable bulls were located for further natural or artificial use. Since December 1962 the average production of herdmates calving within 2 months of a daughter were included in herdmate comparisons.

A method for genetic appraisal of bulls was based on number of herds represented by natural or A.I. daughters to determine the *predicted average* of his progeny. This was changed to a *predicted difference* between expected production of his daughters as compared with herdmates. The formula for calculating *predicted difference* between production of daughters and herdmates was changed in May 1965 to:

$$\text{Predicted difference} = \frac{\text{Number of daughters}}{\text{No. of daughters} + 12} \times \left\{ \begin{array}{l} \text{Adjusted} \\ \text{daughter} \\ \text{average} \end{array} - \begin{array}{l} \text{Breed} \\ \text{average} \end{array} \right\}$$

Daughters of natural service sires were evaluated on their own performance versus herdmates. On A.I. progeny, performance of daughters versus herdmates and of paternal half-sisters were included in calculating the index. As the number of daughters increases, variations in production due to environment become smaller. Hence the *predicted differences* represent the expected deviations of bulls' A.I. progeny from herdmates in herds producing at levels near the breed average.

Installation of a new electronic system (IBM 360-40) in June 1967 enabled incorporating improvements to change summarizing sires to presumably more representative estimates of genetic merit of dairy bulls. Age correction factors vary with season of calving and regional environmental conditions with resulting milk and fat yields. Six 2-month seasons appeared to reduce bias due to season of calving with relation to production. Variations in milk varied with those for butterfat.

Because of varying environmental influences, *predicted differ-ence* was adjusted with relation to the number of daughters and their records, the herdmates, and the number of herds involved. Appropriate weighting factors take into account the number of daughters, records per daughter, and the number of herds involved. A repeatability factor will indicate the relative confidence attributable. However, few bulls produce viable semen long enough for it to be available for long-term computations to be made (stores of frozen semen would be exceptions).

Continuous production records enabled owners to recognize and build transmitting cow families within herds. J. C. McDowell found from DHIA records that the better producing cows were from the higher producing dams on the average.

DHIA records were searched for the possibility of locating the leading cows for possible breeding value, assuming that they would be among the top 2 percent of daughters of a bull. Criteria for selection were: (a) registered animals; (b) daughters sired by bulls summarized in July or November 1963 (and later) with five or more A.I. daughters; (c) cows with one or more lactations with herdmate data; and (d) cows with index values for milk yield exceeding arbitrary requirements assumed for breeds. An index was computed for the cow's production and that of A.I. paternal half-sisters and was weighted according to number of lactations. The method of computation was thought to account for genetic influence on milk yield, or the cow's true breeding value.

Weightings used for computing an index were:

Number			Weights		
Cows' records	Paternal sisters	Cow		Paternal sisters	Relative emphasis
1	5	0.24		0.19	0.8
1	25	0.22		0.49	2.2
1	200	0.20		0.74	3.7
2	5	0.33		0.17	0.5
2	25	0.31		0.43	1.4
2	200	0.29		0.66	2.3
4	5	0.41		0.15	0.4
4	25	0.38		0.38	1.0
4	200	0.36		0.59	1.6

As an example, assume the cow's one lactation was 12,641 pounds of milk, and the adjusted average production of five paternal half-sisters amounted to 11,137 pounds of milk. If the average deviation of the five half-sisters from their herdmates was 361 pounds of milk, the computation would be:

$$\text{Index} = 0.24(12,641 - 11,137) + 0.9(361) = 492$$

The index of 492 would indicate that probably this cow was above the average but was not outstanding.

The first search of DHIA records was for the leading 2 percent of cows for possible true breeding value. Those from daughters of bulls summarized in July or November 1963, as above, included 72 Ayrshire cows with milk index levels of 1,995 pounds; 127 Brown Swiss with 1,586 pounds; 408 Guernsey cows, 1,312 pounds; 2,392 Holsteins, 1,711 pounds; 386 Jerseys, 1,189 pounds; and two Milking Shorthorns with milk index levels of 1,350 pounds. Indexes of these cows ranked in the highest 2 percent for computed genetic levels for milk production. Such evaluations are published three times yearly.

COOPERATIVE BULL ASSOCIATIONS

Owners of small herds in Denmark in 1874, and owners around Zurich, Switzerland, in 1887 organized associations or breeding syndicates for cooperative ownership and use of dairy bulls. Registered bulls in the bailiwick of Guernsey were required to stand for public service under certain conditions. The Michigan Agricultural College assisted farmers to form a cooperative bull association in 1908. Through the state extension services and Joel E. Winkjer of the Dairy Division, Bureau of Animal Industry, USDA, 36 such organizations were active in 14 states in 1917 with 1,158 members owning 189 registered dairy bulls. Members formed three or more blocks located conveniently, each block using one bull for 2 years. Bulls were rotated between blocks. Purchases of bulls and association expenses were assessed on a per-cow basis. A peak of 306 bull associations operated with 63,564 cows in 1943.

Less than one-fourth of dairy bulls in the United States were registered purebreds in 1908. Few of them were retained until their daughters freshened. Cooperative bull association members were encouraged to keep production records in DHIA or otherwise.

BREEDING DAIRY CATTLE

During the feudal period Walter of Henley advocated against crossing cows with bulls of a type larger "than the land would support." Improvement of cattle became important soon after enclosure of land enabled individual owners to control matings of their animals, and to store feeds for use in seasons of scarcity.

There was a prejudice against mating closely related animals. Bakewell gained prominence by breeding "the best to the best" irrespective of relationship in order to propagate and improve type. Breeding was an art. Through observation and research it is changing gradually to a science.

Prior to about 1920 all breeding of dairy cattle in the United States was by natural service. Most owners of cows also kept a bull, or had access to one in the neighborhood. Transmitting ability of this animal was important to improvement, since half the hereditary characters were contributed by him. Since bulls often became dangerous, special provision often was made for his care at time of service. *A gentle bull never should be trusted; a temperamental bull will not be.* The practice became common in large herds to house the bull separately. A breeding chute, with a gate that swung into the pen, allowed the bull access to the cow; by swinging the gate inward, the cow could be removed without entering the bull's pen. Too often an unproved young bull was used for two years, then slaughtered before his daughters gave indication of his transmitting ability.

The breeding committee of the American Dairy Science Association established differences in average gestation periods existing between the dairy breeds, as follows:

Breed	Number of gestations	Average period (days)
Ayrshire	1,039	278.7
Brown Swiss	1,548	290.8
Guernsey	576	284.0
Holstein-Friesian	5,548	278.9
Jersey	3,118	279.3

H. W. Norton, Jr. summarized 11 studies (10,476 gestations) with

Holstein-Friesians in which average periods were reported between 276.7 and 281 days, some differences being due to inclusion of terminal dates. The 1,106 heifer calves in Norton's random sample were carried 280.4 days; 959 male calves for 281.6 days; and 26 sets of twins for 275.6 days.

Cows were slaughtered at the Illinois station at intervals after calving. The uterus returned to normal size and shape in 30 days, but the mucous lining required 20 to 30 more days for complete recovery.

Heifers attain the age of puberty at 6 months or younger in Guernseys and Jerseys, and at older ages in the large breeds that take more time to mature. A low plane of nutrition, or a deficiency in essential nutrients, delays the onset or disrupts regularity of the reproductive functions.

Failure to conceive to first service often was an erroneous reason for discarding many bulls. Breeding records of 6,751 reproductive periods with Holsteins at Carnation Farms were analyzed by R. E. Erb and R. A. Morrison. Estrus or "superheat" occurred in 5.6 percent of the cases, the proportion averaging as high as 7.5 percent in the second and third reproductive periods. Superheats were observed in 3.4 percent of cases in an English study.

Breeding records were tabulated of 16,555 conceptions from natural service in 12 station herds. Some 64.3 percent of the animals conceived at first service; 19.9 percent to the second, and 8.5 percent to the third service. A few conceptions occurred at the ninth to fourteenth service. An average of 1.78 natural services were required in eight of the herds. Health of the animals is not known. The Nebraska station found over more than a 30-year period that 24 percent more services were required for heifers under 2 years old than for cows. Health of the reproductive organs and the interval since calving to first service also affect conception.

Hereditary defects occur in the acrosomal cap and in coiled, folded, or broken tails of spermatozoa from several breeds of cattle. These defects may cause low fertility or sterility in certain bulls. Since 1960 the Royal Guernsey Agricultural Society has required that unproved bulls 12 months or older be examined by a veterinary surgeon before being sold.

BREEDING INTERVAL

Cows bred within 50 days after calving had a 31 percent conception rate; 51 to 60 days, 67 percent; 60 to 90 days, 70 percent; and a 76 percent conception rate was observed among cows bred after 90 days from calving. This work at Cornell University and observations at other stations led to recommending delay of first service until 60 days after calving. In another investigation, 50 cows were bred within 60 days after calving and required 2.09 services per conception, 48 percent of them conceiving at the first to fourth services. The second group of cows, bred at 61 to 90 days, required an average of 1.55 services, with 70 percent of conceptions at first service. Cows not bred until 90 days after calving required 1.54 services, and 76 percent conceived at first service.

There is a trend, however, toward breeding at the first estrus after 45 days postpartum for more frequent reproduction and higher average daily milk yield during the cow's productive lifetime. Study with Jerseys at the Florida station showed:

Calving interval	Lacations	Average daily milk yield during lifetime
(days)		(pounds)
385 or fewer	148	21.5
386–445	136	20.0
446–525	81	18.6
More than 525	51	16.1

Breeding efficiency may be reduced slightly because of incomplete involution of the uterus in some cows. The ultimate result was increased milk production past 2 years of age to the close of productive life. Others have obtained similar results recently.

MULTIPLE BIRTHS AND SUPERFETATIONS

Multiple births occur more frequently in dairy cattle than among beef breeds. Gilmore computed a frequency of one set of twin dairy calves in 85 total births. Arnold assembled reports of four dairy breeds in nine station herds, which indicated 3.04 percent of twins in dairy cattle. Hancock commented that the average twinning frequency of 0.6 to 1.0 percent among dairy cattle in New Zealand was very low. Frequency of monozygotic twins was independent of twinning rate. He believed the tendency for a fertilized ovum to divide and form two individuals was conditioned by heredity, since two bulls each had sired two sets of identical twins at intervals of 1 year or more. A recent study of 7,387 calving records at Carnation Farms over a 30-year period showed 4.2 percent of multiple births. Both the sire and the dam contributed significantly to the twinning tendency, according to R. E. Erb and associates.

Instances of superfetation with living calves are few. A grade Jersey cow in the Florida State Hospital dairy herd was bred at successive heat periods to a Jersey and a Holstein bull, and dropped living calves at an interval to the respective services. A 2-year old Jersey at the Tennessee A & I State University dairy farm was inseminated at three successive estrus periods. She dropped a heifer calf on July 18, 1959, and a male calf on September 28, 1959.

ARTIFICIAL BREEDING

An Arab reportedly inseminated a mare successfully in 1322 with semen from a stallion. The Italian physiologist Spallanzani impregnated a bitch in 1780 with semen collected from a dog, and others confirmed the practice later. E. I. Ivanoff, a Russian veterinarian, practiced artificial breeding in 1899 at the Royal stud. He trained many inseminators to breed cattle and sheep before World War I.

King of the Ormsbys 178078 was purchased jointly by Tranquility Farm in New Jersey and Winterthur Farm in Delaware and was stabled alternately at these farms. Before 1920 Superintendent Arthur Danks of Tranquility Farm aspirated semen from a natural service and carried it to the other farm for special matings. This was the first insemination of dairy cattle in America known to O. E. Reed, Chief of the USDA Bureau of Dairy Industry.

As early as 1920 the veterinarian at Carnation Farm in Washington followed a similar practice with semen from the full brothers Matador Segis Walker 148839 and Segis Walker Matador 166136. In 1934 C. L. Cole collected semen from an albino Holstein bull by rectal massage of the ampulae. He used the semen to inseminate an albino heifer which gave birth to an albino calf. He similarly collected semen from one bull in 1937–38 and obtained 98 pregnancies among cows in seven herds near the North Central School of Agriculture at Grand Rapids, Minnesota.

While Enos J. Perry studied in Denmark during 1937, Konsulent

K. M. Andersen took him to visit the first farmers' cooperative artificial breeding association, established on the island of Samsoe in 1936. A year later, Perry patterned a New Jersey cooperative after the Samsoe one. Dr. K. A. F. Larsen, first artificial technician at Samsoe, advised for 2 months in 1938 with the Cooperative Artificial Breeding Association No. 1 in New Jersey, the first farmers' cooperative artificial breeding organization in America. A similar cooperative began operating in Missouri during the same year.

SEMEN COLLECTION AND PRESERVATION

The first collections of bull semen in the Soviet Union were crude— with a sponge or swab. Development of artificial vaginas in Italy and the Soviet Union were an advance in collecting pure semen. Rubber vaginas were made later in Denmark, where one was obtained by the New Jersey Cooperative. The vaginas have since been improved in design and materials.

Extenders were developed in the Soviet Union and America. An egg yolk-phosphate extender was formulated at the Wisconsin station in 1940, followed by an egg yolk-citrate extender at Cornell in 1941. *Heated homogenized milk* or *heated skim milk* was added later to equally good formulations. J. Q. Almquist used discoveries with glycerol to develop an efficient heated milk-glycerol extender, preparation of which is quite detailed. Superior motility of bull spermatozoa over several days was obtained by Connecticut investigators, adding egg yolk to a *heated skim milk*-glycerol extender.

G. W. Salisbury, and later E. L. Willett, found that the most practical extension of bull semen was about 1:100, if it provided 12,000,000 spermatozoa per milliliter or per single service ampule. Paul Phillips found that additions of harmless coal-tar dyes were convenient for distinguishing semen of different breeds for field use. Trials at the Connecticut, New York, and Pennsylvania stations established that adding 500 and 1,000 units of penicillin and streptomycin per milliliter increased the fertility levels of bovine semen. F. I. Elliott *et al.,* found in 1961 that also adding polymycin B sulfate increased the protection against vibriosis possibly present in the fresh semen when collected.

Nonreturn rates of 30, 60, and 90 days in cows inseminated have been accepted as practical measures of relative effectiveness of semen from different bulls in artificial service. These rates tend to approximate the proportion of females settled by semen from various collections.

EVALUATING QUALITY OF SEMEN

Several methods were developed to evaluate quality of bovine semen. The only reliable measure of semen quality is breeding efficiency or conception rate. Volume of ejaculate and numbers of spermatozoa estimated by photoelectric colorimeter are partial measures of a bull's capacity for semen production. Estimates of semen quality are based on color, fluidity, and microscopic examinations for motility and abnormal or dead spermatozoa. The first ejaculate often is discarded due to contamination from urine which injures the spermatozoa.

Bulls with low breeding efficiency sometimes have produced semen with an undesirable bacterial count. Studies in Denmark and elsewhere have found that vibriosis and some other diseases can be transmitted with semen from infected bulls. Additions of the antibiotics previously mentioned tend to nullify these factors and have increased conception rates of some bulls significantly.

Frozen semen was stored in three forms: in ampules, straws, or pellets. The methods have been modified, based on field trials. R. Cassou used 1 ml. straws of frozen semen at the Artificial Insemination Center at L'Aigle, Orne, France, in 1967. Two extenders were used in England and Wales. One extender consisted of egg yolk 20 percent, glycerol 7 percent, 73 percent of a 2.9 percent sodium citrate solution, plus the antibiotics penicillin and streptomycin. The second extender consisted of skim milk heated one hour (to 97° C. or 207° F.) 79 percent, egg yolk 10 percent, glycerol 11 percent. Fructose was added at 1.25 gm./100 ml., with the two antibiotics mentioned. The extended semen was allowed to equilibrate at least five hours before being frozen in ampules or straws for storage above liquid nitrogen in tanks. Medium-sized straws (0.5 ml.) were used in France since 1966, to deliver about 14,000,000 live sperms at insemination.

For freezing pellets, the fresh semen was extended 1:4 with a diluent of egg yolk 20 ml., glycerol 4.7 ml., 75.3 ml. of an 11 percent solution of lactose W/M, and 500 units each of penicillin and streptomycin per milliliter. Tubes containing diluted semen were placed in a 250 ml. beaker of water at 30° C. (86° F.) for five hours to equilibrate. The diluted semen then was frozen by dropping 0.07 ml. from a calibrated Pasteur pipette into depressions 2 mm. in diameter by 1 mm. deep on solid block carbon dioxide (dry ice). In two minutes these frozen pellets were transferred with cooled forceps into split aluminum tubes submerged in a liquid nitrogen storage tank. Before insemination the frozen pellet is reconstituted in 3.2 percent sodium citrate solution for immediate use.

The bull's identity was imprinted on each ampule in the method described above.

PHYSICAL EXAMINATION OF BULLS

Major attention has been given to selecting bulls for artificial service. The manager, a bull committee, bull analyst, and technicians continually seek desirable bulls. Several organizations budget one cent per first service to obtain detailed health and reproductive examinations of prospective bulls. Many bulls were selected as "proved" sires at first. The supply of desirable bulls was limited. Associations sampled selected young bulls for early proving. Some 1,002 bulls entered the studs under 3 years old in the 10-year period 1942–51. Of these, 317 were eliminated at an average age of 5½ years while still fertile. There were 230 bulls removed before 5 years old for natural causes, such as low conception rate. Some 45.5 percent survived for selection of those worthy of continued service. This is a higher proportion of good bulls than occurs in natural service, based on the proportion of bulls proved desirable in herds under DHIA supervision.

TIMING INSEMINATIONS

Stages of estrus were observed by rectal palpation with four breeds of cows in the Nebraska station dairy herd. Inseminations were timed with relation to estrus cycle. Conception rates were: at start of estrus, 44 percent; middle of estrus, 82.5 percent; end of estrus, 75; 6 hours after estrus, 62.5; 12 hours after estrus, 32; 18 hours after estrus, 28 percent. The highest conception rates resulted from inseminations between 6 hours after the start and 6 hours after the end of estrus or "standing heat." These were from 12 to 20 hours after the beginning of estrus. These optimum times were confirmed by the Kentucky, New York, Wisconsin, and other stations. Three sets of twins were produced during the Nebraska investigations: a set of identical females from a single ovum, and 2 sets of mixed twins when 2 ova were released at a time.

Greater lifetime production was found from 10,679 lactations in England when cows freshened every 14 to 15 months, as shown in Table 21.3.

TABLE 21.3
RELATION OF CALVING INTERVAL TO LIFETIME PRODUCTION INDEX OF COWS
IN ENGLAND

Calving intervals (days)	Number of records	Yield index (%)
300–359	3,322	92.8
360–379	2,382	100.0
380–419	2,931	103.6
420–459	1,112	104.6
460–499	488	104.0
More than 500	444	102.8

Pregnancy diagnoses following over 14,000 artificial inseminations in Wisconsin usually were made at 30 to 49 days after service, between August 13, 1945, and September 30, 1947. The respective and overall conception rates are listed in Table 21.4.

TABLE 21.4
CONCEPTION RATES AT SUCCESSIVE INSEMINATIONS IN ARTIFICIAL
BREEDING IN WISCONSIN, 1945–47

Order of insemination	Services	Classed as fertile	Conceptions (%)	Conceptions overall (%)
First	8,621	4,690	54.4	
Second	3,463	1,756	50.7	
Third	1,443	671	46.5	
				82.6
Fourth	631	241	38.2	
Fifth	308	113	36.7	
Sixth	152	46	30.3	
Seventh	75	21	28.0	
Eighth and later	78	13	16.7	
Services	14,771	7,551		86.4

From the Wisconsin observations, a breeder may be justified in having an animal examined for abnormalities when she failed to conceive from the first three services. If treatment were not indicated, the animal could be culled as soon as feasible unless she was especially valuable in a breeding program. More effort might be justified in the latter case.

PLACEMENT OF SEMEN

Louisiana and South Carolina workers tested the placement of semen in 2,014 first services in the field and with 103 controlled heifers. No significant differences were observed in nonreturn and conception rates between placement of semen in the os uterus, body of the uterus, or the uterine horns. Superheats sometimes occur, however. Ten cows previously diagnosed pregnant in Illinois, and then inseminated through the cervix into the uterus, were found on slaughter at 64 to 185 days to be carrying dead fetuses that were disintegrating. L. E. Casida and N. L. VanDemark concluded from separate investigations that the second and later inseminations should not be placed farther than the midcervical region.

Pregnancy examinations were made in Wisconsin at 34 to 50 days after insemination over a 5.5 year investigation. Some 690 pregnancies were diagnosed as positive by palpation of the amniotic vesicle. Return of estrus or observed abortion was evidence of embryonic and fetal mortality. Abortions from disease were excluded from the study. Some 44 pregnancies (6.38 percent) terminated before 239 days in gestation. It was estimated from other work that about 40 percent of all potential young in dairy cattle are lost by 60 to 90 days after breeding, nearly half being embryonic deaths.

Kentucky workers found 23 mummified fetuses among 1,509 pregnancies of 504 cows. Statistical analyses concerning the affected cow- and bull-families indicated that the condition might be due to a sex-linked recessive lethal gene.

PRESERVING FROZEN SEMEN

The short life of fresh semen and waste in maintaining field supplies entailed much loss to breeding organizations. Fresh semen could not be held long. Selective matings were difficult in breeders' herds because of discontinuous semen supplies.

Dr. Prevost froze frog spermatozoa (in testes) at −8 to −10° C., at Geneva, Switzerland, in 1839. The spermatozoa returned to motility when thawed gradually. P. Manteguzza of Italy froze human spermatozoa successfully to −17° C. in 1866. Spermatozoa in frog semen were killed when frozen in pond water but survived the temperature of liquid air (−147° C.) in sugar solution.

C. Polge and L. E. A. Rowson diluted bull semen with an equal volume of egg-yolk–citrate buffer at 28° C. and cooled it to 4° C. for four hours. An equal volume of citrate buffer containing 20 percent glycerol was added at 5° C. and held overnight, then cooled slowly for 45 minutes to −79° C. and stored in dry ice and alcohol. After the semen was thawed, 38 cows were inseminated promptly. Thirty were diagnosed pregnant in 6 weeks. These trials in 1951–52 were conducted jointly by the National Institute for Medical Research and the Cambridge and District Cattle Breeders' Society in England. Three heifer calves were born in June and July 1970 from Dairy Shorthorn semen that had been frozen for 16 years by the Cambridge A.I. Center. The American Breeders' Service developed a technique of freezing bull semen to −320° F. (liquid nitrogen). They cooperated with the Linde organization to perfect insulated equipment to store and distribute frozen semen at this temperature. Spermatozoa that tolerate freezing deteriorate slowly when held at constant temperature. However, not all semen withstands present freezing techniques for some unknown reasons. Hereditary defects of spermatozoa include defective acrosome or cap on the sperm head. The tails may be broken, twisted, or recurved.

Proper freezing allowed semen to be accumulated and maintained in storage and reduced losses of semen in the field greatly. Fewer bulls were needed to supply the industry, and interstud exchanges of semen gave wide use of desirable sires. The greatest impact was the ability to make selected matings under contract between desirable bulls and transmitting cows of seedstock quality.

Artificial breeding of dairy cattle increased greatly in the United States after 1938. Small organizations were established in several states. Some failed because per-cow costs were too high. Many problems were solved by experience and research. Small organizations consolidated with operating economies. The demand for quality bulls increased interest in more production records by breeders and commercial dairymen. Cooperative study of DHIA records by extension dairymen and the USDA Bureau of Dairy Industry located many bulls with desirable daughters. Physical examinations were established for bulls prior to acquisition. Some bulls "proved" in individual herds were unsatisfactory in A.I. service later. Too few desirable mature bulls were obtainable. Sire analysts traveled widely to find desirable mature bulls in natural use, select young

TABLE 21.5

EXPANSION OF ARTIFICIAL BREEDING IN THE UNITED STATES[a]

Year	Studs	Bulls	Bulls per stud	Dairy cows bred	Average per bull
1939	7	33	4.7	7,399 [b]	228
1945	67	729	10.9	360,732 [b]	495
1950	97	2,109	21.7	2,619,555	1,245
1955	79	2,450	31.0	5,413,874	2,210
1960	62	2,544	41.0	7,144,679	2,808
1965	46	2,316	50.3	7,879,982	3,402
1969	31	2,275	73.4	8,209,444 [c]	3,608

a. From *Dairy-Herd-Improvement Letter* 46(4), July 1970.
b. Cows enrolled.
c. Of these numbers, 384 beef bulls inseminated 694,916 dairy and beef cows. Not included were 3,230 swine and 235 goats. About 40 percent of the dairy bulls were under restricted progeny test. Some 52.5 percent of dairy cows and heifers were bred artificially during 1969.

sires from good families for sampling, mate under contract, and evaluate transmission of bulls in current A.I. use.

Evaluation of semen, methods of extending it, use of antibiotics in extenders, techniques of handling, storage and distribution, and use of semen by trained technicians contributed toward economies in every phase of artificial breeding. Observation of cows and reporting to technicians as to the hour cows were first seen in heat improved timing of inseminations, raised the nonreturn rates of bulls, and gave higher breeding efficiencies. The increase in artificial breeding among dairy cows in the United States is shown in Table 21.5.

The average useful tenure of desirable dairy bulls in the studs increased 1.72 years in 1939–47. Some 1,577 dairy bulls removed during 1960–64 had entered A.I. service at an average age of 4.43 years (range from 1 to 15 years). They lasted an average of 4.19

years in the studs of the United States and Canada. Twenty eight percent were active at 10 years or older, and a few were past 16 years old at last semen collection.

IDENTIFYING PARENTAGE

Records of parentage of each artificially conceived animal registered depend on integrity of the owner of the cow at time of insemination, as well as on identity of the bull producing the semen and reliability of the inseminator. Investigations at the Ohio, Wisconsin, and other stations found that the red blood corpuscles bore specific antigens identifiable by laboratory tests whereby questions between probable parents could be decided. These factors now are " . . . interpreted on a genetic system by arranging factors into phenogroups within eight definite distinct systems."

Each registered bull producing frozen semen for artificial service on a within-herd, between-herds, or in a semen-producing business (A.I. stud) under contract with the Purebred Dairy Cattle Association is required to have his blood type recorded with the respective breed association. Over 250 phenotypes have been identified in the "B" system of the eight systems recognized in bovine blood. Routine service identifications of parentage were begun at Ohio State University in 1948. The Serology Laboratory, School of Veterinary Medicine at the University of California at Davis, working under contract with the Purebred Dairy Cattle Association since 1955, handled a limited number of blood-typing samples on a definite schedule. Arrangements for blood-typing service were made in advance at a fee through the breed association, following instructions of the respective association. In case of doubtful parentage, blood samples of the dam and progeny in question were taken in treated sample tubes under witness, packed, and transported by specific routines. This method was used to identify which of two blood-typed bulls was the sire of the animal. The method has been recognized legally by a Canadian court.

DISEASE CONTROL

The dairy cattle industry has benefitted from prevention, reduction, or eradication of certain communicable diseases.

Contagious pleuropneumonia of cattle was eradicated from the United States by combined state and federal action following outbreaks from Europe in 1843, 1847, and later. The first Friesian cattle in Massachusetts were destroyed by the Commonwealth in stamping out the outbreak in 1859. Final eradication in this country took place in 1892; in Great Britain in 1898, and earlier in Holland.

Foot and mouth disease (aftosa) attacks all cloven-hoofed animals. The disease was eradicated by slaughter of animals, quarantine, and disinfection of premises following outbreaks in 1870, 1880, 1884, 1902, 1908, 1924, and 1925. The disease broke out at the National Dairy Show in 1914. Under a federal court injunction, show cattle were quarantined at a nearby fair grounds until all danger passed. Some cattle died, but many prominent animals survived.

A joint Mexican-United States commission eradicated an outbreak on Mexican soil during the 1950's by a combination of vaccination, quarantine, and slaughter. Annual preventive vaccination is being practiced in some countries. Aftosa still occurs in parts of South America and Europe. Sanitary regulations, quarantine, and inspection have protected against reinfection by cattle, forages, hides, packing materials, and ship's garbage from countries with known outbreaks. Opening the St. Lawrence seaway in 1959 extended United States and Canadian shorelines necessary to be guarded against entrance of diseases.

Robert Koch of Germany discovered the bacillus causing *tuberculosis* in 1882. Three tests to diagnose its presence—interdermal, opthalmic, and thermal—were used to detect and eradicate the disease from individual herds. A campaign of area eradication began in May 1917. The last counties in the United States were declared modified accredited areas (less than 0.5 percent of reactors eliminated on last test) by November 1940. Slaughter of cattle under federal inspection revealed less than 0.02 percent of reactors in 1940. The goal is total eradication of the disease as a public health measure.

Programs of disease control and eradication have been conducted as economic and public health measures. Purebred dairy cattle imported from Europe and animals for stocking zoological gardens undergo quarantine before release. The ports are closed to animals having such diseases as aftosa, tuberculosis, or others. The livestock industry has benefitted by elimination of these diseases through reduced losses and extension of useful lifespan of breeding animals.

Anthrax outbreaks have been eradicated by slaughter, burning carcasses, and quarantine of premises. The anthrax bacillus is a spore-former that has remained virulent in soil over 40 years under favorable environment and moisture.

Bloat (hoven) is not a disease. It is mainly an excessive accumulation of frothy foam in the rumen fluid from feeding on succulent young legumes. In emergencies, it formerly was necessary to release the gases by a trochar-and-canula puncture into the rumen, leaving the canula for several hours for gases to escape. Medication often was of little avail in the past. Allowing cattle some dry hay prior to grazing lush legume pastures made them less susceptible to bloat. In 1965 workers at the Kansas station found that giving cattle 5 to 10 gm. of poloxalene (bloat guard) twice daily reduced surface tension of rumen fluids and allowed less stable foam to accumulate in the rumen fluids. Bloat thus can be prevented for up to 12 hours, except with a few large cattle.

Brucellosis. Professor B. L. Frederick Bang of Denmark discovered the *Brucella abortus* bacillus in 1896, which is one cause of contagious abortions. Three strains include also *B. abortus suis* of swine and *B. abortus melitensis* of goats. These are communicable to man, causing undulent and Malta fevers. An eradication campaign began in July 1934 on an individual certified-herd plan. State and federal agencies look to eradication of the disease by an area testing plan.

Cattle tick fever is caused by a protozoan, the alternate hosts of which are two species of ticks and cattle. Deer, horses, and mules sometimes carry and spread these ticks to other pastures. These ticks were eradicated in zones covering over 729,852 square miles in 16 states by dipping cattle systematically, beginning in 1906. Countries to the south yet are infested.

Hydrophobia. Several mammals develop lethal hydrophobia (rabies) from a bite by a rabid animal. The French investigator Louis Pasteur's most notable contribution was inoculation against the virus communicated in the saliva of a rabid animal. He attenuated the virus contained in the medulla oblongata that connects the brain with the spinal cord by drying it in sterile air. A succession of progressively more virile injections caused the body to build up resistance against lethal action of the virus. Immunization of 9-year old Joseph Meister, bitten by a rabid dog, was completed July 16, 1888. Pasteur thus demonstrated immunization against that fatal infection was possible if done before advanced stages of the disease.

Johne's disease (paratuberculosis), recognized since 1895, involves the mucous membranes and intestinal lymph glands of cattle. A Johnin test similar to the tuberculin test is applied to detect infected animals. Some states are cooperating with the USDA in testing programs and in elimination of infected animals.

Rinderpest, a critical virus disease of cattle, has been kept out of the United States by embargo, but it still occurs in many tropical regions and parts of the Orient. The infection can be carried in hides, improperly sterilized animal by-products, and materials infected by contact.

Smallpox. On May 14, 1796, the English physician Edward Jenner successfully immunized his son against smallpox by treating him with the attenuated virus from a cowpox lesion on the hand of the milkmaid Sarah Newlme. Jenner initiated the practice still used in immunization against several common diseases of man and livestock, such as blackleg, leptospirosis, and several other infections.

JUNIOR DAIRY ACTIVITIES

Junior extension activities began with corn clubs, to which pig and calf clubs were added. Calves were purchased by business men or others, and placed for 1 year with selected calf club boys or girls to feed, manage, and keep records. The calves were exhibited and sold at auction at the end of the year. Original cost of the calf was remitted to the owner, while any difference went to the club member. A committee of the American Dairy Science Association recommended in 1922 that henceforth the club member own the calf, and keep records at least through the first lactation. Calf clubs then became 4-H Dairy Clubs. Some prominent breeders began to build their herds as 4-H Dairy Club members. Some vocational agricultural schools foster a junior dairy project along the same plan.

Judging at the junior dairy shows first followed the traditional system of a first, second, and third prize to the three top animals irrespective of quality, condition, and showing. Under such a plan many animals were out of the money in large groups. This method of judging was changed to the Danish system whereby animals were divided according to quality or score into blue, red, and white ribbon groups. Every animal in these groups was recognized.

Fitting and showmanship contests and the project record books have been educational features of the junior dairy project. Competition begins with local clubs, and the better entries advance to district, state, and national junior dairy shows.

A 1-day district Black-and-White Show was initiated at Richmond, Utah, by the Holstein western fieldman and Utah extension service in 1915. Jersey breeders followed with parish shows in the midwest in 1928. Brown Swiss canton shows began in 1938.

Feeding and Management

Classroom instruction on care and feeding of dairy cattle formerly was based on successful farm experience. European feeding standards were interpreted by W. A. Henry, and later standards were assembled by his successor F. B. Morrison in the textbook *Feeds and Feeding*. They assembled analyses and nutritive values of American forages and grains.

Controlled feeding trials by T. L. Haecker determined the digestible crude protein, fat, and carbohydrates needed to maintain a dairy cow in the stable. He found that the nutrients above maintenance requirement needs for milk production varied according to the butterfat content and composition of milk. J. L. Hills and associates confirmed Haecker's findings for maintenance and milk production, and added an allowance during advancing gestation. H. P. Armsby determined the energy value of protein, nitrogen-free extract, crude fiber, and fat—with separate animals in a large-animal calorimeter. From these values he attempted to compute the energy value of other feeds from their chemical analyses. J. T. Reid et al. found the maintenance requirements of grazing cows to be about 40 percent above those in the stall (11.3 instead of 7.9

pounds of total digestible nutrients for a 1,000-pound cow). This allowed for the work of grazing.

C. H. Eckles found that a good cow and a mediocre cow digested feeds equally well. The requirements for nutrients to maintain the body were in proportion to live weight. The better cow utilized surplus nutrients for greater milk production in proportion to her ability. He found that the amounts of protein and energy used to develop the bovine fetus were too small to measure by the usual feeding trials. Work begun in Missouri on the feed requirements of growing heifers was completed by A. C. Ragsdale. Eckles investigated requirements for maintenance and growth of dairy calves and the needs of dairy cows for phosphorus.

Long-time calcium and phosphorus balance trials at the Michigan, Pennsylvania, and Vermont stations and at Beltsville found that milk cows stored these minerals during the dry period. Cows drew on these reserves when the daily intakes were inadequate for high milk production. Iodine in trace amounts was found to prevent goiter in young animals by the Montana, Washington, and Wisconsin stations. Needs of cattle for the trace mineral elements—iron, copper, and cobalt—to overcome and prevent nutritional anemia and for respiratory and other functions were investigated at the Florida station. Other stations confirmed results in these fields and adapted them to local feeding conditions.

Eckles directed early investigations on the vitamin requirements of cattle. This led S. I. Bechdel and associates to discover that bacteria synthesize the water-soluble vitamins in the rumen. Also L. M. Thurston found that the diet of calves need not supply vitamin C. However, vitamin A had to be supplied by the feeds, according to I. R. Jones. This field was reviewed by the Subcommittee on Dairy Cattle Nutrition of the National Research Council.

Hormonal relationships to udder development and milk production have been reviewed by the Committee on Animal Nutrition and have been presented in textbooks on animal physiology and milk secretion.

Successive subcommittees of the National Research Council's Committee on Animal Nutrition summarized investigations on the recommended nutrients and the nutrient requirements of dairy cattle. The early reports applied to practical feeding practices. The third report, in 1956, dealt with *minimum requirements* based on controlled investigations in confinement with analyzed feeds of good quality. It allowed no margin for average variations in composition of forages and concentrates, or for work by grazing animals under farm conditions. Margins of 10 to 15 percent above the minimum amounts of digestible crude protein and total digestible nutrients (the equivalent basis for digestible fats, carbohydrates, and protein) were recommended by T. L. Haecker, J. L. Hills, C. H. Eckles, and other workers in their recommended feeding standards. Their recommendations allowed for differences observed between selected analyzed feeds and average feeds, particularly forages, under general farm conditions. J. T. Reid's recommendations on maintenance needs of cows grazing on pastures, cited previously, was based on the Morrison standard at the "recommended" or higher level, which included the margins by the above investigators.

A nationwide survey in 1941–44 found that butter (80 percent fat) contained an average of 15,000 I. U. of vitamin A per pound. A quart of milk (4 percent fat) averaged 1,540 I. U., or 716 per pound of milk. Vitamin A contents varied seasonally with intakes of quality forages. Workers at Purdue University determined that Guernsey cows maintained their bodies and produced butterfat of maximum vitamin A potency on intakes of 200 ml. of carotene daily. Requirements for reproduction were found at the Oklahoma station to increase at the sixth month of gestation, and were practically double the needs at other times. The report of the Subcommittee on Dairy Cattle Nutrition in 1966 adjusted allowances for vitamin A on a practical basis to provide for an average vitamin A potency in milk.

The fourth revised edition (1971) of *Nutrient Requirements of Dairy Cattle* was presented in terms of energy and in metric units. Requirements were listed separately for growth of small and large breeds of heifers, of young bulls, for veal calves, and for mature bulls and cows. Allowances during the last two months of gestation provided for gain in body condition before calving. Requirements for milk production varied with butterfat percentages. For high milk production, the carotene allowances were suboptimal, and could retard reproduction. Allowances were estimated for sixteen major and minor mineral nutrients, vitamins A, D, and E. The B-complex vitamins were listed for calves before rumen synthesis begins to supplement their intake in feeds.

Milk Secretion

Fundamentals of udder anatomy, function, and milk secretion are separate studies, dealt with in a technical textbook, *Physiology of Lactation*, 5th edition, by V. R. Smith, expanding earlier assemblage by D. Espe.

An important function of colostrum—the first milk produced after parturition—is to provide the newborn calf with antibodies against many common bacteria. This function long has been recognized. More recently, B. Campbell and W. E. Petersen determined that the udder could acquire other bacteria from the nursing calf and soon form antibodies against them. This discovery may bear on production of special milks for therapeutic use in the future. Investigations are continuing on this function of the mammary gland.

Cows dependent mainly on pastures during drouths in New York and Wisconsin gave milk low in protein contents. Eckles and Palmer found that cows underfed on total digestible nutrients yielded milk low in casein content. E. B. Powell reported that fineness of the leafy forages, as well as restricted intakes of them, decreased the percentage of fat in cow's milk. Field observations and feeding trials in Florida confirmed the need for long leafy forages in quantity for fat synthesis in the udder. Underfeeding energy at 85 or 75 percent lowered protein contents of milk, independently of heredity and management of the cows.

The cow is a ruminant, adapted to utilize high-quality leafy forages efficiently. H. A. Herman determined that growth stage of *Lespedeza sericea* affected digestibility inversely to its lignin content. During 14 years of rumen fistula studies, O. T. Stallcup of Arkansas confirmed that lignin incrusting other nutrients impeded the rate of passage and digestibility of stalks as contrasted with the leafy portions of corn silage.

Florida workers found that mainly small feed particles passed from the rumen into the omasum. Rhythmic contractions crushed softened particles still finer between the laminal papillae. Much of the finest material soon disappeared by solution and digestion in the abomasum.

HEREDITARY CAPACITY LIMITED BY ENVIRONMENT

Early attempts to influence heredity in cattle were mentioned in Genesis 30:25–43. The herdsman Jacob tried to increase his pay in spotted animals from Laban's flocks and herds. Robert Bakewell

FIG. 21.1. Johann Gregor Mendel, German-born Austrian monk, discovered the independent assortment of dominant and recessive genes by cross-pollinating peas in the Augustinian monastery garden at Brünn, Austria (now Brno, Czechoslovakia).

demonstrated use of inbreeding to concentrate some desirable characters in his Longhorn cattle and Leicester sheep. The Austrian monk Johann Gregor Mendel discovered the principles of inheritance through independent assortment of genes while working with hand-pollinated strains of peas in the monastery garden at Brünn, Austria (now Brno, Czechoslovakia). See Figure 21.1.

Colonel Zadock Pratt observed a century ago that production of cows could be improved by even a single cross to one of the improved breeds. Grading up experiments at the Iowa Station (Table 21.2) and others substantiated this fact.

John Gosling, Kansas City meat cutter, demonstrated at Farmers'

Week programs in 1915 that muscling and spareness of fleshing (dairy temperament in part) were inherited breed characters. See Figure 15.5.

The effect of feeding, good care, and management has been demonstrated many times. Transfers of cows between herds under different management have been reflected in milk and butterfat yields. E. L. Anthony stated that transfers of outstanding animals did not include the environment that enabled good cows to produce up to their hereditary capacities.

The combined effect of improved hereditary capacity, better feeding, and management are evidenced by the increasing average production of cows—from 215 to 483 pounds of butterfat in DHIA between 1906 and 1970 (Table 21.1). The impact of breeding artificially to good sires contributed toward this improvement. In 1969 some 1,911 selected bulls in artificial breeding studs were used to inseminate an average of 3,608 cows per bull (Table 21.5). Dairy cows will become even more efficient.

PUREBRED DAIRY CATTLE ASSOCIATION

Good representative cows of five dairy breeds were displayed on the Borden Company's rotolactor at the New York World's Fair in 1939. In 1940 Henry Jeffers, inventor of the rotolactor, suggested to seven representatives of the breed associations that a joint interbreed organization might cooperate on mutual problems in addition to the public relations exhibit. The group organized temporarily until the Purebred Dairy Cattle Association was formed at Peterborough, New Hampshire, five months later. Membership comprised three representatives from each of five dairy breed organizations—usually the president, secretary, and a breeder. The breeds shared operating obligations in proportion to the number of registered animals in the previous three years. Merits of purebred dairy cattle were promoted with a Court of Dairy Queens (six daughters of a sire of each breed that transmitted production) at the National Dairy Show in 1940 and 1941. An essay contest on the purebred sire drew 2,816 contestants.

The PDCA drew on specialists in different dairy fields to serve on committees and cooperate with other organizations. Projects included the Unified Score Cards for dairy cattle, uniform classes for dairy cattle at county and state fairs, facilities for photography at fairs, and unified rules for production testing, artificial breeding, and use of frozen semen. Some uniformity was attained in type classification of dairy cattle. Diseases were studied with relation to regulations for interstate movement of animals. Recommended sales practices at public auctions and private treaty were incorporated into a recommended code of sales ethics. Interests were interwoven with committees of the American Dairy Science Association on breed relations, supervision of production testing, animal breeding, health, and others. The PDCA now provides the Roll of Honor certificates for meritorious production in DHIA, sponsored earlier by the National Dairy Association.

State livestock associations existed early in several areas, and PDCA groups since have organized in many states.

A Research Council was founded with cooperation of the American Dairy Science Association, with subcommittees in production, reproduction, disease control, and dairy type. This resulted to some extent in unified methods of analyzing sires and dams, subject to revisions. Blood typing of dairy cattle was fostered at the Universities of Ohio, Wisconsin, and California, with service in verification of parentage now under a separate contract, based on blood antigen determinations. All bulls in artificial service under contract with the PDCA now are required to have blood types recorded with the respective breed associations.

A uniform scorecard for judging Junior Showmanship Contests was developed.

The Eckles Club

Eleven former graduates who had studied under Dr. C. H. Eckles, lunched together while attending the National Dairy Show at Springfield, Massachusetts, in 1916. They decided to wire "The Chief" who had inspired them toward their careers. Thus the Eckles Club began and expanded. Later "The Chief" gave or wrote a short message to these former students. (Like: "Offering cheese in convenient-size pieces might increase acceptance." It has.) Doctor Eckles was straightforward and humble about his contributions to the dairy industry. The members of the Eckles Club had each contributed in some way to the dairy industry.

Members of the Eckles Club included 17 Chairmen and 36 staff members of Dairy Departments, 16 Extension dairy specialists, four Deans or Directors at Colleges of Agriculture, 11 leaders in the USDA Bureau of Dairy Industry, two Presidents and four Secretaries of dairy breed associations, and many leaders in their respective fields in the dairy industry.

Dr. Eckles helped to found the American Dairy Science Association and served four years as its President. Former students served a total of 16 years as president. Dr. Eckles helped to establish the *Journal of Dairy Science* and was an associate editor thereafter. Eckles Hall was dedicated in his honor at the University of Missouri on November 3, 1939, and a street on the St. Paul campus of the University of Minnesota was named in recognition of his service there.

The pledge given by Eckles Club members at each annual meeting was: "I believe that dairying is a branch of agriculture which promotes the best interest of this, my country, and I pledge myself to its proper development through research and practice. All this I promise that I may prove worthy of the example of my teacher, counselor and friend: Clarence Henry Eckles."

After Mrs. Eckles passed away, the Eckles Club voted to disband.

It is with humble appreciation that this book is dedicated to "The Chief" as one of the greatest leaders who contributed most significantly toward a better dairy industry.

Dairy Shrine Club

Honor to living persons and recognition of former leaders who made outstanding contributions to any part of the dairy industry were goals for founding the Dairy Shrine Club. Conception of the organization originated with a few men in 1947 and was completed in Minneapolis in June 1949. It crystallized around the creation of a graduate student scholarship in honor of Dean H. H. Kildee on his retirement from administrative duties at Iowa State College on June 30, 1949. The scholarship was to be awarded to the high man or alternate in judging all dairy cattle breeds at the National Intercollegiate Dairy Cattle Judging Contest at the National Dairy Cattle Congress.

Portraits of living Guests of Honor unveiled during the years included:

Dean H. H. Kildee, Iowa State College	1949
William Henry Jeffers, Walker Gordon Company	1950
Charles L. Hill, Guernsey breeder, importer, and President	
of the National Dairy Association	1951
Fred Pabst, Holstein breeder	1951
Dr. E. V. McCollum, food value of milk	1952
E. S. Estel, Manager, National Dairy Cattle Congress	1953
Joe P. Eves, Judge of dairy cattle	1954
Karl B. Musser, American Guernsey Cattle Club	1955
H. W. Norton, Jr., Holstein-Friesian Association of America	1955
John S. Ames, Langwater Farm Guernseys	1956
Alfred W. Ghormley, Carnation Company	1957
Fred S. Idtse, Brown Swiss Cattle Breeders Association	1958
Harold J. Shaw, Holstein breeder	1959
Harry A Strohmeyer, Jr., livestock photographer	1959
James Cash Penney, Guernsey breeder	1960
Dr. W. E. Petersen, teacher and investigator	1960
Dr. C. F. Huffman, teacher and investigator	1961
Elbert H. Brigham, Jersey breeder	1962
Roger R. Jessup, dairy leader	1962
Otto H. Liebers, Guernsey breeder	1962
M. S. Prescott, *Holstein-Friesian World* editor	1963
William D. Hoard, Jr., *Hoard's Dairyman* business manager	1964
Harold E. Searles, Dairy Extension leader	1965
Warren Kinney, Brown Swiss breeder	1966
W. D. Knox, *Hoard's Dairyman* editor	1967
Glen Lake, dairy farmer and milk marketing	1968
Dr. J. L. Lush, teacher and investigator	1969
Dr. Earl Weaver, teacher and investigator, judge	1970
Lawrence O. Colebank	1971

Portraits of prominent dairy pioneers are hung in the Pioneer Room, and volumes contain sketches of their contributions. These men won acclaim as administrators, auctioneers, authors, bacteriologists, breeders, dairy editors, educators, importers, industrialists, inventors, herd managers, judges, organizers, research workers, manufacturers of dairy equipment and dairy products, publishers, breed association officers, dairy extension specialists, teachers, and others in closely allied dairy fields. Each person made some permanent contribution leading to advancement of the dairy industry. Persons recognized from other countries were Bernard Bang, Gustav

De Laval, Robert Koch, Johann Gregor Mendel, and Louis Pasteur. One of the founders wrote:

> Probably the Dairy Shrine Club would never have been founded had it not been for the lasting impressions and inspiration these leaders of the past left with those with whom they shared some phase of life. It may have been as a teacher, a breeder, a coach, an advisor or just the example set by men of profound purpose and devoted application. For "example works stronger and quicker in the minds of men than precedent."

The Dairy Shrine Club portrait gallery, livestock pictures, and sketches of contributions in albums serve as an historical record of cattle, people, and accomplishments in the dairy industry. Secretary-Treasurer Arthur W. Nesbitt, 901 Janesville Avenue, Ft. Atkinson, Wisconsin 53538, will reply concerning the current location of the displays.

REFERENCES

Adametz, L. 1898. Studien uber Bos (brachyceros) europaeus, die wilde Stamform der Brachyceros Rassen des europaischen Hausrindes. Z. *Landwirtsch.* 46:269–320.

Agr. Res. Adm. 1943. Federal legislation, rulings, and regulations affecting the state agricultural experiment stations. Off. Expt. Sta. *USDA Misc. Publ. 515.*

Almquist, J. O. 1946. The effect of coaltar dyes used for semen identification on the livability and fertility of bull spermatozoa. *J. Dairy Sci.* 29:815–20.

———. 1951. A comparison of penicillin, streptomycin and sulfanilimide for improving the fertility of bulls of low fertility. *J. Dairy Sci.* 34:819–22.

———. 1959. An efficient low cost artificial breeding program using liquid semen. *Guernsey Breeders' J.* 104:24–26, 42.

———. 1959. Efficient low-cost results using milk-glycerol diluent. *A. I. Digest* 7(8):11–14, 27.

Armsby, H. P. 1917. *The nutrition of farm animals.* Macmillan, New York.

Arnold, P. T. Dix, and R. B. Becker. 1936. Influence of the preceding dry period and of mineral supplementation on lactation. *J. Dairy Sci.* 19:257–66.

———. 1953. Dairy calves: Their development and survival. *Florida Agr. Expt. Sta. Bull. 529.*

———. 1954. Building a dairy herd. *Florida Agr. Expt. Sta. Bull. 576.*

———. 1958. Dairy cattle and their care. *Florida Agr. Expt. Sta. Bull. 599.*

Babcock, S. M. 1890. A new method for the estimation of fat in milk. *Wisconsin Agr. Expt. Sta. Bull. 24.*

———. 1896. The relation between milk solids and yield of cheese. *Wisconsin Agr. Expt. Sta. 18th Ann. Rept.*, pp. 100–19.

Barrett, C. R., C. A. Lloyd, and R. A. Carpenter. 1948. Order number of inseminations and conception rate. *J. Dairy Sci.* 31:683.

Bechdel, S. I., C. H. Eckles, and L. S. Palmer. 1926. The vitamin B requirement of the calf. *J. Dairy Sci.* 9:409–38.

Bechdel. S. I., et al. 1928. Synthesis of vitamin B in the rumen of the cow. *J. Biol. Chem.* 80:231–38.

Becker, R. B. 1953. American contributions to better dairy cattle. *Hoard's Dairyman* 98:736–39.

Becker, R. B., and P. T. Dix Arnold. 1958. The destiny of "sampler" bulls. *J. Dairy Sci.* 41:736.

Becker, R. B., and L. W. Gaddum. 1937. The comparison of limonite effective and ineffective in correcting "bush sickness" in cattle. *J. Dairy Sci.* 20:737–39.

Becker, R. B., and P. C. McGilliard. 1929. A suggested simplification of the present system of official testing with breeds of dairy cattle. *J. Dairy Sci.* 12:337–50.

Becker, R. B., and C. J. Wilcox. 1969. Hereditary defects of spermatozoa. An illustrated review. *A. I. Digest* 17(12):8–10.

Becker, R. B., S. P. Marshall, and P. T. Dix Arnold. 1963. Anatomy, development, and function of the bovine omasum. *J. Dairy Sci.* 46:835–39.

Becker, R. B., W. M. Neal, and A. L. Shealy. 1931. Salt sick: Its cause and prevention. *Florida Agr. Expt. Sta. Bull. 231.*

———. 1933. Effect of calcium deficient roughages upon milk production and welfare of dairy cows. *Florida Agr. Expt. Sta. Tech. Bull. 262.*

Becker, R. B., C. J. Wilcox, and W. E. Pritchard. 1966. Crampy—Progressive posterior paralysis in mature cattle. *Florida Agr. Expt. Sta. Bull. 639.*

Becker, R. B., W. G. Kirk, George K. Davis, and R. W. Kidder. 1953. Minerals for dairy and beef cattle. *Florida Agr. Expt. Sta. Bull. 513.*

Becker, R. B., C. F. Simpson, L. O. Gilmore, and N. S. Fechheimer. 1964. Genetic aspects of actinomycosis and actinobacillosis in cattle. *Florida Agr. Expt. Sta. Tech. Bull. 670; Ohio Agr. Expt. Sta. Res. Bull. 938.*

Becker, R. B., C. J. Wilcox, W. A. Krienke, L. E. Mull, and E. L. Fouts. 1965. Subnormal milk—Cause and correction. *Florida Agr. Expt. Sta. Bull. 692.*

Bowling, G. A. 1942. The introduction of cattle into colonial North America. *J. Dairy Sci.* 25:129–54.

Brewster, J. E., R. May, and C. L. Cole. 1940. The time of ovulation and rate of spermatozoa travel in cattle. *Amer. Soc. Animal Prod. Proc.*, pp. 304–11.

Bur. Dairy Ind. 1937. List of sires proved in Dairy Herd Improvement Associations 1933–37, arranged by breeds. Div. Dairy Herd Improvement Investigations. *USDA Misc. Publ. 277.*

———. 1945. Vitamin A. in butter. *USDA Misc. Publ. 571.*

Burgess, T. D. 1953. The relationship between 30-, 60-, 90- and 120-day nonreturns to service in the artificial insemination of dairy cattle in Ontario. *Canadian J. Agr. Sci.* 33:261–64.

Campbell, B., M. Sarwar, and W. E. Petersen. 1957. Diathelic immunization—a maternal-offspring relationship involving milk antibodies. *Science* 125:932–33.

Casida, L. E., and W. G. Venske. Observations of reproductive processes in dairy cattle and their relationship to breeding efficiency. *Amer. Soc. Animal Prod. Proc.* 29:221–23.

Cole, C. L. 1938. Artificial insemination of dairy cattle. *J. Dairy Sci.* 21:131–32.

Cole, L. J., and S. V. H. Jones. 1920. The occurrence of red calves in black breeds of cattle. *Wisconsin Agr. Expt. Sta. Bull. 313.*

Copeland, Lynn. 1965. Type and useful lifetime production. *Guernsey Breeders' J.* 115(9):637–38.

———. 1965. A study of Guernsey bulls. *Guernsey Breeders' J.* 116(8):541.

Deaton, O. W., Durward Olds, and D. W. Seath. 1959. A study of some possible genetic causes of mummified fetuses in dairy cattle. *J. Dairy Sci.* 42: 312–14.

Downie, A. W. Jenner's cowpox inoculation. *Brit. Med. J.* 2:251–56.

Eckles, C. H. 1916. Nutrient requirements to develop the bovine fetus. *Missouri Agr. Expt. Sta. Res. Bull. 26.*

Eckles, C. H., and T. W. Gullickson. Nutrient requirements for normal growth of dairy cattle. *J. Agr. Res.* 42:603–16.

Eckles, C. H., and L. S. Palmer. 1916. Influence of plane of nutrition of the cow upon the composition and properties of milk and butter fat. *Missouri Agr. Expt. Sta. Res. Bull. 25.*

Eckles, C. H., and O. E. Reed. 1910. A study of the cause of wide variation in milk production by dairy cows. *Missouri Agr. Expt. Sta. Res. Bull. 2.*

Eckles, C. H., R. B. Becker, and L. S. Palmer. 1926. A mineral deficiency in the ration of cattle. *Minnesota Agr. Expt. Sta. Bull. 229.*

Eckles, C. H., T. H. Gullickson, and L. S. Palmer. 1932. Phosphorus deficiency in the rations of cattle. *Minnesota Agr. Expt. Sta. Tech. Bull. 91.*

Ellenberger, H. L., J. A. Newlander, and C. H. Jones. 1932. Calcium and phosphorus requirements of dairy cows. *Vermont Agr. Expt. Sta. Bull. 342.*

Elliott, F. I., et al. 1962. The use of polymyxin B sulfate with dihydrostreptomycin and penicillin for the control of "vibrio fetus" in a frozen semen process. *A. I. Digest* 10(2):10–13.

Erb, R. E., and E. M. Holtz. 1958. Factors associated with estimated fertilization and service efficiency of cows. *J. Dairy Sci.* 41:1541–52.

Erb, R. E., and R. A. Morrison. Effect of mummified fetuses on the prolificacy of Holsteins. *J. Dairy Sci.* 40:1030–35.

————. Estrus after conception in a herd of Holstein-Friesian cattle. *J. Dairy Sci.* 41:267–74.

Erb, R. E., L. R. Anderson, P. M. Hinze, and E. M. Gildow. 1960. Inheritance of twinning in a herd of Holstein-Friesian cattle. *J. Dairy Sci.* 43:393–400.

Foote, R. H., and R. W. Bratton. 1950. The fertility of bovine semen in extenders containing sulfanilimide, penicillin and polymycin. *J. Dairy Sci.* 33: 544–47, 842–46.

Forbes, E. B., et al. 1935. The mineral requirements of milk production. *Pennsylvania Agr. Expt. Sta. Tech. Bull. 319.*

Fosgate, O. T., and G. R. Smith. Prenatal mortality in the bovine between pregnancy diagnosis at 34–50 days post-insemination and parturition. *J. Dairy Sci.* 37:1071–73.

Gerasimova, A. A. 1940. Duration of heat and time of ovulation of the cow. Problemy Zhivotnovodstva. *Animal Br. Abstr.* 8:32.

Gray, V. 1966. The sire analyst—a dedicated man. *Guernsey Breeders' J.* 118: 10–11, 32.

Greenham, L. W. 1959. Methods of control and the reason for Great Britain's persistency in the slaughter policy. *Ayrshire Cattle Society's J.* 31(2):99–100, 112, 121.

Haecker, T. L. 1907 and 1914. Investigations in milk production. *Minnesota Agr. Expt. Sta. Bulls. 79 and 140.*

Hancock, John. 1954. Studies in monozygotic twins. *New Zealand Dept. Agr. Animal Res. Div. Publ. 63.*

Hansel, William. 1959. Embryo mortality problem in dairy cattle. *Jersey J.* 6(22):38–39.

Hauge, S. M., et al. 1944. Vitamin A requirements of dairy cows for the production of butterfat of high vitamin A value. *J. Dairy Sci.* 27:63–66.

Helmer, L. G., E. E. Bartley, and R. M. Meyer. 1965. Bloat in cattle. *J. Dairy Sci.* 48:575–79.

Herman, H. A. 1971. Registered bull market isn't dead. *Hoard's Dairyman* 116:640.

Hoxie, Jane. 1923. *Solomon Hoxie, a biography by his daughter.* Little &. Ives Co., N.Y.

Huffman, C. F., C. W. Duncan, C. S. Robinson, and L. W. Lamb. 1933. Phosphorus requirements of dairy cattle when alfalfa furnishes the principal source of protein. *Michigan Agr. Expt. Sta. Tech. Bull. 134.*

Hurst, V. 1953. Site of semen deposition as related to fertility in dairy heifers. *J. Dairy Sci.* 36:577.

Huston, C. R. 1965. A. I. in long pants. *Guernsey Breeders' J.* 116:14–15, 25.

Ivanoff, E. T. 1922. On the use of artificial insemination for zootechnical purposes in Russia. *J. Agr. Sci.* 12:244–56.

Jenner, Edward. 1798. *Inquiry into the cause and effects of variolas vaccinae.*

Jones, I. R., C. H. Eckles, and L. S. Palmer. 1926. The role of vitamin A in the nutrition of dairy calves. *J. Dairy Sci.* 9:119–39.

Kalkus, J. W. 1920. Goiter and associated conditions. *Washington Agr. Expt. Sta. Bull. 156.*

Kidder, H. E., et al. 1954. Fertilization rates and embryonic death rate in cows bred to bulls of different levels of fertility. *J. Dairy Sci.* 37:691–97.

Kildee, H. H., and A. C. McCandlish. 1916. Influence of environment and breeding in increasing dairy production. I. *Iowa Agr. Expt. Sta. Bull. 165.*

Knight, C. W., et al. The relation of site of semen deposit to breeding efficiency of dairy cattle. *J. Dairy Sci.* 34:199–202.

Knoblauch, H. C., E. W. Law, and W. P. Meyer. 1962. State Agricultural Experiment Stations. A history of research and policy procedures. *USDA Misc. Publ. 904.*

Kuhlman, A. H., and W. D. Gallup. 1942. Carotene requirements of dairy cattle for conception. *J. Dairy Sci.* 25:688–89.

Loosli, J. K., E. E. Bartley, W. P. Flatt, N. L. Jacobson, C. H. Noller, and M. Ronning, with the cooperation of P. W. Moe. 1971. Nutrient requirements of dairy cattle. 4th rev. ed. *Natl. Res. Council Publ.*

Loosli, J. K., R. B. Becker, C. F. Huffman, N. L. Jacobson, and J. C. Shaw. A report of the Subcommittee on Dairy Cattle Nutrition. 1966. 3rd rev. ed. *Natl. Res. Council Publ. 1349.*

Luyet, R. J., and E. L. Hodapp. 1938. Revival of frog's spermatozoa vitrified in liquid air. *Proc. Soc. Expt. Biol. Med.* 39:433–34.

McCandlish, A. C., L. S. Gillette, and H. H. Kildee. 1919. Influence of environment and breeding in increasing milk production. II. *Iowa Agr. Expt. Sta. Bull. 188.*

McDowell, J. C. 1925. Cow testing associations, and stories the records tell. *USDA Farmers' Bull. 1446.*

McDowell, J. C., and J. B. Parker. 1926. Better cows from better sires. *USDA Dept. Circ. 363.*

Marx, G. D., and C. L. Cole. 1966. Protective ability of bovine milk and bovine serum antitoxin in mice. *J. Dairy Sci.* 49:718–19.

Meigs, E. B., and T. E. Woodward. 1921. The influence of calcium and phosphorus in the feed on the milk yield of dairy cows. *J. Dairy Sci.* 4:185–217.

Mendel, Gregor. 1866. Versuche uber Pflanzenhybriden. *Verhand.' Naturforsch. Vereines in Brunn.* 4:3–47.

Miller, R. H. 1962. A new method of comparing sires. *Hoard's Dairyman* 107: 816–17.

———. 1968. A cow index. *Jersey J.* 15(9):17–21.

Milovanov, V. R. 1938. Isskusstueneye Cesememis Selcke-Khosia venny Jiveenuky. [Artificial insemination in farm animals.] Moscow, Selkkozgiz.

Morrison, F. B. 1956. *Feeds and Feeding.* 22nd ed. Ithaca, N.Y.

Nalbandov, A., and L. E. Casida. 1942. Ovulation and its relation to estrus in cows. *J. Animal Sci.* 1:189–98.

Neal, W. M., and C. F. Ahmann. 1937. The essentiality of cobalt in bovine nutrition. *J. Dairy Sci.* 20:741–53.

Neal, W. M., R. B. Becker, and A. L. Shealy. 1931. Natural copper deficiency in cattle rations. *Science* 74:418.

Nelson, J. B. 1961. Normal immunity reactions of the cow and the calf with reference to antibody transmission in colostrum. *Missouri Agr. Expt. Sta. Res. Bull. 532.*

Nelson, R. E. 1961. Utilization of blood typing. *Jersey J.* 12(17):40.

Norton, H. W., Jr. Gestation period for Holstein-Friesian cows. *J. Dairy Sci.* 39:1617–21.

Olds, Durward. 1959. Insemination timing and conception. *Guernsey Breeders' J.* 104:22–23, 42.

Perry, E. J. 1939. *Among the Danish farmers.* Interstate, Danville, Ill.

———. 1968. *The artificial insemination of farm animals.* 4th ed. Rutgers Univ. Press, New Brunswick, N.J.

Phillips, P. H. 1945. A practical method of coloring semen for identification purposes. *J. Dairy Sci.* 28:843–44.

Pickett, R. M., W. A. Cowen, A. F. Fowler, and D. C. Gosslee. 1960. A comparison of motility of bull semen extended in egg-yolk-citrate-glycerol and skimmilk-egg-yolk-glycerol. *A. I. Digest* 8(2):19.

Pirtle, T. R. 1926. *History of the dairy industry.* Mojonnier Bros., Chicago.

Plowman, R. D., and R. T. McDaniel. 1967. Changes in DHIA Sire Summary procedures. *Jersey J.* 14(18):20–21.

Polge, C., and L. K. A. Rowson. 1952. Fertilizing capacity of bull spermatozoa after freezing at −79° C. *Nature* 169:626–27.

Powell, E. B. 1938. One cause of fat variations in milk. *Proc. Amer. Soc. Animal Prod.* Pp. 40–47.

Prevost, Dr. 1840. Recherches sur les animalcules spermatique. *Compte rend. hebromadaires des secances* (Academy of Science). 11:907–8.

Radot, Rene Vallery. 1928. *The life of Pasteur.* Doubleday, Garden City, N.Y.

Ragsdale, A. C. 1934. Growth standards for dairy heifers. *Missouri Agr. Expt. Sta. Bull. 336.*

Reid, J. T. 1956. Nutrition and feeding of dairy cattle. *J. Dairy Sci.* 39:735–63.

Roman, J., C. J. Wilcox, R. B. Becker, and M. Koger. 1969. Tenure and reasons for disposal of artificial insemination dairy sires. *J. Dairy Sci.* 52:1063–69.

Ronning, M., E. R. Berousek, A. H. Kuhlman, and W. D. Gallup. 1953. The carotene requirements for reproduction in Guernsey cattle. *J. Dairy Sci.* 36: 52–56.

Russell, W. A. 1966. Practical thoughts on handling dairy bulls. *Brown Swiss Bull.* 45(1):11–13, 40, 45, 72–73.

Salisbury, G. W. 1946. Fertility of bull semen diluted 1:100. *J. Dairy Sci.* 29:695–97.

Salisbury, G. W., and N. L. VanDemark. 1961. *Physiology of reproduction and artificial insemination of cattle.* Reinhold, New York.

Schalk, A. F., and R. S. Amadon. 1928. Physiology of the ruminant stomach (bovine). Studies of the dynamic factors. *North Dakota Agr. Expt. Sta. Bull. 216.*

Smith, V. R. 1959. *Physiology of lactation.* 5th ed. Iowa State Univ. Press, Ames.

Sorensen, Edward. 1938. *Kunstig Saedoverfording hos Huspattedyrene.* Royal Agr. and Vet. College, Copenhagen.

Stallcup, O. T. 1965. Some reasons why high quality forage is important. *Guernsey Breeders' J.* 115:794–95.

Stormont, Clyde. 1966. What blood typing tells us. *Hoard's Dairyman* 111: 1151, 1182.

Sullivan, J. J., F. I. Elliott, D. E. Bartlett, D. M. Murphy, and C. D. Kurdas. 1966. Further studies on the use of polymyxin B sulfate with dihydrostreptomycin and penicillin for control of vibrio fetus in frozen semen. *J. Dairy Sci.* 49:1569–71.

Swanson, E. W., and H. A. Herman. 1944. The digestibility of lespedeza hay and ground Korean lespedeza seed by dairy heifers. *J. Dairy Sci.* 27:263–68.

Sykes, J. F., F. H. Andrews, F. W. Hill, F. W. Lorens, J. W. Thomas, and C. F. Winchester. 1953. Hormone relationships and applications in the production of meat, milk and eggs. *Natl. Res. Council Publ. 266.*

Thurston, L. M., C. H. Eckles, and L. S. Palmer. 1926. The role of the antiscorbutic vitamin in the nutrition of calves. *J. Dairy Sci.* 9:37–49.

Trimberger, G. W. 1948. Breeding efficiency in dairy cattle from artificial insemination at various intervals before and after ovulation. *Nebraska Agr. Expt. Sta. Res. Bull. 153.*

———. 1954. Conception rates in dairy cattle from services at various intervals after parturition. *J. Dairy Sci.* 37:1042–49.

Trout, G. M. 1956. Fifty years of the American Dairy Science Association. *J. Dairy Sci.* 39:625–50.

True, A. C. 1928. The history of agricultural extension work. *USDA Misc. Publ. 25.*

———. 1929. The history of agricultural education in the United States, 1785–1925. *USDA Misc. Publ. 36.*

———. 1937. The history of the agricultural experimentation and research in the United States, 1607–1925. *USDA Misc. Publ. 251.*

Turner, C. W. 1934. The functional individuality of the mammary glands of the udder of the dairy cow. *Missouri Agr. Expt. Sta. Res. Bull. 211.*

USDA. 1942. Keeping livestock healthy. *USDA Yearbook of Agriculture.* GPO, Washington, D.C.

———. 1956. Animal diseases. *The Yearbook of Agriculture.* GPO, Washington, D.C.

VanDemark, N. L. 1952. Time and site of insemination in cattle. *Cornell Vet.* 42:215–22.

VanDemark, N. L., and G. W. Salisbury. 1950. The relation of post-partum breeding interval to reproductive efficiency in the dairy cow. *J. Animal Sci.* 9:307–13.

VanDemark, N. L., G. W. Salisbury, and L. E. Boley. 1952. Pregnancy interruption and breeding technique in the artificial insemination of cows. *J. Dairy Sci.* 35:219–23.

Van Slyke, L. L. 1896. Effect of drouth upon milk production. *New York (Geneva) Agr. Expt. Sta. Bull. 105.*

Virtanen, Artturi. 1966. Milk production of cows on protein-free feed. *Science* 155(3744):1603–14.

Weaver, Earl, C. A. Matthews, and H. H. Kildee. 1928. Influence of environment and breeding in increasing milk production. Part 3. *Iowa Agr. Expt. Sta. Bull. 251.*

Werner, G. M., L. E. Casida, and I. W. Rupel. 1939. Best time for insemination of cows. *Wisconsin Agr. Expt. Sta. Bull. 446,* p. 15.

Wilcox, C. J., K. O. Pfau, R. E. Mather, and J. W. Bartlett. 1959. Genetic and environmental influences upon solids-not-fat content of cow's milk. *J. Dairy Sci.* 42:1132–46.

Willett, E. L. 1950. Fertility and livability of bull semen diluted at various levels to 1:100. *J. Dairy Sci.* 33:43–49.

———. 1956. Developments in the physiology of reproduction of dairy cattle and in artificial insemination. *J. Dairy Sci.* 39:695–711.

Winkjer, Joel L. 1917. Cooperative bull associations. *USDA Yearbook of Agriculture 1916,* pp. 311–19.

Government Acts

1862. An Act donating public lands to the Several States and Territories which may provide for the benefits of agriculture and the mechanic arts.

1955. A Bill. To consolidate the Hatch Act of 1887 and the laws supplementary thereto relative to the appropriation of Federal funds for the support of agricultural experiment stations in the States, Alaska, Hawaii and Puerto Rico.

Hoard's Dairyman

1968. English dairy herds hard hit by foot-and-mouth disease. 113:79.

1969. Young sire programs provide bull power. 114:71, 105.

Jersey J.

1960. Young Jersey matron does double duty. 7(4):66.

1965. Editorial. A. I. Issue. 12(17):27.

1965. Summary of blood typing requirements. 12(17):58.

Other Publications

1964–70. Methods of freezing semen. Milk Marketing Board. *Report of the Breeding and Production Organization* 15:102–4, 1964–65; 18:123–35, 1967–68; 20:32–34, 1969–70.

1965. DHIA Cow Performance Index List. Registered progeny of DHIA sires, August and September, 1964. *Dairy-Herd-Improvement Letter.* ARS–44–154.

1965. Genetic appraisal of sires. Genetic appraisal of cows. *Dairy-Herd-Improvement Letter.* ARS 44–161. Vol. 41(4).

1966. Requirements governing artificial insemination of purebred dairy cattle, Effective July 1, 1966. Purebred Dairy Cattle Association.

1967. Unified rules for DHIA (in addition to standard DHIA rules). *Brown Swiss Bull.* 42(2):75–79.

1968. Automated unit tests milk electronically. *Dairy Herd Management* 5(7):8–9.

CHAPTER 22

SUMMARY

THIS PART of *Dairy Cattle Breeds* is an analysis divided into 21 units based on the preceding breed chapters. These sections outline and amplify the practices and principles used in good dairy herds. They are adapted for use with herds of registered dairy cattle maintaining permanent herd records. Certain units apply for dairymen raising replacement heifers. They were assembled particularly for 4-H and FFA members. The units were based on the practical experience of many persons, including Dr. C. H. Eckles, Dr. H. H. Kildee, Professor C. S. Plumb, Lynn Copeland, J. B. Fitch, C. R. Gearhart, Guy E. Harmon (my former "boss"), Arthur B. Klussendorf, James W. Linn, and P. C. McGilliard. These men were among many who contributed to the art and science of improving dairy cattle.

Instructors may arrange the sequence of units to accommodate local events, such as shows, official type classifications, field trips to breeders' herds, or to purebred sales. Since some individuals have "eyes that see not and ears that hear not," ingenuity of the leader in presenting such events will give them the values of actual experience to breeders or students. Preparation should be made before a field trip. Discussion of observations will give them educational advantages.

Dr. H. H. Kildee selected certain photographs for illustration, reviewed the entire manuscript critically, and suggested that "The Score Card" be developed as a standard for the improvement of dairy cattle.

Selection has contributed toward improvement of cattle, since herdsman Laban attempted to influence breeding characteristics of animals in Jacob's herd and flocks. Robert Bakewell improved his Longhorn cattle and Leicester sheep by four means. He selected animals that approached his ideal; mated the best animals together irrespective of relationship; and leased out or sampled males and then used those that proved successful. Finally, animals that failed to approach his ideal were culled closely. His methods apply today.

There is much satisfaction in working with good animals, trying to improve and increase their numbers. People in business and public life have developed some leading herds of dairy cattle. Their herds were a source of satisfaction or a hobby, when under efficient management of skilled caretakers.

Because generations of dairy cattle are of short duration, students should keep up to date with their selected breed journals. They should be conversant with selected articles in the breed journals and in assigned technical publications. A detailed report of a preferred breed may conclude classroom instruction.

There are three classes of dairy cattle. Few registered dairy animals are of seedstock quality, capable of improvement at the top level. More registered dairy cattle multiply the desirable kind. These two classes provide the breeding stock that maintain the inherited qualities of good commercial dairy cattle through natural service or artificial breeding. The third class includes the good commercial producers. Animals of all three classes should be expected in every leading herd. Continued study and culling are essential to good management. These are among the units condensed in the Summary, with supporting references.

WHAT IS A PUREBRED?

Cattle originally were brought to the western hemisphere from Europe before breeds were developed and recognized as such. Some animals undoubtedly were good, and others less so. Attempts were not made to keep different kinds separated. Cattle of a community ran at large or were herded, as on the Boston Commons. Little opportunity was afforded for improvement by individual owners.

When enterprising persons and Agricultural Societies recognized the greater productive value of kinds of livestock developed in small communities in Europe, they began to import and breed them as pure lines. Breeders organized into societies to publish herdbooks that would be reliable records of parentage or pedigree. Animals meeting all provisions acceptable for entry became known as purebreds. Such breed organizations had avowed objectives. That of the Brown Swiss Cattle Breeders Association is typical. The preamble to the Constitution stated:

> We, the breeders of Brown Swiss Cattle, recognizing the importance of a trustworthy Herd Book, that shall be accepted as a final authority in all questions of Pedigree, and desiring to secure the influence and co-operation of those who feel genuine interest in jealously guarding the purity of this stock, do unite in forming an Association for the publication of such a Herd Book and for such other purposes as may conduce to the successful breeding of these cattle, and do therefore adopt the following Constitution . . .

A purebred animal then is one whose purity of breeding has met the standards established for entry in the herdbook of the respective organization. Proper records must have been kept of sire and dam, breeder, owner, transfer of ownership, breeding records, color markings, tattoo, or other standards established by the herdbook organization. The animal must have been properly identified and bear the recognized characteristics of the breed. A registration certificate bearing the pedigree facts shall have been granted such animal in the name of the breeder and owner by the breed organization.

Any transfer of title shall have been inscribed by the breed secretary for each change of ownership. If a female is bred to a sire under different ownership, the service record shall have been accounted for on forms prepared by the breed organization and bear the recognized legal signature of the sire's owner or authorized agent.

The registration certificate is a legal paper, the same as title to an automobile or a deed to land. Any changes on the certificate must be made only by the respective breed association office over signature by the breed secretary.

When dairy cattle possess the desirable characteristics of seed-stock animals in their native countries, they may be recorded (now only females) as Foundation Stock in the respective herdbooks.

Colonel Zadock Pratt observed in 1861 that a cow by a purebred sire attained milking ability above that of common cows. Dr. H. H. Kildee continued such a demonstration at Iowa State College in 1909—milk production was increased through three generations. The Kansas, Oklahoma, and South Dakota stations conducted less extended demonstrations.

The American Milking Shorthorn Society's official grading-up plan involves type inspection and acceptable production through four generations, whereby acceptable fifth generation cows may be registered. The Dairy Herd Improvement Section of the USDA pointed out the unavailability of many DHIA records for sire analyses because of failure to identify the sire. An industry committee recommended steps to use DHIA records for Identity Enrollment of cows approved for conformation. The first Brown Swiss cow and two desirable heifers were approved for Identity Enrollment during the Dairy Production Conference at the University of Florida in 1969. If their female progeny meet production and type standards in turn, they may become fully registered and recognized as purebreds along with the progeny of registered parents.

Understanding of the Constitution and By-Laws of a preferred breed may be acquired as a laboratory exercise.

Application for Membership in a Dairy Cattle Registry Association

Provisions are made in the Constitution and By-laws or regulations concerning membership and the duties of officers of a herdbook organization. Rules outline the methods of conducting business, the major interest of which is eligibility and registration of approved animals of the breed.

Membership in a purebred dairy cattle association is a privilege and not a right. An application for membership must have been endorsed by one to five members, depending on the particular breed organization, and a fee submitted therewith. Integrity of the applicant is investigated before the Board of Directors considers the application. Membership is attained when the applicant meets all requirements and is accepted or approved by the Board of Directors.

Three forms of membership are recognized, depending upon the organization: (a) annual in one organization at a small fee; (b) term membership for individuals, partnerships, corporations, or institutions; and (c) life membership for individuals. A member becomes inactive upon failing to transact business with the association within a definite time. Membership is not transferable.

Tenure of membership depends upon integrity and business relationships of the member. Continuance of membership is a privilege granted by the association and can be cancelled by the Board for cause after giving the member notice and opportunity for a hearing on infringement of rules or bylaws or for conduct unbecoming a member. If membership is cancelled, a definite time is given for disposal and transfer of all registered cattle. Registration certificates can be voided upon abuse or misrepresentation by the member or by failure to abide by rules of the organization.

Each member files his legal signature, or that of an authorized agent to represent him, which is recognized in all transactions dealing with cattle records or business of the organization. Only this signature(s) is recognized in transactions conducted by him or his agent with this breed.

BREEDING RECORDS

Breed registry associations require complete records to be kept of breedings, calvings, and identity of each animal in herds of registered dairy cattle. The same records are useful in commercial dairy herds. They are simple and kept easily. Dairy breed journals and supply houses sell several types of useful record forms. A simple form for commercial herds may be written as a daily diary in a notebook, as in Table 22.1. A barn breeding record on heavy cardboard may be posted with pertinent dates concerning each cow on separate lines.

Some forms have spaces for recording estrus periods before the cow has been fresh long enough to be bred, allowing a person to anticipate future heat periods and note any irregularity. The notebook

TABLE 22.1
A Suggested Permanent Breeding Diary

Cow's name or herd number	Date of service	Barn name of bull	Calving date	Sex of calf	Herd number, ear tattoo, or neck chain
No. 3	1–3–70	Stan	rebred		
No. 4	1–7–70	Stan	10–18–70	M	killed
No. 1	1–10–70	Oxford	rebred		
No. 3	1–24–70	Stan	11–1–70	F	UF-21
No. 5	1–25–70	Oxford	11–3–70	F	UF-22
No. 1	1–31–70	Oxford	11–4–70	M	killed
No. 2	4–11–70	Stan			

BARN BREEDING RECORD
List of cows

	No. 1	No. 2	No. 3	No. 4	No. 5
Last calving date	11–2–69	2–1–70	9–25–69	8–31–69	9–28–69
Condition at calving	clean	clean	clean	retained	clean
In heat, before bred					
First date	11–29–69	2–28–70	11–23–69	11–8–69	11–30–69
Second date	12–18–69	3–19–70	12–14–69	11–29–69	12–18–69
Third date				12–19–69	1–6–70
Fourth date					
Dates bred					
First service	1–10–70	4–11–70	1–3–70	1–7–70	1–25–70
Bull	Oxford	Stan	Stan	Stan	Oxford
Second service	1–31–70		1–24–70		
Bull	Oxford		Stan		
Third service					
Bull					
Fourth service					
Bull					
Due to calve	11–12–70	1–21–70	11–5–70	10–16–70	11–6–70
To turn dry	9–30–70	12–8–70	9–23–70	8–31–70	9–25–70
Date calved	11–4–70		11–1–70	10–18–70	11–3–70
Sex of calf	M		F	M	F
Tattoo of calf	killed		UF-21	killed	UF-22

should have a key to the name and registration number of herd sires and of registered females.

Because such a diary is consecutive, the herdsman can see at a glance the cow that may calve next, so as to allow her a 4- to 8-week dry period before calving time. It aids in anticipating approximate calving time, so that the animals may be separated and watched occasionally if assistance or veterinary care is needed.

Each calf needs to be identified permanently with an ear tattoo, color sketch, photograph, or other appropriate method shortly after birth. A record form on which all information on each animal is entered on a single page during its lifetime aids greatly in herd management. Permanent breeding records are useful on many occasions —in registration of progeny, in production testing, health examinations, when selling animals, and in direct management of each animal. Each student should become fully acquainted with the use of herd records.

A more complete and useful barn breeding record, by Extension Dairyman C. W. Reaves at the University of Florida, allows for entry of ten facts concerning a cow and her reproductive history. This record form provides for the following entries: cow's name and number; date of last calving; condition after calving; heat periods prior to breeding (three spaces); dates when bred (spaces for five services, and bull used); date due to calve; date to turn dry; date calved; sex of calf; and ear tag or tattoo of calf.

BREEDING EFFICIENCY

Regular reproduction is important in economical milk production, as suggested by the axiom—"No calf, no milk." Many factors affect breeding efficiency. These include breed or heredity, state of nutrition, health of animal, involution of the uterus after calving, time from previous calving to first service, and interval between calvings. Several factors have been investigated relating to calving efficiency. Some investigators measured breeding efficiency in percentages, basing 100 percent as a calf every 365 days, or number of calves produced per 12-month period after the animal reached 2 years old.

There is a difference between a 60- to 90-day nonreturn rate and a conception rate based on positive diagnosis of pregnancy by rectal palpation. Superheats are not uncommon. Repeated natural breedings under such circumstances have resulted at rare intervals in a cow delivering a second calf about 3 weeks after birth of the first calf, or in abortion of an 8-month fetus at time of natural calving. An instance has been recorded of a cow bred to a Holstein bull and to a Jersey bull at the same estrus period that delivered maternal twins by the separate sires.

The instructor may wish to divide the students into groups to analyze the permanent records of the college dairy herd over certain years for separate factors.* Problems may include the following: age of male and female with relation to conception rate, interval from calving to date of first service with relation to number of services per conception in natural breeding, inseminations per conception in artificial breeding, relation of calving interval to average daily milk yield between calvings, and other problems. Students may be assigned selected references and then report their findings to the class.

Applications for Registration and Transfer of Ownership of Purebred Dairy Animals

Applications are made for registration of one solid-colored purebred animal (Brown Swiss or Jersey) and of one broken-colored animal (Ayrshire, Guernsey, or Holstein-Friesian). Note is made of the sex, color sketch or tattoo, sire and dam (with registration numbers), single or twin birth, naturally horned or polled, date of service and date of birth, whether the result of artificial insemination with fresh or frozen semen, and pasture breeding or natural service. Owner of sire and owner of dam at time of service and at birth of the calf are entered only with an *authorized legal* signature. The fee is determined for a member or nonmember with relation to time from date of birth. Payment must accompany an application for registration or for transfer of ownership.

The name and registration number of the animal are entered

*P. T. Dix Arnold and R. B. Becker. 1953. Dairy calves. Their development and survival. Florida Agr. Exp. Sta. Bull. 529.
P. T. Dix Arnold and R. B. Becker. 1956. Building a dairy herd. Florida Agr. Exp. Sta. Bull. 576.

with the *authorized* name and correct mailing address of the buyer. The seller's name, address, and *legal* signature are required on the application for transfer form.

If the animal is a nonpregnant female represented as being sold open, the words "Not served" are inscribed across the face of the service statement. If bred, date of service, name, and registration number of the service sire are entered over the recognized signature of the bull's owner or authorized agent. If pasture-bred, inscribed are the dates between which the female was on pasture with the bull, his full name, and his registration number.

When artificially inseminated, the official artificial breeding receipt shall be attached. It shall bear the name and registration number of the bull, name, registration number, and full tattoo identification of the female (letters and numerals), whether or not the female was identified from the registration certificate, and signature of the authorized technician. All writing shall be in ink or indelible pencil and signature shall be written. The artificial insemination receipt shall bear the date. If the female was bred previously, the number of times and the last previous date and sire are recorded on the receipt.

If frozen semen was used, the words "frozen semen" shall be indicated on the artificial breeding receipt.

Errors in Applications for Registration

Receipt of incomplete or inaccurate applications from breeders is a problem of breed registry associations. Correction of these inaccuracies involves correspondence, extra labor, and delays in issuing correct certificates. Charles H. Bohl made a survey of inaccuracies in applications for registration received by the American Jersey Cattle Club. Bohl reported: "We found that 14 percent of all the applications for registration received [in June 1956] were returned. The breakdown of the results is listed [in Table 22.2]."

Some 12.68 percent of applications received during 2 weeks by the Holstein-Friesian Association required completion or corrections. Other associations could have made similar tabulations. Most of these errors and delays would have been avoided if the applicants had rechecked all entries on each application before forwarding it to the breed registry office.

TABLE 22.2
ERRORS IN APPLICATIONS FOR BREED REGISTRATIONS

Reasons returned	Number	Number returned	Percent of all registrations
Signatures			
Wrong person	90		
Omitted	46		
Not countersigned	40		
No legal authority to sign	23		
Printed	17	216	3.7
Fee, not sufficient money		117	2.0
Omissions			
Tongue color	36		
Switch color	30		
Body color	19		
Other	13		
Birth date	12		
Sex	2	112	2.0
Artificial insemination receipt			
No. A.I. receipt	43		
Not identified	22		
No tattoo	13		
Completed in pencil	9		
Wrong A.I. receipt	5		
No bull name and number	4		
Altered	3		
No A.I. signature card on file for technician	3	102	1.8
Altered			
Tattoo	20		
Birth date	18		
Description	9	47	0.8
Conflict of calving		47	0.8
Tattoo			
Omitted	17		
No letter	17		
No number	6		
Wrong tattoo	5	45	0.8
Abnormal gestation		24	0.4
Completed in pencil		14	0.2
Others, including applications depending on other applications, immature breeding, HIR or ROM conflicting, and animal already on record, etc.		85	1.5
Total		809	14.0

MILK AND BUTTERFAT RECORDS

Walter of Henley mentioned cows kept for the dairy in England over three centuries ago. Later writers cited the amount of milk, butter, or cheese a good cow ought to yield in a season.

The first public milking trial was sponsored for a 5-day period by the Duke of Atholl at Ayr in 1860. He purchased the winner, whose milk was measured daily after the next calving. She yielded over 15,000 pounds of milk in that lactation.

Colonel Zadock Pratt of Duchess County, New York, reported the average production of about 50 native cows between 4,355 and 5,209 pounds of milk yearly during 1857 to 1861. He mentioned that even one cross to an improved breed would increase the milk from the first-cross cows.

Solomon Hoxie of the Dutch Friesian Cattle Breeders Association divided that organization's herdbook into two parts: the Pedigree Register and the Main or Advanced Register. Cows were entered in the Pedigree Register if the sire and dam traced solely to Friesian cattle in Friesland or North Holland. Registered animals were promoted to the Main or Advanced Register when they qualified in milk production and/or by scoring and body measurements.

Dr. S. M. Babcock developed a simple rapid method of analyzing milk for percent of butterfat in 1890 at the Wisconsin Agricultural Experiment Station. It was used publicly first in milking trials during the World's Columbian Exposition in 1893. Dr. Nicholas Gerber developed a rapid practical test in Switzerland in 1888, which was published in 1891. The Babcock and Gerber tests are used to determine butterfat percentage of milk on farms and in commercial dairy plants and to determine butterfat yields of dairy cows. A new (alkali) TeSa test was used first in 1958. The light-interception Milko-Tester method, developed in Denmark, is being used on milk samples assembled in some district laboratories.

Six plans have been employed to record production of dairy cows in commercial herds: private records, owner-sampler, weigh-a-day-a-month, DHIA records with grade and registered dairy cows, official testing (Advanced Registry, Register of Merit, Record of Merit,

Register of Production), and now the Dairy Herd Improvement Registry with registered dairy cows.

Private records are kept by the owner or his employees. They may be simple or elaborate, as desired by the owner. There is no way to assure the public of their authenticity.

Two unofficial methods now are in use. For the owner-sampler method, owners keep the milk weights of cows and send milk samples to the local supervisor for butterfat tests. The second method is the weigh-a-day-a-month plan, in which the owner keeps individual milk weights on the middle day of each month, and total feed weights for the herd. Each month he sends the weights to the County Agent's office, where for a small fee, an employee computes the records and returns them to the owner for herd management. Records obtained by these methods are not considered official by breed organizations.

Cow testing association (milk control) records began in Denmark in 1895. Helmer Rabild helped to organize the Cow Testing Association in Michigan in 1905. The name was changed later to Dairy Herd Improvement Association. Under this plan, several farmers organize and employ a supervisor trained by the Extension Dairyman to keep individual records of production and feed consumption. These records are computed from one day's milk yield and butterfat test each month during the year. The supervisor reports concentrate consumption per cow and forage on a herd basis. The records enable feeding each cow according to production; allow analyses of herd sires by daughter-dam or herdmate comparisons; aid in culling a herd (selecting heifers from the better cows for herd replacements); and contribute to other herd management practices.

Machine calculation of milk and butterfat yields was begun by Lyman H. Rich in Utah. Other states added feeding and additional records, with improved electronic computations. Several states joined in establishing regional processing centers to service DHIA records. Information was expanded to include (a) birthdate, sire, and dam, (b) calving, breeding, and dry dates, (c) taped estimate of body weight after calving, (d) milk, butterfat, and possibly protein or solids-not-fat (protein-lactose-mineral) yields, (e) days in milk, dry, and in gestation during the lactation, and (f) value of product, and income over feed cost.

Concentrate consumption is on an individual basis. Pasture, corn silage, grass silage, other succulents, and dry forages are reported on a herd basis and allotted in proportion to estimated body weights to the individual cows. Feed intakes and needs are computed on an estimated energy basis in therms. As of 1966 practically all DHIA records are computed electronically at state or regional centers. The records cover business management of the dairy herd in more detail than formerly. Labor is recorded on a herd basis, and milk production per worker-day is computed.

If a laboratory is desired on DHIA methods and results, a 19-cow record report for the year perhaps could be analyzed on a class basis. Methods of supervision would be obtainable from the supervisor's *Manual*. Questions might be asked concerning breeding problems, time for first insemination, dates to turn cows dry, if any cows should be culled and their order, source of replacement heifers, and similar management problems. Students might be asked to supervise a monthly test with five cows and make the report. None but a complete report should be accepted.

SUPERVISION OF DHIR RECORDS

Official production records of registered dairy cows on a 305- or 365-day basis are under joint supervision of three parties: the owner of the cow, the State Superintendent of Official Testing, and the breed registry organizations. An electronic computing laboratory jointly serves the State Superintendent and the breed registry organization. Recording methods have been modified; for example, selective testing has been replaced with electronic computation of testing of all individual cows in a herd.

Each party has certain responsibilities for the record of each cow. The owner is responsible for normal care and management of the cows. He applies to the breed registry organization for permission to test his milking herd for a year, listing the cows and paying the fees. He joins the local DHIA and notifies the State Superintendent of Official Testing. He pays for the monthly test supervision in the prescribed manner. The DHIR is a DHIA test with additional breed organization requirements. He receives a form from the breed organization with the list of cows on test for the first month and a form from the Computing Center for the supervisor's use in subsequent months. He provides registration certificates (or photostatic copies of them) for the supervisor to identify each cow by color markings, ear tattoo, ear tag, or other permanent identification indicated on the certificate. He supplies dates of calving, breeding, turning dry, or leaving the herd. He is responsible for board, lodging, and transportation of the supervisor under conditions set out by the local DHIA, and provides a place for the supervisor to work. (There are central testing laboratories in some areas.)

The breed organization grants the owner permission to test the registered cows in the herd if the owner is in good standing and registration records correct. The breed organization receives a 305-day and a complete lactation record card from the Computing Center at completion of the records. The breed organization issues final certificates and publishes the records in breed publications. The records may be used in determining Approved Sires, Dams, and other breed promotion programs.

The State Superintendent of Official Testing is responsible for the training and reliability of the men and women who supervise a 1-day test each month. He also supervises one or more surprise tests yearly, following the latest *Unified Rules for Official Testing* approved by the Purebred Dairy Cattle Association and the American Dairy Science Association. These uniform rules help to maintain public confidence in the records in the country. Each supervisor is supplied with the *Uniform Rules for Official Testing,* accurate milk scales, sample bottles, Babcock or TeSa glassware and tested thermometer, and dividers for reading the butterfat tests. If Milkometers are used for milk weights, the supervisor checks their accuracy by one of the approved methods.

An approved supervisor visits the herd each month. He identifies each cow from the registration certificate or photostatic copy by color markings, ear tattoo, ear tag, or other permanent identification. He sees each cow milked, weighs the milk in pounds and tenths of pounds (or reads the meter), and enters the weight on the owner's and his records. He samples the milk and determines

U. S. DEPARTMENT OF AGRICULTURE
AGRICULTURAL RESEARCH SERVICE

INDEX NO.

LIFETIME HISTORY OF INDIVIDUAL COW

BARN NAME		TATTOO NUMBER	EARTAG NUMBER

REGISTRATION NAME		REGISTRATION NUMBER

BREED	DATE OF BIRTH	OFFICIAL TYPE CLASSIFICATION	CHECK IF COW IS PROGENY OF ARTIFICIAL INSEMINATION ☐

SIRE		REGISTRATION NUMBER OF SIRE

DAM		INDEX NUMBER	EARTAG OR REGISTRATION NUMBER OF DAM

TEST DAY MILK WEIGHTS

LACT NO.	DATE FRESH			MONTH OF LACTATION											
	MO.	DAY	YEAR	1	2	3	4	5	6	7	8	9	10	11	12
1															
2															
3															
4															
5															
6															
7															
8															
9															
10															

LACTATION PRODUCTION SUMMARY

LACT NO.	TYPE OF RECORD	AGE OF FRESHENING (Months)	WEIGHT WHEN FRESH	DAYS DRY BEFORE CALVING	FIRST 305 DAYS						COMPLETE LACTATION				
					DAYS CARRIED CALF	DAYS MILKED 3 TIMES	MILK	%	BUTTER FAT	DAYS IN MILK	LACTATION/LIFETIME TOTALS			SUPVR'S INITIALS
											MILK	B.F.	INCOME ABOVE F.C.	
1														
2														
3														
4														
5														
6														
7														
8														
9														
10														

DHIA Form 1057
Jan 1959

CALVING RECORD

LACT NO.	DATE OF CALVING			CALF'S NAME & EAR TAG NUMBER	SEX OF CALF	SIRE OF CALF	DISPOSAL OF CALF
	MO.	DAY	YEAR				
1							
2							
3							
4							
5							
6							
7							
8							
9							
10							

BREEDING RECORD

LACT NO.	DATE OF CALVING			DATE BRED	SIRE USED	DATE BRED	SIRE USED	DATE BRED	SIRE USED	DATE BRED	SIRE USED	REMARKS
	MO.	DAY	YEAR									
1												
2												
3												
4												
5												
6												
7												
8												
9												
10												

HEALTH AND VETERINARY RECORD MAINTAINED BY OWNER

DATE	DISEASE, AILMENT, OR TEST	TREATMENT AND REMARKS

☆ U. S. GOVERNMENT PRINTING OFFICE : 1959 O - 496412

the butterfat percentage of each milking during the 24 hours. Protein or solids-not-fat (protein-lactose-mineral) may be determined if the local DHIA has made such a provision and if the owner desires it. The supervisor enters these records on the monthly report form and signs and forwards it promptly to the regional center for electronic computation. The Computing Center reports all cows exceeding surprise test requirements to the State Superintendent, who orders surprise tests when required (see below).

	Surprise Test Production Requirements				
	A			B	
Breed	Milk	Fat		Milk	Fat
	(pounds)			(pounds)	
Ayrshire	22,500	800		25,000	875
Brown Swiss	22,500	850		25,000	875
Guernsey	19,000	800		21,000	900
Holstein	25,000	900		27,500	1,000
Jersey	17,000	800		19,000	900

If after 90 days in lactation a cow under DHIA supervision produces equal to or exceeds the projected 305-day mature equivalent production in Column A, a surprise check test with a preliminary dry milking will be ordered by the State Superintendent of official Testing, usually by a different supervisor. If the projected production exceeds that in Column B, a second such surprise test shall be made before the 305th day of lactation. Surprise tests may be in lieu of or in addition to the regular monthly test. The breed organization pays only for surprise tests beyond those indicated; the owner pays for all others.

Emphasis on evaluation of milk is changing gradually to include protein and other milk constituents. Certain protein-binding dyes—amido black and orange G—or a formol titration method are used to estimate protein contents of milk. Such methods are confirmed periodically by Kjeldahl nitrogen determinations in some European countries.

Specific gravity is determined by a lactometer under controlled temperature in foam-free milk. With the specific gravity and butterfat test, the solids-not-fat content is computed from a standard table. N. S. Golding devised a series of plastic beads with varied densities for rapid field determinations of solids-not-fat in milk samples. These methods are used in different states.

The Herd Improvement Registry or Herd Test was handled by the same three parties before modification of the rules and report forms.

If desired, each student may acquaint himself with the latest *Unified Rules* and the supervisor's *Manual.* It may include supervision and report of five or more cows on DHIR (DHIA) test. Only a complete report should be accepted by the instructor.

Burke, J. D. 1959. The accuracy and usefulness of central processing DHIA feed records. *J. Dairy Sci.* 42:942.

Corwin, A. R., F. R. Allavie, Jr., and S. N. Gaunt. 1960. Comparison of methods for determining solids-not-fat and protein content of milk. *J. Dairy Sci.* 43:1888.

Golding, N. S. 1959. A solids-not-fat test for milk using density plastic beads as hydrometers. *J. Dairy Sci.* 42:899. See also 41:227 (1958).

Krienke, W. A. 1959. Tests for evaluating the milk plasma phase of whole milk. *J. Dairy Sci.* 42:899-900.

Washburn, R. G. 1959. A progress report on the TeSa reagent test for butterfat. *J. Dairy Sci.* 42:942.

ANALYZING OR PROVING SIRES

The influence of a sire upon production of his daughters may be analyzed by several methods. The influence of feed and care (environment) upon production of milk, butterfat, and solids-not-fat often overshadows heredity to a considerable extent. It is desirable when a breeder evaluates the transmitting ability of a sire that the environmental factors be similar throughout the lactation periods of daughters and their dams, or their probable influence considered. A cow's production is affected by age and size at calving, previous feed and care, condition of the animal, kind and quality of feed, number of milkings daily, days carrying a calf during the lactation, and skill of the caretaker. Maximum production is limited by heredity.

Lynn Copeland analyzed many Jersey records and concluded that at least ten unselected daughters should be tested when measuring the transmitting ability of the sire. Other workers confirmed this conclusion. With bulls in artificial use and their daughters under differing environments, a comparison of at least 30 daughter-dam pairs has been recommended.

The instructor will decide whether to use (a) actual production records at corresponding lactations, or (b) records computed to a uniform age basis. Each system has advantages and disadvantages. Many cows attain their maximum milk and butterfat production in the third to fifth lactation, other lactations having lower yields. Records computed to a uniform 305-day $2\times$ mature equivalent basis are inflated and hence may mislead students concerning actual yields to be expected. Such records do not consider environmental factors other than age, nor do they allow for idiosyncracies of cows in single lactations. Actual records of daughters and dams in corresponding lactations can be standardized to 305-day length. They are not inflated but involve small errors due to age differences, condition of the animal, mature size, season of calving, precocity, and other factors.

Age correction factors have been computed for the different breeds of cows with DHIA records, and by the breed associations with official records. They may be used for machine calculation when facilities permit. The Holstein-Friesian Association of America research department stated in general that "the mature equivalent record is intended to indicate what a young cow would have produced under the same feeding, management and environment had she been a mature cow [six to eight years] when the record was made." Actual records will be used for simplicity in the examples below.

Several methods have been applied to compare milk and butterfat production of a bull's daughters with that of their dams. Among them are (a) arithmetical averages of the records of daughters and dams; (b) equal parent index; (c) regression index with modifications; (d) arrow chart; (e) spot graph; and (f) a comparison of production of a bull's daughters with that of stablemates in the same season, year, and age. Each method has its proponents.

Most bulls in artificial service have daughters in many herds. Machine computations have been developed to compare production by herdmates based on breed-year-season averages, and on daughter-herdmate differences (subtracting the breed average production in the recent 2-year period). (Dairy-herd-improvement letter, Volume 41, Number 8. 1965.)

An example of a sire analysis is shown in Table 22.3 using actual 305-day records of ten cows and of their respective dams milked twice daily in their first lactations. These records were not corrected for age.

Since the formulas for regression index involve the inclusion of average production of the entire herd and breed average, and formulas differ, the instructor may choose one applicable to the breed involved. The method in current use may be obtained from the breed association.

An arrow chart comparing the production of daughters with their respective dams provided a convenient way of visualizing these re-

TABLE 22.3
EVALUATION OF SIRES ON PRODUCTION OF DAUGHTERS

A. Comparing corresponding lactations of daughters and dams

	Daughters				Dams			
Animals	Age (yr., mo.)	Milk (lbs.)	Test (%)	Fat (lbs.)	Age (yr., mo.)	Milk (lbs.)	Test (%)	Fat (lbs.)
1	2, 2	11,240	3.4	382	2, 5	13,470	3.3	445
2	2, 7	9,725	3.6	350	2, 6	10,120	3.4	344
3	2, 5	8,705	3.5	305	2, 7	9,890	3.4	336
4	2, 4	13,320	3.6	479	2, 5	12,050	3.3	398
5	2, 3	12,610	3.5	441	2, 5	11,725	3.4	399
6	2, 2	10,465	3.4	356	2, 4	8,690	3.5	304
7	2, 6	9,745	3.5	341	2, 5	11,170	3.6	402
8	2, 5	11,430	3.6	412	2, 1	10,725	3.5	375
9	2, 4	14,120	3.5	495	2, 5	13,470	3.3	445
10	2, 1	13,285	3.4	452	2, 9	11,415	3.4	388
Average	2, 4	11,465	3.5	401	2, 5	11,273	3.4	384

						Number of daughters excelling dams		
	Lactations	Milk (lbs.)	Test (%)	Fat (lbs.)		Milk (lbs.)	Test (%)	Fat (lbs.)
Sire A								
Daughters	10	11,465	3.5	401		6	7	7
Dams	10	11,273	3.4	384				
Difference		+192	+0.1	+17				

B. Equal parent index

	Milk (lbs.)	Test (%)	Fat (lbs.)
Average production of daughters	11,465	3.5	401
Increase or decrease	+192	+0.1	+17
Sire A's equal parent index	11,657	3.6	418

lationships. Records of Sire A's daughters are compared with those of their dams in the arrow chart shown in Figure 12.3. Production of the dam is gauged on the base line above the abscissa, while those of the daughters are measured on the ordinate at the left of the chart. At the point of the dam's production, the length of the arrow shaft indicates the amount and direction upward or downward whereby the daughter's production differs from that of her dam. Some artificial breeding associations have used this method.

A spot graph has been used by the *Netherlands Herd Book* to depict the production of daughters of Friesian bulls with their respective dams. It is illustrated in Figure 12.4. Production of the dam is gauged on the base line and that of the daughter on the ordinate at the left of the chart. A dot indicates the point at which a line parallel to the base from any daughter's production level on the ordinate will intersect a vertical line above the dam's production on the base. If the general field of dots points upward and to the right, the sire has made some improvement of his daughters over their dams. However, if the general field should point upward and to the left, the trend has been toward a decrease.

Progressive improvements in methods of analyses were presented in the *USDA Dairy-Herd-Improvement-Letter* 38(4), April 1962, and 41(8), November-December 1965. They were based on comparisons of milk, test, and butterfat production of a sire's daughters with those of parentally unrelated herdmates at the same period. In May 1965 "Predicted Difference" from the breed average replaced "Predicted Average" for A.I. sires, computed by the formula:

$$\text{Predicted difference} = \frac{\text{No. of daughters}}{\text{No. of daughters} + 20} \times \left\{ \begin{array}{l} \text{adjusted} \\ \text{daughter} \\ \text{average} \end{array} - \begin{array}{l} \text{breed} \\ \text{average} \end{array} \right\}$$

This was nearer to a representative method of estimating breeding value of A.I. dairy bulls.

In 1968, five dairy breed associations and the American Milking Shorthorn Society adopted the USDA Sire Summaries for bulls in artificial service. The sire summaries gave the average production of daughters on a 305-day $2\times$ mature equivalent basis in comparison with an adjusted average production of herdmates calving in the same year and season. This was presented as a "predicted difference" in pounds of milk. An estimated repeatability percentage increased as daughters produced under the environment of a larger number of herds. The sire summaries were recomputed each January, May, and September as additional daughters' records became available from registered and known-grade daughters of a registered bull.

This method of analysis is available to study and understand, but is beyond average laboratory facilities to carry through the mechanics of calculation.

Anonymous. 1962. New DHIA Sire Record. *USDA Dairy-Herd-Imp.-Letter* 38(4).
Anonymous. 1965. Improvements in Methods of Sire Evaluation. *USDA Dairy-Herd-Imp.-Letter* 41(8).
Kendrick, J . F. October 1953. Standardizing Dairy-Herd-Improvement-Association Records in Proving Sires. *USDA BDI Inf.* 162.
Plowman, R. D., and B. T. McDaniel. 1967. Changes in USDA Sire Summary Procedures. *Jersey J.* 14(18):20–21.
Schooley, Ray R. 1968. The new Jersey Sire Award program. *Jersey J.* 15(8):26–27, 31, 97.

Cow Families

A cow produces few calves during her lifetime; the average is three to four in many herds. Bull calves are not contributions, unless they are raised for breeding. The true value of a cow in a breeding or commercial herd cannot be determined fully in one generation. Good herds are built upon cow families.

The class may work as a group on the cow family laboratory. Complete herd records are desired for this project. Each student is assigned to study two cows that were acquired or dropped their first calves at least 10 years earlier. Records of each cow and her progeny should be tabulated through three generations, including each calf dropped, its disposal, number of lactations, and average production of each female descendant.

Data for each animal should be assembled on a chart. The chart should indicate the total number of male and female calves dropped in each cow family in the three generations, number of lactations, average production per lactation and *overall average daily production* since first calving, and number of females alive in the herd. Reasons for loss or removal of each female in the cow family should be stated. Herdmate comparisons may be computed from the assembled records.

The instructor may desire to discuss rate of turnover, causes of losses of cows, maintenance of herd numbers, and improvement. (See *Florida Agr. Exp. Sta. Bull.* 576). What rate of increase or decrease in the total number of cows has occurred? Could the herd be culled to advantage? Are there desirable surplus animals that could be sold as breeders? Would additional females need to be purchased to maintain herd numbers? Such facts should be gained from this project.

Cow INDEXES

Search has been made of DHIA records for registered cows sired by bulls summarized in July 1963 and later. Daughters of non-A.I.

TABLE 22.4

PROPORTION OF BULLS PROVED IN 1961 THAT MAINTAINED OR INCREASED PRODUCTION OF THEIR DAUGHTERS[a]

| | | | | Average production | | | | | |
| | | | | Dams | | | Daughters | | |
Butterfat production of dams	Number of sires	Maintained or raised production	Percentage that raised production	Milk (lbs.)	Test (%)	Fat (lbs.)	Milk (lbs.)	Test (%)	Fat (lbs.)
Over 650	13	1	7.7	18,479	4.1	764	14,065	4.1	574
625–649	15	2	13.3	15,569	4.1	632	14,083	4.0	568
600–624	50	11	22.0	15,789	3.9	609	14,430	3.9	559
575–599	81	21	25.9	15,419	3.8	586	14,447	3.8	555
550-574	204	52	25.5	14,939	3.8	560	14,915	3.8	531
525–549	432	116	26.9	14,222	3.8	536	13,353	3.8	507
500–524	737	273	37.0	13,637	3.7	511	13,050	3.8	495
475–499	1,122	458	40.8	13,047	3.7	486	12,638	3.8	479
450–474	1,545	726	47.0	12,383	3.7	462	12,147	3.8	460
425–449	1,673	859	51.3	11,666	3.7	437	11,511	3.8	440
400–424	1,406	753	53.6	10,876	3.8	413	10,846	3.9	412
375–399	1,143	646	56.5	10,063	3.9	388	10,124	3.9	392
350–374	799	506	63.3	9,473	3.8	363	9,756	3.9	381
325–349	442	303	67.8	8,462	4.0	338	8,881	4.1	361
300–324	228	150	68.6	8,096	3.9	314	8,557	3.9	337
275–299	131	101	77.1	7,373	3.9	289	8,015	4.0	319
Under 275	74	65	87.8	6,646	3.8	255	7,981	3.9	311
Totals	10,095	5,043	49.9	11,535	3.8	436	11,377	3.8	437

a. Computed from *Dairy-Herd-Improvement Letter* 37(9):3, October 1961.

bulls were evaluated solely on differences in milk production from herdmate comparisons. For A.I. progeny, performance of their paternal half-sisters also was included in computing an index of production. About 2 percent of cows in each breed that exceeded herdmates' production by an arbitrary high standard were selected. Such cows were assumed to include those of good breeding value.

OFFICIAL TYPE CLASSIFICATION

Type classification by scoring was one of the requirements which qualified Dutch-Friesian cattle for advancement from the Pedigree Register to the Main or Advanced Register of the *Dutch-Friesian Herd Book*. It was carried into the Holstein-Friesian Association in 1885. The program was not developed fully, and Solomon Hoxie recommended that it be discontinued during a depression. The Ayrshire Advanced Registry and Jersey Register of Merit had a provision for bulls to be scored along with the milk and butterfat qualifications of their daughters. Both organizations later discontinued scoring.

The Holstein-Friesian Association developed "true type" models of a bull and cow. Could the educational benefits of the show ring be extended to record permanently the type of individual animals in entire herds? Official type classification was developed to fit this need, beginning in 1929. Standards were established so that cows of milking age and bulls past 3 years old could be classified by an association classifier as Excellent, Very Good, Good Plus (added later), Good, Fair, or Poor. Other breed associations developed similar programs.

The Brown Swiss Cattle Breeders' Association developed "breakdown" ratings in 1943 for the anatomical subdivisions of the animal, similar to divisions of the Unified Dairy Score Card for dairy bulls and cows. Other associations adopted this system with slight differences in details. Permanent records of the type ratings are published for use in breed programs for herd improvement.

The Holstein-Friesian Association engaged G. W. Trimberger to analyze their method of type classification. F. W. Atkeson's plan, based on the Unified Dairy Score Card, was expanded in 1967 to include ratings on stature, head, front end, back, rump, hind legs, feet, fore udder, rear udder, udder support and floor, udder quality, and teats. This classification plan was integrated with the USDA sire summaries of production in 1968 into a sire guide—Registered Holstein Sire Performance Summaries—available three times yearly for active bulls with ten or more daughters. Students should become acquainted with this advancement.

One laboratory assignment should be attendance at an official type classification of a registered dairy herd. If the owner permits, the class may benefit by analyzing the effects of at least two sires on the breakdown scores of their daughters. What characteristics did each bull transmit best? Where could a breeder obtain a sire to overcome specific weaknesses in the next generation of cows? When an official classification cannot be observed, a substitute laboratory may deal with all daughters of two bulls in the college herd. Daughter-dam pairs should be included in their reports, if possible. Do any of the cows show hereditary defects? An alternate laboratory may be an analysis of type transmission by two Brown Swiss, Holstein-Friesian, or other bulls from records published by the breed association in the latest Sire Performance Summaries.

Shows and Their Functions

The earliest fairs were said to have been held in China. Fairs or markets were granted charters during the feudal period in Europe, setting times and places for livestock, goods, or merchandise to be displayed for barter or sale. Itinerant merchants exhibited their wares and farmers brought specified products to trade. Assembly on market day had educational values to inspire improvement.

Competitive shows for livestock were an outgrowth of fairs and markets of the feudal period. The first competitive shows were sponsored by agricultural societies interested in improvement of animals. Much improvement in the conformation of dairy cattle has been attributed to the influence of shows. How does this come about? How are shows operated, and what are the benefits?

Shows serve several functions. They establish standards of excellence for animals; bring good stock before the public; create desire for good animals; and inform people where they may be obtained. Ideals are established. Breeders compare the conformation of their animals with those of other breeders, and they recognize good and poor qualities by comparison. This gives incentive for further improvement through selection of breeding stock.

This laboratory deals with operation of shows for dairy cattle. They are sponsored by agencies or organizations in an area interested in improving agriculture, in display of goods and products, and incidentally in expanding the benefit through sales. Breeders desiring to enter animals at shows must comply with rules concerning registration of animals, health, entry fees, care of animals on the fair grounds, their display in the show, and their return home when released from the show. Large fair associations prepare a printed catalog and calendar of events for the public. Many interested persons attend the show, visit with exhibitors, and witness the judging. Magazines and newspapers report results of the show.

Animals are grouped by breed into classes of similar ages and sex to be judged. In 4-H and FFA shows the animals are rated into blue ribbon, red ribbon, and white ribbon groups according to relative merits against a standard, and every meritorious animal receives an award. At other shows, animals are judged according to comparative merits as first prize, second prize, third prize animal, and so on, among those exhibited. The judge is privileged to withhold a prize if in his judgment animals present in the show ring are unworthy of a higher award. Associations sometimes withhold first premium moneys when only one exhibitor shows animals in a class.

SHOW CLASSIFICATIONS

The show classifications group animals on exhibition by age and sex into separate classes to be judged. Shows held during the summer and autumn of 1970 were divided into male and female classes according to age, as follows:

Calf, born after June 30, 1969, and over 4 months old
Junior Yearling, born between January 1, 1969, and June 30, 1969

Senior Yearling, born between July 1, 1968, and December 31, 1968

2-year-old, born July 1, 1967 to June 30, 1968

3-year-old, born July 1, 1966, to June 30, 1967

4-year-old, born July 1, 1965, to June 30, 1966

Bulls past 3 years old had no separate classification. They were considered old enough to be evaluated upon the winnings of their progeny, or to be shown as 3-years-old or over.

A special class for best uddered cow was judged while the animals wore blankets over their entire bodies except the flanks and udders. Some shows provided for a best udder rating in each class of cows.

First-prize animals in the Calf and Yearling classes competed for the Junior Champion Bull and Junior Champion Female ribbons. First-prize animals in the bull and cow classes 2 years and older competed for the Senior Champion Bull and Senior Champion Female, respectively. Grand Champion Female and Grand Champion Bull were judged among the Junior and Senior Champions.

The group prizes are more coveted than an individual championship, since they are often more representative of a breeder's achievement than a possible prize winner that may have been purchased. Group prizes in 1970 were for the following:

Get of Sire—Four animals of either sex, not more than two being bulls. The sire must be named. Some shows had a Junior Get of Sire class consisting of yearlings or younger, and a Senior Get of Sire among animals 2 years and older.

Produce of Dam—Two animals of either sex over 4 months old, the progeny of one cow. The dam must be named.

Dairy Herd—Four cows in milk, owned by the exhibitor.

Premier Exhibitor—A banner was awarded to the exhibitor that won the most money in open single classes on animals owned and exhibited by himself.

Premium Breeders—A banner was awarded to the breeder of animals that won the most money in open single classes. This included animals bred by him even though exhibited by other owners.

Additional ribbons and trophies often were awarded by breed associations or donors, under special rules.

Shows and exhibitions uphold high ethical standards. For the 1970 shows a rule provided that premiums would be withheld if regulations had not been complied with, or if deception or fraud had been practiced or attempted.

Some rules differed slightly among fairs. These applied to tuberculin and brucellosis tests, calfhood vaccination and certificates therefor, date for entry of animals before the show, entry fees, milking out time prior to show day, uniforms for attendants and showmen, and responsibilities of exhibitors.

Many larger shows held a Herdsmens' Contest. Judges inspected the herds and observed conduct of the herdsmen with regard to care and display of their animals and attractiveness of the exhibit during the show. Suitable personal trophies were awarded the winner.

THE KLUSSENDORF MEMORIAL TROPHY

More than passing mention should be made of the Klussendorf Memorial Trophy. The National Dairy Cattle Congress premium list described the trophy thus:

> This trophy commemorates the character and friendliness of the master dairy showman Arthur E. ("Art") Klussendorf. It is awarded to the cattleman in any of the dairy breeds who excells in Endeavor, Ability and Sportsmanship. The award is based on the record throughout the show season and, while announcement is made at the National Dairy Cattle Congress, the recipient need not necessarily be showing there.

A representative appointed by each of the five dairy breed associations determined by observing throughout the season the man best qualified in each breed. These representatives and the breed judges at the National determine the man best qualified. The trophy was awarded first in 1937.

An incident will illustrate why many people respected Art Klussendorf. He showed a good heifer, well fitted and trained, at a state fair early in the show season. A young breeder brought an outstanding heifer among other animals for his first show. The heifer was "in the rough," led by an old stable halter. Art recognized a good heifer and his feelings were hurt when the judge placed Art's better prepared animal over this pearl-in-the-rough. Art learned that the beginner planned to take his animal to the next state fair. He asked the man to help show the groups, and in return helped him to clip the head, ears, and tail, polish the horns, and clean the hoofs. Art showed him how to lead and display the animal, and loaned him a russet show halter to use. The sequel? Art had the satisfaction of seeing the outstanding heifer receive the blue ribbon that was merited. Art's good heifer stood in second place.

This spirit of fair play, love of good animals, and good sportsmanship was an ingrained part of Art Klussendorf's life. After his death, many friends joined Harry Strohmeyer, Guy E. Harmon, dairy editors, judges, former associates, and friends to contribute funds for the Klussendorf Cup as a rotating trophy. A smaller replica becomes permanent property of the winner each year.

Fine people work with good cattle.

THE SCORE CARD

At the close of the first elective course on Dairy Cattle Breeds, Dr. H. H. Kildee arranged a tour of four prominent dairy herds. Typical of his teaching ability and diplomacy, he announced that a class of four animals would be judged at each herd, and someone would be asked to explain his placings.

In giving the reasons on placing No. 1 first, say: "I placed No. 1 over No. 2 because she excelled in breed character of head, strength and straightness of topline, body capacity, and rear udder attachment. However, No. 2 surpassed her in udder capacity. When it comes to the bottom pair, give your reasons for placing No. 3 over No. 4. *Do not say* that you placed No. 4 last for lack of certain characteristics. The owner recognizes those points, but that animal may be better than most of you will ever own."

That was a lesson in dairy type, and in diplomacy.

After Dr. Kildee reviewed the *Dairy Cattle Breeds* manuscript, he suggested that *The Score Card* be developed. He had served on the Purebred Dairy Cattle Association's committee that drafted the unified dairy cattle scorecards, which apply to each dairy breed.

Standards of ideal conformation for dairy bulls and cows were needed when competitive shows were established by agricultural societies in Scotland, the Channel Islands, and elsewhere. Ideals in the minds of some leaders were accepted by the public in the beginning. William Aiton, the historian, made drawings and described the ideal Ayrshire bull and ideal Ayrshire cow in Scotland. These sketches, approved by the Kilmarnock Farmers' Club, were published eventually in 1811. George Garrard, under patronage of the Board of Agriculture, sketched the pictures of typical bulls and cows of several breeds in England. His drawings and descriptions were assembled in a portfolio in 1800 to 1803.

A committee of seven persons drafted a scale of points in 1817 that was used by the first Agricultural Society on the Island of Guernsey to evaluate the best bulls for public service. Bulls were selected each year and judged according to this standard. Cows and heifers competed in the early Island shows.

In 1833 the (Royal) Jersey Agricultural and Horticultural Society appointed five cattle dealers as a committee to draw up scorecards for the bull and cow of that breed. They selected the best bull and cow on the Island for head, forequarters, and barrel, and the best for rear quarters and mammary development. The committee, serving under officers of the Society, drafted scales of points based on the characteristics of these animals. These standards were accepted for judging animals at the Island shows.

The State Agricultural Society in New York desired fixed standards to evaluate excellence in improved breeds of cattle. Their committee—Lewis F. Allen, C. N. Bement, E. P. Prentice, Francis Rotch, and George Vail—met October 17, 1843, during the sixteenth annual fair. The objectives were to determine: " . . . those forms, qualities and properties which most conduce to intrinsic values; and also that the distinctive characteristics of each separate breed may be as closely defined as possible."

Committees were appointed to draft points of excellence for Ayrshires, Devon, Durham, Hereford, and native stock. George Randall of New Bedford drew up a scale of points and a description for Ayrshires, which were accepted by the Society on January 3, 1844. This was the first Ayrshire scale of points in America. Cattle had been shown in a single group, but soon the entries were separated and judged according to breed.

After herdbook societies were organized for registrations, each society accepted scorecards drafted overseas or appointed a committee to establish scorecards for the breed. These scorecards were revised, based on experience and improvements among the better animals. Sometimes the points for a particular part were increased to emphasize need for improvement or to recognize estimated relative importance. Size of animal, udder conformation, feet and legs, and skin secretion have been emphasized by different associations.

UNIFIED SCORE CARDS

The Purebred Dairy Cattle Association appointed 12 breed representatives, assisted by three leading judges and instructors in the United States and five men from Canada to draft unified scorecards for the dairy bull and cow. John S. Clark served as chairman. Coaches and teachers of dairy cattle judging were surveyed for criticisms and suggestions. Forty score cards were assembled. Comments and score cards were a background for study.

A model dairy cow was prepared, based on a Holstein-Friesian cow's picture, adding Brown Swiss horns, a Guernsey udder, Jersey color, and Ayrshire style. Anatomical parts were labeled with standard names. A similar dairy bull was prepared. There was no thought of modifying breeds to the same standard. Some characteristics are alike in all dairy breeds. Characteristics of each breed were listed separately as to general description, color, horns, and size. The model bull and cow of each breed were reproduced in colors. A list of defects included suggested discriminations. It stated that "breed characteristics should be considered in the application of this score card." Order of observation of the score card's subdivisions aids in observing animals systematically.

The copyrighted unified scorecards are presented by permission of the Purebred Dairy Cattle Association.

A resolution adopted by the American Dairy Science Association in 1963 stated:

> Whereas the Type Committee recognized the value of type to the purebred dairy cattle breeder and the commercial dairy-

man and the improvements being made in production-type relationships in classification and the showring, the Type Committee wished to re-emphasize the importance of:

A. Udder conformation, attachments, and quality
B. Dairy character
C. Size, scale and substance
D. Feet and legs.

A recent survey of their members suggested that the points be reallocated by increasing the points on the mammary system, feet, and legs, and removing the mammary veins as a subcategory.

FITTING AND TRAINING DAIRY CATTLE FOR SHOW OR SALE

The instructor may wish to divide this subject into more than one laboratory, terminating in a show based on improvement of the animal and on showmanship.

Good management in preparation of a show herd is a year-round business. Cows need to be bred for their calves to be dropped as early as possible within the respective show age groups—after July 1 or after January 1. Thus the calf may show to advantage with regard to size in the group. A cow to be shown should have calved shortly before the show, yet with sufficient time for any congestion to be out of the udder and for the barrel to show "fill."

Dairy cattle should be thrifty, not thin. On the other hand, excessive fat causes thick necks, coarse shoulders, and a patchy tailhead. The "happy medium" stage should be observed. Judges discriminate against excessive fatness in dairy animals. They also object to seeing an animal stunted from inadequate nourishment.

CLIPPING THE ANIMAL

If the hair is long and coarse, the animal may be clipped all over, but this should be done in warm weather and at least 8 or 10 weeks before the first show. This allows the hair to grow sufficiently for the customary trimming a few days before the show. Photographs of show cattle in breed journals illustrate the "customary trim" for the breed. A light blanket sometimes is used to hasten shedding of loose hair, mellow the skin, and keep the hair clean.

The head and ears usually are trimmed to the "halter line." The tail is trimmed *above* the switch and blended smoothly at the tail-

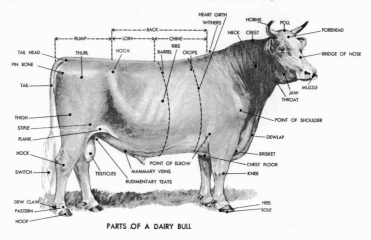

PARTS OF A DAIRY BULL

DAIRY BULL UNIFIED SCORE CARD

Breed characteristics should be considered in the application of this score card

Perfect
Score

Order of observation

1. General Appearance   45

(*Attractive individuality with masculinity, vigor, stretch, and scale, harmonious blending of all parts, and impressive style and carriage. All parts of a bull should be considered in evaluating a bull's general appearance*)

BREED CHARACTERISTICS—(see page 510)   15

HEAD—clean cut, proportionate to body; broad muzzle with large, open nostrils; strong jaws; large bright eyes; forehead, broad and moderately dished; bridge of nose straight; ears medium size and alertly carried

SHOULDER BLADES—set smoothly and tightly against the   15
body

BACK—straight and strong; loin, broad and nearly level

RUMP—long, wide, and nearly level from HOOK BONES to PIN BONES; clean cut and free from patchiness; THURLS, high and wide apart; TAIL HEAD, set level with backline and free from coarseness; TAIL, slender

LEGS AND FEET—bone flat and strong, pasterns short and   15
strong, hocks cleanly moulded. FEET, short, compact, and well

rounded with deep heel and level sole. FORE LEGS, medium in length, straight and wide apart, squarely placed. HIND LEGS, nearly perpendicular from hock to pastern from the side view, and straight from the rear view

2. Dairy Character 30

(*Angularity and general openness, without weakness; freedom from coarseness*)

NECK—long, with medium crest and blending smoothly into shoulders; clean cut throat, dewlap, and brisket. WITHERS, sharp. RIBS, wide apart, rib bones wide, flat, and long. FLANKS, deep and refined. THIGHS, incurving to flat, and wide apart from the rear view. SKIN, loose, and pliable

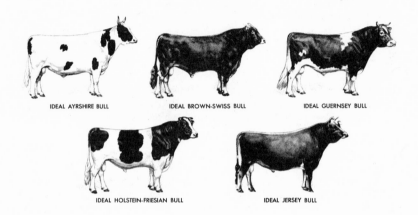

IDEAL AYRSHIRE BULL   IDEAL BROWN-SWISS BULL   IDEAL GUERNSEY BULL

IDEAL HOLSTEIN-FRIESIAN BULL   IDEAL JERSEY BULL

3. Body Capacity 25

(*Relatively large in proportion to size of animal, providing ample capacity, strength, and vigor*)

BARREL—strongly supported, long, and deep; ribs highly and widely sprung; depth and width of barrel tending to increase toward rear   12

HEART GIRTH—large and deep, with well sprung fore ribs blending into the shoulders; full crops; full at elbows; wide chest floor   13

Subscores are not used in breed type classification.

Total   100

BREED CHARACTERISTICS

AYRSHIRE

Strong and robust, showing constitution and vigor, symmetry, style and balance throughout.

COLOR—Light to deep cherry red, mahogany, brown, or a combination of any of these colors with white, or white alone, distinctive red and white markings preferred, black or brindle objectionable. Red markings usually a deeper shade on the bull than on the cow.

SIZE—A mature bull in breeding condition should weigh at least 1850 lbs.

HORNS—Inclining upward, medium size, medium length and tapered toward tips. No discrimination for absence of horns.

GUERNSEY

Size, strength and vigor, with quality and character desired.

COLOR—A shade of fawn with white markings clearly defined. Skin should show golden yellow pigmentation. When other points are equal, a clear (buff) muzzle will be favored over a smoky or black muzzle.

SIZE—A mature bull in breeding condition should weigh about 1700 lbs.

HORNS—No discrimination for absence of horns.

JERSEY

Strong and vigorous. Size and ruggedness with quality desired.

COLOR—A shade of fawn, with or without white markings.

SIZE—A mature bull in breeding condition should weigh about 1500 lbs.

HORNS—Incurving, refined, medium length and tapering toward tips. No discrimination for absence of horns.

BROWN SWISS

Strong and vigorous, but not coarse. Size and ruggedness with quality desired. Extreme refinement undesirable.

COLOR—Solid brown preferred, varying from light to very dark. White or off-color spots are objectionable. Males with any white or off-color markings, or with white core in switch do not meet color standards of the Brown Swiss breed, and shall be so designated when registered. Pink noses and light streaks up the side of the face objectionable.

SIZE—The minimum weight for mature bulls should be about 2000 lbs.

HORNS—Tending to incline slightly forward, of medium length, not coarse, tapering toward tips. Polled animals not barred from registry. No discrimination for absence of horns.

HOLSTEIN

Strong masculine qualities in an alert bull possessing Holstein size and vigor.

COLOR—Black and white markings clearly defined. Color markings that bar registry are solid black, solid white, black in switch, black belly, black encircling leg touching hoof head, black from hoof to knee or hock, black and white intermixed to give color other than distinct black and white.

SIZE—A mature bull in breeding condition should weigh at least 2200 lbs.

HORNS—No discrimination for absence of horns.

EVALUATION OF DEFECTS

In a show ring, disqualification means that the animal is not eligible to win a prize. Any disqualified animal is not eligible to be shown in the group classes. In slight to serious discrimination, the degree of seriousness shall be determined by the judge.

EYES
  1. Total blindness: Disqualification.
  2. Blindness in one eye: Slight discrimination.
  3. Cross-eyes: Slight discrimination.

WRY FACE
  Slight to serious discrimination.

CROPPED EARS
  Slight discrimination.

PARROT JAW
  Slight to serious discrimination.

SHOULDERS
  Winged: Slight to serious discrimination.

TAIL SETTING
  Wry tail or other abnormal tail settings: Slight to serious discrimination.

LEGS AND FEET
  1. Lameness—apparently permanent and interfering with normal function: Disqualification.
    —apparently temporary and not affecting normal function: Slight discrimination.
  2. Bucked knees: Slight to serious discrimination.
  3. Evidence of arthritis, crampy hind leg: Serious discrimination.
  4. Boggy hocks: Slight to serious discrimination.

LACK OF SIZE
  Slight to serious discrimination.

TESTICLES
  Bull with one testicle or with abnormal testicles: Disqualification.

OVERCONDITIONED
  Slight to serious discrimination.

TEMPORARY OR MINOR INJURIES
  Blemishes or injuries of a temporary character not affecting animal's usefulness: Slight discrimination.

EVIDENCE OF SHARP PRACTICE
  Animals showing signs of having been operated upon or tampered with for the purpose of concealing faults in conformation, or with intent to deceive relative to the animal's soundness: Disqualification.

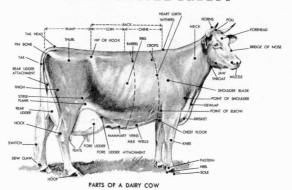

PARTS OF A DAIRY COW

DAIRY COW UNIFIED SCORE CARD

Copyrighted by The Purebred Dairy Cattle Association, 1943. Revised, and Copyrighted 1957
Approved—The American Dairy Science Association, 1957

Breed characteristics should be considered in the application of this score card

Perfect
Score

Order of observation

1. General Appearance   30

(*Attractive individuality with femininity, vigor, stretch, scale,
harmonious blending of all parts, and impressive style and carriage. All parts of a cow should be considered in evaluating a
cow's general appearance*)

BREED CHARACTERISTICS—(see page 514)   10

HEAD—Clean cut, proportionate to body; broad muzzle with
large, open nostrils; strong jaws; large, bright eyes; forehead,
broad and moderately dished; bridge of nose straight; ears
medium size and alertly carried

SHOULDER BLADES—set smoothly and tightly against the   10
body

BACK—straight and strong; loin, broad and nearly level

RUMP—long, wide and nearly level from HOOK BONES to PIN
BONES; clean cut and free from patchiness; THURLS, high and
wide apart; TAIL HEAD, set level with backline and free from
coarseness; TAIL, slender

LEGS AND FEET—bone flat and strong, pasterns short and   10
strong, hocks cleanly moulded. FEET, short, compact and well
rounded with deep heel and level sole. FORE LEGS, medium in
length, straight, wide apart, and squarely placed. HIND LEGS,
nearly perpendicular from hock to pastern, from the side view,
and straight from the rear view

2. Dairy Character   20

(*Evidence of milking ability, angularity, and general openness,
without weakness; freedom from coarseness, giving due regard
to period of lactation*)

NECK—long, lean, and blending smoothly into shoulders; clean   20
cut throat, dewlap, and brisket. WITHERS, sharp. RIBS, wide
apart, rib bones wide, flat, and long. FLANKS, deep and refined.

THIGHS, incurving to flat, and wide apart from the rear view, providing ample room for the udder and its rear attachment. SKIN, loose, and pliable

3. Body Capacity   20

(*Relatively large in proportion to size of animal, providing ample capacity, strength, and vigor*)

BARREL—strongly supported, long and deep; ribs highly and   10 widely sprung; depth and width of barrel tending to increase toward rear

HEART GIRTH—large and deep, with well sprung fore ribs   10 blending into the shoulders; full crops; full at elbows; wide chest floor

4. Mammary System   30

(*A strongly attached, well balanced, capacious udder of fine texture indicating heavy production and a long period of usefulness*)

UDDER—symmetrical, moderately long, wide and deep, strongly   10 attached, showing moderate cleavage between halves, no quartering on sides; soft, pliable, and well collapsed after milking; quarters evenly balanced

FORE UDDER—moderate length, uniform width from front   6 to rear and strongly attached

REAR UDDER—high, wide, slightly rounded, fairly uniform   7 width from top to floor, and strongly attached

TEATS—uniform size, of medium length and diameter, cylindri-   5 cal, squarely placed under each quarter, plumb, and well spaced from side and rear views

MAMMARY VEINS—large, long, tortuous, branching   2 "Because of the natural undeveloped mammary system in heifer calves and yearlings, less emphasis is placed on mammary system and more on general appearance, dairy character, and body capacity. A slight to serious discrimination applies to overdeveloped, fatty udders in heifer calves and yearlings."

Subscores are not used in breed type classification.

Total   100

IDEAL AYRSHIRE COW   IDEAL BROWN-SWISS COW   IDEAL GUERNSEY COW

IDEAL HOLSTEIN-FRIESIAN COW   IDEAL JERSEY COW

BREED CHARACTERISTICS
AYRSHIRE
Strong and robust, showing constitution and vigor, symmetry, style and balance throughout, and characterized by strongly attached, evenly balanced, well shaped udder.

COLOR—Light to deep cherry red, mahogany, brown, or a combination of any of these colors with white, or white alone, distinctive red and white markings preferred, black or brindle objectionable.

SIZE—A mature cow in milk should weigh at least 1200 lbs.

HORNS—Inclining upward, refined, medium length and tapered toward tips. No discrimination for absence of horns.

GUERNSEY
Size and strength, with quality and character desired.

COLOR—A shade of fawn with white markings clearly defined. Skin should show golden yellow pigmentation. When other points are equal, a clear (buff) muzzle will be favored over a smoky or black muzzle.

SIZE—A mature cow in milk should weigh at least 1100 lbs. "In milk" means normal condition after having been in milk from 3 to 6 months.

HORNS—No discrimination for absence of horns.

JERSEY
Sharpness with strength indicating productive efficiency.

COLOR—A shade of fawn, with or without white markings.

SIZE—A mature cow in milk should weigh about 1000 lbs.

HORNS—Incurving, refined, medium length and tapering toward tips. No discrimination for absence of horns.

BROWN SWISS
Strong and vigorous, but not coarse. Size and ruggedness with quality desired. Extreme refinement undesirable.

COLOR—Solid brown varying from very light to dark. White or off-color spots objectionable. Females with any white or off-color markings above the underside of the belly, or with white core in switch do not meet color standards of the Brown Swiss breed, and shall be so designated when registered. Pink noses and light streaks up the side of the face objectionable.

SIZE—The minimum weight for mature cows should be about 1400 lbs.

HORNS—Incurving and inclining slightly up. Of medium length, lacking coarseness, tapering toward tips. Polled animals not barred from registry. No discrimination for absence of horns.

HOLSTEIN
Rugged, feminine qualities in an alert cow possessing Holstein size and vigor.

COLOR—Black and white markings clearly defined. Color markings that bar registry are solid black, solid white, black in switch, black belly, black encircling leg touching hoof head, black from hoof to knee or hock, black and white intermixed to give color other than distinct black and white.

SIZE—A mature cow in milk should weigh at least 1500 lbs.

HORNS—No discrimination for absence of horns.

EVALUATION OF DEFECTS
In a show ring, disqualification means that the animal is not eligible to win a prize. Any disqualified animal is not eligible to be shown in the group classes. In slight to serious discrimination, the degree of seriousness shall be determined by the judge.

EYES
1. Total blindness: Disqualification.
2. Blindness in one eye: Slight discrimination.
3. Cross-eyes: Slight discrimination.

WRY FACE
Slight to serious discrimination.

CROPPED EARS
Slight discrimination.

PARROT JAW
Slight to serious discrimination.

SHOULDERS
Winged: Slight to serious discrimination.

TAIL SETTING
Wry tail or other abnormal tail settings: Slight to serious discrimination.

LEGS AND FEET
1. Lameness—apparently permanent and interfering with normal function: Disqualification.
—apparently temporary and not affecting normal function: Slight discrimination.
2. Bucked knees: Slight to serious discrimination.
3. Evidence of arthritis, crampy hind leg: Serious discrimination.
4. Boggy hocks: Slight to serious discrimination.

ABSENCE OF HORNS
No discrimination.

LACK OF SIZE
Slight to serious discrimination.

UDDER
1. Blind quarter: Disqualification.
2. Abnormal milk (bloody, clotted, watery): Possible disqualification.
3. Udder definitely broken away in attachment: Serious discrimination.
4. A weak udder attachment: Slight to serious discrimination.
5. One or more light quarters, hard spots in udder, obstruction in teat (spider): Slight to serious discrimination.
6. Side leak: Slight discrimination.

DRY COWS
Among cows of apparently equal merit: Give strong preference to cows in milk.

FREEMARTIN HEIFERS
Disqualification unless proved pregnant.

OVERCONDITIONED
Slight to serious discrimination.

TEMPORARY OR MINOR INJURIES
Blemishes or injuries of a temporary character not affecting animal's usefulness: Slight discrimination.

EVIDENCE OF SHARP PRACTICE
1. Animals showing signs of having been operated upon or tampered with for the purpose of concealing faults in conformation, or with intent to deceive relative to the animal's soundness: Disqualification.
2. Uncalved heifers showing evidence of having been milked: Serious discrimination.

head. With a milking cow, the long hair is trimmed on the inside of the thighs, on the udder, and on *either side* of the mammary veins on the belly.

TRAINING TO LEAD

One may begin training an animal to lead by tying it with a rope halter when the animal is about to be fed. Thus restraint of a halter may become associated with a pleasant experience. If water bowls are not used, animals may be led to water two or three times daily. Leading never should be a tug-of-war ordeal—the animal should want to go.

A homemade halter of the "Scotch two-loop" type may be made from 14 feet of rope. Use quarter-inch-diameter rope for calves, three-eighths inch for yearlings and cows. The longer halter rope allows more freedom for animal and trainer, with less feeling of restraint to the animal.

The leader holds the halter rope in his right hand, leading from the left side of the animal, with *not too snug* a hold. Patience is required always. If the animal hesitates long, the lead rope is changed to the left hand and the person steps to the side and pats the animal on the rump, moving freely when it decides to move. Lessons in leading should be brief and should always precede training to stand and pose. Lessons should never tire the animal.

More patience is required when both voice and hand train the animal to stand and pose at attention. Every trainer has to earn and merit confidence and friendship of his animal. Each animal has a different personality. An understanding love of animals and pride of achievement are attributes of a skilled showman.

HORNS AND HOOFS

Horns are not necessary on show animals, but they require attention when present. The horns are cleaned in advance, usually with sandpaper and emery cloth, and oiled. The final *polish* is done the morning of the show. Hoofs are cleaned. If they need trimming, it should be done by an experienced man at least a month before the show so that the feet will not be tender when walking. Shoe polish may be used on black hoofs.

GROOMING

The skin and hair should be kept clean and free from stains. Grooming gently each day with a *soft* brush helps to mellow the skin, remove loose hair, clean the animal, and accustom it to being handled.

Cattle may need washing in warm weather. Showmen have preferences for kinds of soap. Cheap pinetar soap has been among the favorites. If the skin and white hair are stained, bluing water or a weak bleach sometimes is used. Animals should be bedded well in a clean lot and should be kept clean. The switch often becomes soiled and may require washing frequently with soap.

CARE OF THE SWITCH

A bushy switch is attained by braiding the long hair into small "Dinah" braids the night before the show, then moistening the hair, and wrapping the switch in gauze to keep it clean overnight. Braids are undone the next morning, the hair is combed backward along the tail, and then the tail is "swished" freely to let the hair fall into natural position before entering the show ring.

RING ATTIRE

A dairy showman wears a neatly pressed uniform when leading animals in the ring. Attire at major shows sometimes is prescribed for uniformity. The catalog number of the animal is worn by the showman on neat white cards for benefit of the ringside, clerks, and others.

A "normal" body fill or capacity sometimes is difficult to obtain on the show circuit without silage and fresh green forage. Some showmen use soaked beet pulp for this purpose. Changes in kinds of water to drink often become a problem. Sometimes an exhibitor may try to obtain body fill by withholding drinking water the morning of the show, adding a limited amount of salt to the feed, and watering the animal just before entering the ring. This can be harmful unless handled skillfully. Excessive salt has a laxative action. Too much cold water can cause the hair to stand on end, and even give the appearance of a slack heart girth and pot belly. It is well to avoid excessive salt, and to limit the water to an amount learned by experience. A final light brushing before entering the ring gives the hair a smooth appearance.

RING CONDUCT

The animal should be handled correctly when entering the ring, during the show, and returning to the stall. The showman should watch his animal continually, and follow instructions of the ring attendant and judge courteously. Whatever prize may be awarded, the showman should accept it with a smile. He should be a good sportsman always.

No two animals are alike. Every animal has a different personality which the trainer learns by observation and working with that animal. Training and showing cattle are skills that come from practice and by watching good showmen as they fit, train, and exhibit their animals.

An excellent illustrated reference is *Fitting and Showing Dairy Cattle,* by Jack Spearing (Iowa State University Press, Ames, 1953). Several differing suggestions are based on his skilled experience with other cattle, each of which responds according to its individual personality.

PHOTOGRAPHING DAIRY CATTLE

"A good picture is worth a thousand words." A poor picture of a good animal does it a discredit. The principles of animal photography may help a person to take acceptable pictures.

The animal should be clean and prepared for a picture, clipped as for a show, if necessary, and groomed. Many animals are in condition for pictures when fitted for a show or sale. If a show halter is not available for leading, a neat rope halter may be used. One-quarter-inch rope may be used when making a halter for calves, and three-eighths-inch rope for senior yearlings and cows. The attendant should be in neat attire. The camera portrays an animal as the lens sees it. Retouched pictures of livestock are unethical.

THE BACKGROUND

Many farms and showgrounds have desirable settings in which to pose animals for pictures. The background should have *no straight lines,* as they emphasize by contrast any irregularities in the topline.

Animals with white or light markings along the topline need to be in front of shrubbery, a distant field, or hillside. When silhouetted against the sky, white spots and sky blend indistinguishably and fail to portray the topline accurately.

The foreground may be a lawn, a closely grazed pasture, smooth roadway, or smoothly raked soil. Green shrubbery photographs dark gray, hence dark or black-and-white animals should be distant from it.

ANGLE OF LIGHT

The sunlight should illuminate the side of the animal toward the camera. Usually the sun will be from behind the photographer either from his left or right side. Sunlight should strike the animal from the "eleven o'clock" angle to light up and show depth of flank and details of the udder. Photographs should not be taken under natural lighting between 11:00 A.M. and 1:00 P.M. because shadows fall under the animal during that time. Pictures taken then lack perspective and appear flat.

THE CAMERA

Presentable pictures have been taken with a plain box camera when the light was ideal and atmosphere not hazy. The photographer has more latitude of light and time with an adjustable-focus lens and shutter. One-hundredth of a second exposure time is sufficiently fast for a snapshot to stop ordinary slow motion with an alert posed dairy animal. The shutter opening (aperture) can be adjusted with relation to intensity of light. Practice and experience will guide the photographer in this essential. The farther the lens is from the animal (15 to 25 feet), the less distortion will be evident in the picture.

POSING THE ANIMAL

Study of the better pictures in breed magazines shows that in most instances the animal was posed on level ground, or headed very slightly uphill. Four legs should be visible to the lens. Often the picture was taken from slightly to the rear of a direct side view, giving some indication of fore and rear udder development. Only with a telephoto lens may pictures be taken from the front.

Just before the cameraman snaps the shutter, some person outside the photographic field may attract the attention of the animal.
The cameraman should be patient and alert to snap the shutter at
the right instant. Most animals are not photogenic, so the responsibility rests with the cameraman to catch the picture when the animal is ready.

Human interest pictures differ greatly from animal portraiture,
and seldom are obtained. Sometimes a negative with sharp definition, taken from a distance, can be enlarged satisfactorily as a glossy
print.

Good pictures are essential with registered cattle. Most sales of
purebred animals are made by correspondence as the result of good
advertising. Pictures need to be taken in advance, labeled, and retained until needed.

Each student may be assigned a breed magazine to study and
criticize. He should select three pictures that portray best the desirable characteristics of a bull, a cow, and a yearling heifer.

PEDIGREES AND THEIR EVALUATION

Practice should be given in assembling pedigrees of registered
dairy cattle. A good pedigree should include the name, registration
number, tattoo or earmark, date of birth, name and address of the
breeder, and owner. For a female, the pedigree should include date
of previous calving, current breeding and service sire, official production records and type classification of the individual. Was the
animal calfhood vaccinated against brucellosis? Is the herd accredited as free from tuberculosis?

The production records of all females in the pedigree and type
classification ratings of males and females should appear under the
name of the respective animal, following the accepted practice for
that breed. "Padding" a pedigree with distantly related or irrelevant
statements should be deprecated and pointed out as unethical.

The latest Star Award method of the American Jersey Cattle
Club, based on research with Jersey records by Lynn Copeland,
may be used in comparative rating of potential values in pedigrees.
This Star Award method of evaluation is presented by permission
of the American Jersey Cattle Club in Table 22.5.

TABLE 22.5
SCALE OF POINTS FOR STAR BULL AWARD[a]

Star bull award (maximum credits 38)	Sire (maximum credits 11)	Paternal grandsire (maximum credits 4)
One Star awarded for	Having 10 or more tested daughters averaging—	Ten or more tested daughters averaging—
each 5 credits,	360 lbs. fat earns 1 credit 390 lbs. fat earns 2 credits	420 lbs. fat earns 1 credit 450 lbs. fat earns 2 credits
provided at least	420 lbs. fat earns 3 credits 450 lbs. fat earns 4 credits	Ten or more classified daughters averaging—
one-third are from	480 lbs. fat earns 5 credits 510 lbs. fat earns 6 credits	81% earns 1 credit 82% earns 2 credits
the paternal and	540 lbs. fat earns 7 credits Having 10 or more officially	
one-third from the	classified daughters averaging—	Paternal granddam (maximum credits 2)
maternal side of	81% earns 2 credits 82% earns 3 credits	
the pedigree.	84% earns 4 credits	Three or more tested
For Four Stars or	Having been Starred, may transmit to his son as follows—	progeny averaging— 420 lbs. fat earns 1 credit 450 lbs. fat earns 2 credits
more, at least 3	2 Star Bull transmit 2 credits 3 Star Bull transmit 3 credits	
credits must	4 Star Bull transmit 4 credits 5 Star Bull transmit 5 credits	
be from	6 Star Bull transmit 5 credits 7 Star Bull transmit 7 credits	
classification.	If earned credits exceed those transmitted, only earned credits are re-garded, and vice versa.	

	Dam (maximum credits 15)	
	Having a production record of 390 lbs. fat earns 2 credits	Maternal grandsire and maternal granddam earn
	450 lbs. fat earns 3 credits	credits as above
	510 lbs. fat earns 4 credits	
	Ton of Gold earns 5 credits	
	Classification—	
	Good Plus earns 2 credits	
	Very Good earns 3 credits	
	Excellent earns 4 credits	
	Having 3 or more tested progeny averaging (as for sire, above)	

SOURCE: By permission of the American Jersey Cattle Club.
a. All production records computed to a 305-day 2× milking mature equivalent basis.

If library facilities permit, the instructor may select registered dairy animals for each student to assemble two- or three-generation pedigrees. This should include as much of the information listed above as may be obtainable. As a substitute, it may be desirable to study pedigrees in sale catalogs of two or more breeds.

ADVERTISING DAIRY CATTLE

Many sales of registered dairy cattle are transacted through correspondence, largely from advertising.

A beginner in breeding registered dairy cattle usually has no animals of seedstock quality to sell from the herd. Animals that are not an asset in the herd should go only for direct slaughter. The character of the herd and reputation of the breeder have to be established from an initial unknown status. Hence the outlet for unneeded animals bears serious consideration. A breeder establishes his own reputation for integrity, based on business transactions as well as on the quality of his animals.

After acquiring the foundation animals, the second step in building a herd is to set up a permanent system of herd records for production and reproduction. Type classification is desirable at regular intervals. The production records need to be under a breed program. A desirable goal is to strive consistently toward qualifying for the Progressive Breeders' Register or its equivalent.

If surplus animals are too good for beef yet not average within the herd, some of them may go into some commercial dairy herd through small classified advertisements in a local paper, offering them on their merits.

The best advertising comes from farm news in the reading columns of a paper. Production records and type classification are useful farm news, and may justify a small spot in a state or breed magazine.

Active membership in the local or state breed organization helps to acquaint the new breeder, as well as to spread news of his herd. The first sale of a meritorious animal often may be by consignment to an organized sale. The animal should not be below the average of the herd, but seldom should a breeder sacrifice the best animal from his herd. It is desirable that at least a small advertisement describe such a consignment to the sale.

How much money may be afforded for advertising? This may be estimated as a percentage of the expected sale price, after deducting probable expenses. This advertisement may well be placed in the major magazine used by the sponsoring organization to announce the sale.

Advertising is a regular sales channel for registered dairy cattle. A campaign needs to be planned in advance, based on a percentage of yearly sales, the proportion increasing with quality of the animals. A spot advertisement may offer a single animal. For a herd, there may be periodic, alternate issue, or continuous advertisements, depending upon the volume of business.

Some good advertisements attract attention through commanding position, a distinctive style (farm name, uniform art border in successive issues, etc.), appropriate use of white space with the legend in a few selected words that stand out, a good picture, or other features. There is value in repetition. When a breeder carrying a periodic advertisement has no animal to sell, the space is used for educational values such as an item about the herd sire, a cow family, keeping the herd before the public, farm news, and good will.

Reading columns of the local, state, or breed magazine may carry farm news—production records, sire proofs, cow families, type classification, show winnings, participation in activities, a Progressive Breeders' Award, any recognition, and announcements of sales of good animals and names of the buyers. An attractive homestead and neat farm sign help to distinguish a farm home and herd.

Every inquiry concerning an animal should be answered promptly on neat letterhead correspondence paper. A pedigree and a recent picture often are enclosed to a prospective buyer.

The above background may suggest a laboratory study on advertising dairy cattle. Each student may be assigned an issue of a breed journal to examine and present an analysis of advertisements before the class, in the form of Table 22.6. The number and size of advertisements and the nature of information contained in them may be assembled.

What do advertisements cost in this journal? How many sub-scribers does the journal reach?

Select the best large advertisement and the best small advertise-ment. Explain why each excells.

The instructor may provide each student with information about a particular animal, and ask the students to prepare two types of

TABLE 22.6

ANALYZING CONTENTS OF ADVERTISEMENTS OF DAIRY ANIMALS (SAMPLE)

Contents	Size of advertisement, pages					Size, in inches		
	1	½	⅓	¼	⅛	4	2	1
Photograph of animal	—	—	—	—	—	—	—	—
Production records	—	—	—	—	—	—	—	—
Transmitting ability	—	—	—	—	—	—	—	—
Daughter-dam comparison	—	—	—	—	—	—	—	—
Herdmate comparison	—	—	—	—	—	—	—	—
Predicted difference	—	—	—	—	—	—	—	—
Repeatability, percent	—	—	—	—	—	—	—	—
Type classification	—	—	—	—	—	—	—	—
Show winnings	—	—	—	—	—	—	—	—
Pedigree	—	—	—	—	—	—	—	—
Farm news	—	—	—	—	—	—	—	—
Good will	—	—	—	—	—	—	—	—
Other information	—	—	—	—	—	—	—	—
Distinctive design	—	—	—	—	—	—	—	—
Farm name, address	—	—	—	—	—	—	—	—

advertisements concerning it. He also may ask the student to write a letter answering an inquiry about this animal.

A recent reference is *Selling Purebreds for Profit,* by J. W. Bart-lett, Theodore Prescott, and Allen N. Crissey (*Holstein-Friesian World,* Lacona, New York, 1958).

PUBLIC SALES OF DAIRY CATTLE

Dairy cattle are sold in three kinds of public sales: consignment, reduction, and dispersal sales. Public demand is gauged by sales. They help to publicize new bloodlines, measure relative price levels, and sell to the highest bidder.

Several expenses are incurred in well-conducted auction sales: fitting and preparing animals; training to lead and pose; preparing pedigrees and sometimes obtaining pictures; health tests and certificates; preparing catalogs and advertising; construction or rental of sales and loading facilities; feed, bedding, and labor at the sale; transporting animals; engaging a sale manager, auctioneer, and clerk; arranging meals for the public; transfer of registration certificates to the buyers; return of rented facilities; and final cleaning of the sale area. Other expenses may include meeting prospective buyers at trains, busses, or airports. Some attention is given to comfort, convenience, and welfare of prospective buyers. Owners, consignors of cattle, or an organization sponsoring a sale may employ a sale manager who undertakes preparation of pedigrees and sale catalogs, distributes them to a selected mailing list, assists with the sale order of animals, and "reads" pedigrees or makes announcements at the sale.

A sale committee often sends letters to breeders for consignments. They visit the farms and inspect animals offered. They may accept or reject animals, or suggest that others be consigned. Consideration is given to the proportion of cows, heifers, and bulls consigned, depending upon probable demand and objective of the sale.

The instructor may wish the class to visit one sale, preferably a consignment sale, as a field laboratory. Students should study catalogs in advance, become conversant with terminology, and make relative evaluation of the pedigrees. It is desirable to examine the animals and their pedigrees before the sale begins.

Observations on the sale grounds include:

1. Facilities to receive and accommodate animals.
2. Care and preparation of cattle by consignors.
3. Inspection of the cattle, noting their display to interested persons.
4. Preparation of the order of sale.
5. Accommodations for consignors and the public.
6. Arrangements for lunch on the grounds, or elsewhere.
7. Convenience of the auction stand, sale ring, clerks' facilities, and seating arrangements for the public.
8. Convenience for bringing animals into the sale ring; their display before the public; return to the stalls when sold; arrangement and care of the sale ring.

9. Conduct of the sale manager; any corrections of the printed catalog, announcements from the stand, or by the consignors; conduct
of the auctioneer and ring attendants, and of the sale clerk during
the auction and when settling accounts.

10. Delivery of animals to the buyers, and assistance in loading animals after the sale.

A proper health certificate to enable interstate shipment should
accompany each animal. Sale of a registered dairy animal includes
delivery of the registration and transfer certificate by the seller or
sale management through the breed association to the buyer.

Students may submit a written report of their observations as a
laboratory exercise.

TERMS AND CONDITIONS OF PUBLIC SALES OF REGISTERED DAIRY CATTLE

Sales serve a useful function in sounding out the market price levels
of commercial and registered dairy cattle. The Purebred Dairy Cattle Association adopted a set of suggested sales practices and procedures in 1954, approved by the major dairy breed associations.
The three types of sales recognized are the same as for other public
sales:

1. Dispersal sales in which an entire herd is sold without reserva-
tion. If some member of the firm or an employee plans to bid on
or buy any animal, this fact should be announced publicly before
and/or from the auction stand.

2. Reduction sales, with an explanation.

3. Consignment sales in which names of the breeder and consignor
are listed in a mimeographed or printed sale catalog.

The Purebred Dairy Cattle Association has adopted a Sales Code
virtually as presented on pages 535–538. This code may be revised
later as experience may require.

CULLING A DAIRY HERD

The average annual turnover of dairy cows has been estimated at
20 to 25 percent in herds maintained largely with home-raised replacements. The annual turnover is greater where no heifers are

raised and all replacements are bought. S. A. Asdell assembled records from Extension Dairymen in 17 states, which had been tabulated from 2,792,188 cows on DHIA test at various times between 1932 and 1949. He also assembled the records of turnover among 276,937 cows on DHIA test in the state of New York. The percentages of turnover attributed to various causes are listed in Table 22.7.

TABLE 22.7
REASONS ATTRIBUTED FOR CULLING DAIRY COWS FROM HERDS UNDER DHIA
TEST IN 17 STATES BETWEEN 1932 AND 1949[a]

Reasons for removal	New York (%)	United States(%)
Sold for dairy purposes	5.5	5.1
Low production	7.2	7.3
Udder troubles	4.3	2.5
Abortions	2.2	1.5
Sterility	1.9	1.8
Died	1.1	1.1
Age	0.9	0.6
Other reasons	2.7	1.7
Average yearly turnover	25.8	21.6

a. Asdell, S. A. 1951. Variations in amount of culling from D.H.I.A. herds. *J. Dairy Sci.* 34:529–35.

Losses from abortion decreased from 3.8 percent in 1935 to 0.7 percent in 1949 due partly to the campaign against brucellosis. There is little systematic control of vibriosis and other causes of abortion. Mastitis occurs more frequently from inefficient handling of milking machines than from careful hand milking, which now is almost out of practice. The incidence of mastitis can be reduced by attaching the teat-cups when letdown of milk has begun, and removing them as soon as the last milk is out. This method reduces damage to tender mucous linings of the teats and milk cisterns. Tips of the teats may be dipped in a mild disinfectant solution after removing the teat-cups.

Some low breeding efficiency and some diseases can be reduced by timely attention by a veterinarian. Skill and careful management are important with dairy animals.

Every cow reaches a time when she needs to be replaced. Gus

Heebink, Extension Dairyman in West Virginia, listed factors to be considered in culling, as follows:

Is she a first-calf heifer, producing 30 percent below the average of the herd? Has she produced under 130 pounds of butterfat in the first four months of lactation? Is her 305-day lactation record on a mature equivalent basis below average of herdmates freshening the same year and season? Will she be dry 6 months or longer? Has she had mastitis? Is she positive to brucellosis? What about her health?

Has she a record of breeding trouble? Will it pay to replace her with a higher producer? Will it pay to remove, and not replace her? Is a replacement cow available? How old is this cow? Is she still a regular breeder? Did she have complications at last calving? Has she had milk fever or ketosis? Is space needed for fresh heifers? Is the price of beef average to good? Is she a slow milker? Is she below average in type and udder attachment? Is she going to freshen at an off-season for the milk market?

The instructor may wish to list about 20 cows from the herd, with birth dates, previous and current production, last calving date, breeding dates, reproductive history, health, barn capacity (supply of homegrown feed), and the market for beef animals. Perhaps some of the cows listed may include heifers in first lactation as well as good cows advancing in years. The number of replacements available may be mentioned. Each student should list in order five animals that he would remove, giving the reasons and time of removal. After written reports are submitted, there should be a roundtable discussion in which each student participates.

ARTIFICIAL BREEDING

Artificial breeding is taught as a full-semester course or intensive short course usually. One or a series of laboratories can give a general understanding to breeders or students of dairy cattle by excluding techniques and skills. References for detailed information include:

H. H. Dukes. 1962. *The physiology of domestic animals,* 7th ed. Comstock Publishing Co., Ithaca, New York.

H. A. Herman and F. W. Madden. 1963. *The artificial breeding of dairy cattle. A handbook and laboratory manual.* Lucas Brothers, Columbia, Mo.

J. P. Maule. 1962. The semen of animals and artificial insemination. *Tech. Comm.* No. 15. Commonwealth Bureau of Animal Breeding and Genetics. Farnham Royal, Bucks, England.

J. A. McClean. 1957. The progress of artificial insemination—Its techniques and probable future. *A.I. News* 5(11):21–24.

A. V. Nalbandov. 1958. *Reproductive physiology.* W. H. Freeman, San Francisco.

E. J. Perry. 1968. The artificial insemination of farm animals. 4th ed. Rutgers University Press, New Brunswick, N.J.

G. W. Salisbury and N. L. VanDemark. 1961. *Physiology of reproduction and artificial insemination of cattle.* W. H. Freeman, San Francisco.

Reproductive organs may be studied by several methods. Photographs, diagrams, or a series of photographic slides may be used. Plastic models may be purchased from biological supply houses. Fresh organs may be secured through cooperation with a local abattoir or by dissection from slaughtered animals.

MALE REPRODUCTIVE ORGANS

The male reproductive organs include the scrotum which surrounds and contains the testes. Spermatozoa produced in the testes are stored and matured in the convoluted epididymis. They pass downward through the vas deferens at ejaculation. The seminal vesicles, prostate, and Cowper's glands contribute fluids which dilute and transport the spermatozoa onward to the penis. The penis is controlled in part by the retractor muscle, and contained partly within the sheath. The semen is transmitted through the penis at ejaculation.

METHOD OF COLLECTING SEMEN

Semen is collected twice a week, weekly, or biweekly, according to productivity of the bull. A barren cow, steer, another bull, or a dummy may be used as a mount. Semen is collected in a special artificial vagina at proper temperature at the time of thrust.

When a bull persistently refuses or is unable to mount, an electro-ejaculator may be used rectally to cause ejaculation. The bull should be on *dry ground or on a dry floor,* and the electric current should not be greater than that recommended by the manufacturer. A bull has been electrocuted through oversight of these precautions.

Rectal massage of the ampulae also was a technique for semen collection developed by Dr. F. W. Miller and E. I. Evans, Bureau of Dairy Industry, Beltsville, Maryland (*Journal of Agricultural Research* 48: 941–47. 1934.)

SEMEN EVALUATION

A better method of semen evaluation is needed. The present methods of evaluating semen quality from a bull include observation of yield, color, consistency, pH reading, sperm count, and morphology and motility of spermatozoa. The structure of the heads and tails is observed with specially stained samples under a high-power microscope. Longevity of the sperm in hours and proportion of live sperm are highly important in relation to extension and usability of the semen.

Three types of extenders are in use to dilute the semen. Others are under experimental trial. The egg-yolk citrate buffer has been in general use. The second one includes *heated* homogenized milk or centrifugal skim milk as a constituent. A third extender contains glycerol added by careful technique to semen extended in egg-yolk or *heated milk* base. Approved dyes sometimes are added to distinguish semen from each of the breeds in the field. Addition of 1,000 units of penicillin and of streptomycin per 1 ml. of extender has aided in combating bacteria that are present unavoidably in semen produced under careful manipulation with sterilized equipment.

Glycerol was used in an extender at Cambridge University in 1952 in preparing frozen semen. Several experiment stations investigated its use to prolong livability of fresh semen. A technique was developed in Pennsylvania whereby extended fresh semen, collected twice weekly from selected bulls, could be available in the field daily. All extenders require precise manipulation, especially those containing glycerol and/or heated milk. Elliott and associates developed an extender and process for control of vibrio fetus in frozen semen. Research is needed to find out why semen from some bulls fails to survive freezing.

FEMALE REPRODUCTIVE ORGANS

The female reproductive organs, from the exterior inward, are the vulva, vagina, cervix or os uterus, uterus, right and left uterine horns, fallopian tubes, fimbriated funnels, and right and left ovaries. The broad ligament suspends the uterus and its associated parts loosely in the posterior part of the abdominal cavity.

The first estrus occurs at varying intervals after calving, depending on health of the reproductive organs, nutrition, and other conditions. An ovum is produced within an ovarian follicle and is released after termination of estrus. The ovum is gathered into a fimbriated funnel, and descends through a fallopian tube into a uterine horn.

The estrus interval varies in healthy animals from about 16 to 24 days. At the approach of estrus, the cow smells and attempts to ride other cows. The vulva appears slightly swollen, red, and moist. In true heat, the cow appears nervous and excited, bawls, rides other cows, and stands to be ridden. Milk yield may or may not be affected. True heat usually continues over 16 to 18 hours. After heat the cow smells other cows but refuses to stand to be ridden. A clear mucous discharge comes from the vagina. The optimum time for breeding or insemination is from the middle of estrus to 6 or 8 hours after its termination. So-called "silent heat" in which there is no noticeable manifestation occurs sometimes.

H. O. Dunn tabulated artificial services in New York and found conceptions as follows:

First service after calving	Conceived at first service	Calving to conception
(days)	(percent)	(average days)
Within 50	31	100
51–60	67	75
61–90	70	94
More than 90	76	130

A herdsman needs to observe each cow daily and record the dates of estrus. Inseminations are more effective following an interval of at least 60 days after calving, although the tendency is to shorten this interval slightly. The time of insemination with relation to estrus cycle and placement of semen have been studied intensively.

Duration of estrus varies with individual cows; hence records of successive estrus periods facilitate timing inseminations or services.

FORMS OF SEMEN

Extended fresh semen sealed in 1 ml. ampules was cooled gradually to about 40°F for prompt delivery to technicians. It was usable for over 48 hours when kept cold, but it deteriorated soon thereafter. J. O. Almquist of Pennsylvania developed a method to use *heated milk*-glycerol extender that could be held in refrigeration over four days.

Investigators at Cambridge, England, discovered that semen could be safely frozen and stored. They used an egg yolk-citrate buffer, with a final concentration of 10 percent of glycerol. The extended semen in 1 ml. ampules was cooled slowly to 5°F, then cooled further and stored at −110°F. Many studs in Canada and the United States used this method. The procedures and facilities at the stud and in the field cost slightly more than with fresh semen. However, use of frozen semen reduced waste, permitted stockpiling from individual bulls, and facilitated selective matings. More cows were inseminated per bull in a year than with extended fresh semen.

Much research enabled the American Breeders Service to utilize liquid nitrogen as a refrigerant at −320°F. Special containers are used in the field. The limitations of stockpiling and storage of semen yet are unknown. Some frozen semen has been used successfully after 16 years in frozen storage. Semen frozen with liquid nitrogen now is used almost universally.

ARTIFICIAL INSEMINATION

The cow should be quiet, with no fear of the technician. He must be gentle and considerate. By rectal palpation, the technician grasps the uterine cervix through the intestinal wall; directs the plastic inseminating pipette along the vagina and into the cervix (os uterus), discharging 1 ml. of extended semen into the uterus. If a repeat breeding is necessary at a following estrus, the semen usually is placed intercervically, because of a possible superheat in an already pregnant animal. The spermatozoa pass quickly along the uterus to the horns and fallopian tubes. A period of maturation occurs in the uterine fluids before conception. One spermatozoa may penetrate the ovum and result in conception. Superheats occur in 3.5 to 7.5 percent of cows, and hence precaution is needed in inseminations.

The empty ovarian follicle fills with a special tissue—the corpus luteum or yellow body. The latter persists if conception has occurred, and serves as an important endocrine gland influencing continuing attachment of fetal membranes to the uterine wall (at the cotyledons) during gestation. The ovaries and fetal membranes also produce hormones of value to the female physiology. If conception did not occur, the corpus luteum normally degenerates soon and is resorbed, permitting the ovary to function again.

MEASURING BREEDING EFFICIENCY

A conception rate of 60 to 70 percent nonreturns is expected on first service in healthy cows. The majority of fertile females conceive at one of the first three services. A 60- to 90-day nonreturn rate has been accepted to estimate breeding efficiency of bulls in artificial service. There is a trend to inseminate at the first estrus after the forty-fifth day from calving, for greater lifetime production.

LABORATORY SUGGESTION

A field trip to an artificial breeding stud may be justified *if students have made sufficient advance preparation* to understand the basis of operations and practices. An appointment should be made well in advance of the trip for convenience with the routines of stud operations. Students must know that *bulls are temperamental,* and must conduct themselves quietly as guests of the organization. Also *no smoking* in the barns and laboratories. There is an enormous investment involved, and fire hazards are great in the stables and laboratories.

HEREDITARY CHARACTERS IN DAIRY CATTLE

Hereditary characters of two general classes occur in dairy cattle: simple and complex. Many herds afford material to observe a simple character. Dominant simple characters include polled or naturally hornless, and black haircoat. Other characters conditioned by a single pair of genes in homozygous form are flexed pastern, crampy (neuromuscular spasticity, progressive posterior paralysis, or stretches) in mature cattle, imperfect skin, posterior paralysis in newborn calves, and others.

A brief review of hereditary processes concerns maturation of germ cells to form an ovum (in female) or four spermatozoa (in male), Mendel's principles of dominant and recessive characters, independent assortment of genes from heterozygous parents, and use of the Punnet square to illustrate chance recombination of genes. Mating a black breed carrying the recessive red color gene results as follows:

	♂ B	b
♀		
B	BB	Bb
b	Bb	bb

1 BB—pure for black coat color

2 Bb—black, but carrying red

1 bb—pure for red coat color

Evidence that the dominant gene "covers up" the recessive one may be demonstrated in many herds where dairy cows with wry tails often have daughters appearing normal, and vice versa. Such examples illustrate inheritance of simple recessive characters.

Selected references for review on inheritance of hereditary characters may include:

Atkeson, F. W., F. Eldridge, and H. I. Ibsen. 1944. Prevalence of "wry tail" in cattle. *J. Heredity* 35:11–14.

Becker, R. B. 1933. Recessive coloration in Dutch Belted cattle. *J. Heredity* 24:283–86.

Becker, R. B., and P. T. Dix Arnold. 1949. "Bulldog head" in cattle. *J. Heredity* 40:282–86.

Becker, R. B., and C. J. Wilcox. 1969. Hereditary defects of spermatozoa. An illustrated review. *A.I. Digest* 17(12):8–10.

Becker, R. B., C. J. Wilcox, and W. R. Pritchard. 1961. Crampy or progressive posterior paralysis in mature cattle. Florida Agr. Exp. Sta. Bull. 639.

Becker, R. B., C. J. Wilcox, and C. F. Simpson, and L. O. Gilmore and N. S. Fechheimer. 1964. Genetic aspects of actinomycosis and actinobacillosis in cattle. Cooperative publication—Florida Agr. Exp. Sta. Tech. Bull. 670 and Ohio Agr. Exp. Sta. Bull. 938.

Cole, L. J., and S. V. H. Jones. 1935. The occurrence of red calves in black breeds of cattle. Wisconsin Agr. Exp. Sta. Bull. 313.

Gilmore, L. O. 1950. Inherited non-lethal anatomical characters in cattle. A Review. *J. Dairy Sci.* 33:147–65.

Gregory, P. W., S. W. Mead and W. M. Regan. 1944. Hereditary congenital lethal spasms in Jersey cattle. *J. Heredity* 35:195–200.

Some hereditary characteristics may be combinations of one or more characters. The more complex characteristics include body capacity and size, breed character, dairy character (light natural fleshing), strength and straightness of legs and feet, milk composition (percentages of butterfat, casein, chlorides, lactalbumin, color-carotene hydrolysis), peak daily milk yield, persistency of lactation, shoulder attachments, teat placement, shape and size, udder attachments and texture, type of spermatozoa, multiple births, and other characteristics that are less evident. Oklahoma workers summarized a report on heritability and repeatability estimates dealing with some of these characteristics.

A practical laboratory may deal with some complex characteristics. Cows in the herd may be assembled in groups according to their sires, or into cow families. Resemblances can be distinguished within separate groups attributable to the parent. Major changes within a breed or herd generally require more than one generation of selective matings to prepotent sires. Herds seldom retain enough individuals alive to demonstrate such slow changes. Some may be demonstrated for analysis from records of successive type classifications under the breakdown system. Feeding and management allow animals to develop within the limits of potential hereditary capacity. Some of the slow changes within the breeds were indicated in a popular article: "American Contributions to Better Dairy Cattle," *Hoard's Dairyman* 98(17):736–39 (Sept. 10, 1953).

A well-planned laboratory on inheritance in a dairy herd can inspire students and breeders with the possibilities of planned breeding for livestock improvement.

SALES CODE

The Sales Code approved June 24, 1966, by The American Jersey Cattle Club is substantially that adopted by the Purebred Dairy Cattle Association on January 21, 1966, with exception of breed association names. It applies to sales of registered dairy cattle by private treaty and public auction.

SALE POSITIVE

The highest bidder shall be the buyer. In event of dispute, the auctioneer will recognize the person from whom he accepted the last bid, and open the bidding to those involved. Every animal is pledged to absolute sale, unless withdrawn for just cause, and by-bidding is prohibited. Every animal must be transferred to the new owner and may not be transferred back to the seller within a period of one year, except at the discretion of the breed association.

ERRORS

A sincere effort shall be made to correct any errors noted in the catalog; such announcement from the auction stand will take precedence over printed matter in the catalog. All data in the catalog is correct to the best knowledge, but the sponsors or sales manager do not assume any personal responsibility for errors.

WARRANTIES

The seller warrants clear title to the animal and right to sell same. Unless otherwise noted in the sales catalog, or announced from the auction stand, each animal is sold as sound. Prospective bidders are cautioned to pay attention to all announcements from the auction stand.

REGISTRATION CERTIFICATES

Every animal is recorded (or in the event of baby calves, will be recorded) promptly in the herd registry of the breed association. A certificate of registry with recorded transfer will be furnished by the seller to the purchaser free of charge.

BREEDING GUARANTEE

Bulls. Any bull is sold to be able to serve and settle, after reaching 14 months of age, and having semen that will have 20 percent of its sperm cells progressively motile after 24 hours of storage at 25° to 50°F. Should any bull not meet these requirements the buyer shall within 4 months from date of sale or within 4 months after such bull has reached the age of 14 months, report that fact in writing to the seller and arrange to return the bull to him, who shall have 4 months from date of receiving the bull in which to prove him as meeting these requirements before refunding the purchase price. In the case of proven bulls, additional information relative to terms of sale and guarantee may be appropriate.

Females. Should any female not pregnant when sold fail to become pregnant after being bred to a bull known to be a breeder and after having been treated by a licensed veterinarian, the buyer shall within 4 months from date of sale or within 4 months after such female shall have reached the age of 16 months, report that fact in writing to the seller and arrange to return such female to him, who shall have 4 months from date of receiving this animal in which to prove her as pregnant before refunding the purchase price. Any cow having freshened normally within 90 days previous to date of sale will thereby be considered a breeder. Females bred when sold are not thereby guaranteed to be pregnant; moreover, pregnant females are not guaranteed to deliver normal calves or carry calves full time.

Above warranties cease when animal is disposed of by the original purchaser. Transportation charges on all animals subject to adjustment shall be paid by the shipper.

Responsibility

The seller is the responsible party for all recommendations and warranties. The sales manager, auctioneers, and the sponsors are acting as agents only and not responsible in any way, but they will endeavor to protect the interest of both buyer and seller.

The seller, or anyone associated in the sale, shall not be responsible in case of accident resulting in injury to anyone attending the sale.

Health of Animals

All animals shall be from herds that are T.B. Accredited and Bang's Certified; or from Accredited and Certified free herds, with negative herd tests within 12 months; or animals accepted from herds with a negative test within 3 months.

In addition, all animals shall have a negative (tube test for

Bang's) within 30 days, except calfhood vaccinated under 18 months of age, for both brucellosis and tuberculosis.

Each animal enters the sale with an official health certificate approved by the state veterinarian of the state from which the animal originated, showing status of the herd and date of last negative herd test for both brucellosis and tuberculosis.

Calfhood vaccination shall be recorded on the health certificate, showing date and age of vaccination, or individual calfhood vaccination certificate furnished. Unvaccinated animals should be so identified throughout the sale catalog.

ABSENT BIDDERS

While it is preferable that prospective bidders attend the sale in person, for the convenience of those not present, any of the following men are well qualified to act and will gladly represent you, and should be addressed _____.

Any persons handling bids for absentee buyers act only as agents of the buyer and should be under no personal liability. Bids may be handled by ————.

PENALTIES

Violations of this Code shall be considered as an unethical sales practice and shall subject the violator to such penalties as may be imposed by (The American Jersey Cattle Club).

TERMS OF SALE

Cash: Unless other arrangements are made with the owner or sponsor in advance of the sale. Parties unknown to the seller who desire to pay by check should furnish bank statement or certified check. No animal to be removed until terms of sale are complied with.

INDEX OF NAMES

SUBJECT INDEX

Acts (laws), 189, 291, 292, 401, 404, 469, 475; Adams, 436; Hatch, 436, 475; Morrill, 436; 475; Purnell, 436; Research and Marketing, 436; Smith-Hughes, 437; Smith-Lever Extension, 436

Advertising dairy cattle, 522–24

Age conversion factors, 341

Agricultural education, 436, 437, 474

Agricultural extension, 474

Agricultural research, 474

Agricultural societies, 67–70, 87, 105, 192, 193, 293, 294, 352, 390, 391, 404, 405, 435, 441

Agriculture: in Ayr, 72; in England, 48–50, 342–43; in Switzerland, 134–37; on Guernsey, 185; in the Netherlands, 228, 248; on Jersey, 287, 301, 305, 308; in Denmark, 386, 404

Aiton's surveys, 75, 79, 83

Alderney cattle, 78, 204, 288, 289, 309, 320, 349

All-Jersey milk, 341

American Dairy Science Association, 262, 279, 379, 439, 440, 445, 459, 466–67, 474

American Guernsey Cattle Club, 205

American Jersey Cattle Club, 312

American Red Danish Association organized, 421

Analyzing (proving) sires, 97, 210, 493–97

Antibiotics, 469, 471, 473

Approved, Ayrshire milk (Scotty-Milk), 130

Approved breeding bulls, 142

Approved dams, 97, 98, 122

Approved sires, 97, 120

Artificial breeding, 100, 101, 129, 130, 146, 169, 172, 199, 200, 249, 250, 277, 278, 284, 306, 335, 361, 362, 383, 385, 406, 416, 425, 440, 448–56, 470, 474, 475